AF885310

Understanding the World Around Through Simple Mathematics

From Credit Card Interest, Basal Metabolic Rate To Earthquake Magnitude

M. Kemal Atesmen

2010

INFINITY PUBLISHING

ISBN 978-0-7414-6311-1

Printed in the United States of America

Published January 2011

INFINITY PUBLISHING

Toll-free (877) BUY BOOK
Local Phone (610) 941-9999
Fax (610) 941-9959
Info@buybooksontheweb.com
www.buybooksontheweb.com

Table of Contents

Preface

Mathematics is a rich tool that helps one solve problems in healthcare, economics, physical sciences, engineering, and helps us to understand the World around us. One starts to learn mathematics in early life by simple numbers, additions, subtractions, and multiplications. One moves on to algebra, geometry, logic, calculus, statistics and probability, vectors, complex numbers, and to differential equations in high school and in college. The challenge is to be able to apply all what was learned to the problems we face in our daily lives. This book uses different mathematical tools that we learned in high school and in college to solve in detail one hundred everyday problems.

In Section 1 of this book, detailed solutions to everyday problems that we face using basic functions such as logarithmic, exponential, square root, trigonometric, and other power functions, are provided. In Section 2, series solutions to everyday problems using arithmetic progression, geometric progression, and Taylor series, are detailed. In Section 3, differential and integral calculus such as rate change, derivatives, definite integrals, and numerical integration are exercised to solve everyday problems. In Section 4, statistics and probability principles such as distributions, expected values, control charts, permutations, and combinations are used to solve everyday problems. Since personal computers are an inseparable part of our lives, in Section 5, spreadsheets are used to solve everyday problems.

Section 1 provides fourteen everyday problems that are solved using basic linear equations. These problems range from a farmer's fertilizer need for his lawns and trees to head and tail wind effects to airplane travel time. Section 1 also tackles twenty three everyday problems that are solved using basic non-linear equations. These problems range from credit card interest, basal metabolic rate, world oil production, spider silk strength, earthquake magnitude in Richter scale to wind chill factor.

Section 2 provides solutions to seven everyday problems using series. These problems range from payments for a fixed interest rate loan, life cycle cost of a car to tidal forces between two bodies.

Section 3 provides solutions to thirty five everyday problems using differential and integral calculus. Optimizations problems, finding maximum and minimum of governing functions, range from finding minimum piping costs through a ranch, maximizing viewing angle that covers height of a television screen to maximizing the volume of a open top square box. Integration solutions to everyday problems are detailed for jerk, acceleration, speed, and distance for a track athlete, for pressure changes in atmosphere and below sea level, and for volume of a spherical candle holder. Numerical integration solutions are used for motion of a falling object, for star temperatures, for Sun protection factor (SPF), and for surface area of a leaf. First and second order differential equations are solved for loss of memory by time, for population growth and decline, for cooling of tea with and without milk, and for wire tension in musical instruments.

Section 4 provides solutions to fifteen everyday problems using statistics and probability. Statistical analysis is done on a person's weight, on monthly expenses, on baseball players, and on mid-term grades of a class. Probability analysis problems range from betting on soccer and baseball games, uptime for sputtering equipment in a manufacturing environment, to finding percentage of patients without high glucose and cholesterol levels.

Last Section 5 is for spreadsheet applications. There are six everyday problems that use spreadsheets for their solutions. These applications range from a large party drink requirements, carbon footprint, home energy usage, nutrition facts to foreign currency exchange.

Author's wish is to stimulate the reader to use simple tools of mathematics and spreadsheets to solve everyday problems that are encountered everyday and to understand the World around us. This book will enhance the reader's judgment about common issues we encounter daily and make his or her decisions more accurately.

M. Kemal Atesmen
Santa Barbara, California

Acknowledgements

The idea on this book was solidified and supported by members of my family, colleagues, and friends. Every day of my personal and professional life, I came across an issue that needed mathematical tools to solve. This book is a combination of problems that emerged from questions and discussions with members of my family, colleagues, and friends.

One day a good friend of my son, Beau, asked me how many bottles of liquor he should buy for his upcoming large party which led to a spreadsheet solution to "Party Drinks" everyday problem? In another occasion a discussion regarding the meaning of a company name "Velocity11" led to the calculation of Earth's escape speed.

My wife Zeynep used to ask me about the monthly payments for a fixed interest rate home loan and how much we will gain from taxes for home loan interest deduction. I went back to basic mathematics tools and wrote a simple program that can be used to answer her questions.

During a thunder storm, I was watching the lightning and listening to the thunder with my son Omer. He asked me how far is the location of the lightning? A square root function derived in thermodynamics gave me the speed of sound in air and answer to his question.

I would like to dedicate this book to members of my family, colleagues, and friends who asked me challenging questions throughout my grownup life and gave me the distinct opportunity to use my engineering and physics knowledge along with basic mathematics tools to provide answers to them.

M. Kemal Atesmen
Santa Barbara, California

Introduction

Everyday World problems in areas of healthcare, economics, and physics are solved in detail by using different mathematical tools that can be categorized into five sections. Section 1 uses basic linear and non-linear equations and their solutions to achieve the desired results. Section 2 uses series concepts to attack the problems from the given information. Section 3 utilizes differential and integral calculus tools. Section 4 focuses on statistics and probability principles. Section 5 uses spreadsheet formations to solve everyday problems that have a large number of inputs, calculations, and outputs.

Main focus of Section 1 is to be able to set up the basic linear and non-linear equations from the given information for a specific problem. Once the equations are setup, the unknowns are determined, all the conditions are met, and solutions to an everyday problem fall out with ease.

Section 1 provides solutions to daily economic issues around us. These economic issues cover compounding interest on credit card debt, oil production projections, purity of gold bar, cost benefit of cooking at home or buying cooked food, required fertilizer nutrients, minimizing cost of fertilizer mixing, movie theater audience mix, learning curve, office cleaners' options, depreciation in value of a car, optimum house wall insulation and heating costs, different raffle payment choices, equilibrium point in supply and demand, adjusting cookie diameter for profit, buying a common lounge TV, a typist's base and bonus pays, and distribution of bonuses.

Section 1 also provides solutions to daily health issues that affect us. These health issues are basal metabolic rate, wind chill factor, heat stress, temperature scales, World track records, water vapor in saturated air, and relative humidity.

Section 1 also tackles physics of daily issues around us. These physics problems cover head and tail wind effects to airplane travel time, volume of ice below sea surface, rowing a boat in a river, spider silk strength, earthquake magnitude – Richter scale, disk drive areal density increase, intensity of sound, brightness of stars, Venturi effect, speed of sound, and surveying to find height of a mountain.

Main focus of the Section 2 is to be able to set up series solutions to everyday problems from the given information.

Section 2 provides solutions to some of the daily economic issues around us. These economic issues cover payments for a fixed interest rate loan, life cycle cost of owning a car, profits from book sales, manufacturing cost versus hours, and saving money.

Section 2 also tackles two physics problems of daily issues around us. These physics problems are tidal forces between two bodies and center of mass of vehicles and how they affect rollover accidents.

Main focus of the Section 3 is to be able to solve everyday World problems using differential and integral calculus tools. Differential calculus is used to find minimum or maximum of a function. First order differential equations are solved using applicable boundary and initial conditions. Solution to second order differential equation, the wave equation, is used to detail string tension in musical instruments. Standard tables are used for differentiation and integration. In several cases where closed form integration is difficult, numerical approximation is utilized. To understand the sensitivities of a dependent variable to its governing independent variables, differential calculus is used.

Section 3 provides solutions to daily economic issues around us. These economic issues cover optimizing piping costs through a ranch, shortest ladder over a wall, maximizing area of a rectangular building on an odd shaped lot, minimizing material cost of a can, maximizing viewing angle that covers height of a TV screen, population growth or decline, optimizing magazine sales price, maximizing volume of a right circular cylinder in a sphere, income, consumption, and savings relationships, and maximizing volume of an open top square and rectangular box.

Section 3 also provides solutions to daily health issues that affect us. These health issues are spread of contagious diseases, loss of memory by time, bacterial growth, Sun protection factor (SPF), and sensitivity of drag forces on a bicycle cyclist.

Section 3 tackles nineteen physics problems of daily issues around us. These physics problems are distance, speed, acceleration, and jerk relationships for a short distance track runner who has aspirations to set the World record, stopping distance for a vehicle during braking, motion of a falling object, pressure changes below sea level and in atmosphere, cooling of tea with and without adding milk, a basketball player's hang time, radioactive carbon dating, inductor and resistor in series applied to a disk drive write head, emptying a water tank with no inflow and with a constant inflow, Earth's escape speed, evaporation of a drop, star temperatures, string tension in musical instruments, cooling of an electronic chip, acceleration in G's, surface area of a leaf, volume of a candle holder, and speed of a revolving light.

Main focus of the Section 4 is to solve fifteen everyday problems using statistics and probability principles.

Three of the everyday problems focused on control charting of different types of data and finding the average, median, standard deviation, lower and upper control limits. In addition to control charting, normalized distribution function is utilized to determine desired probabilities. Expected value concept for discrete random events is introduced in several examples. Frequency distribution and cumulative frequency distribution plots are also utilized. Binomial distribution function that applies to events with only two outcomes is applied to several problems. Standard deviations in a set of independent events are treated. One problem focused on accuracy and precision of a measurement system along with its repeatability, reproducibility, and gage capability. A 2x2 full factorial design of experiments is detailed by introducing the additive and interaction effects of independent variables on the dependent variable. Addition rule of probability for mutually exclusive and for mutually non-exclusive events is exercised. Joint probability of independent events is used in several problems. Permutations of possible outcomes of an event and percentage combinations are utilized in solving some of the everyday problems.

Most of the everyday problems in Section 4 fall into the economy category. These are monthly living expenses, expected value of an investment, egg size improvement in a farm, distribution of mid-term grades, duration of operation for a sputtering cell, optimum curing conditions for an adhesive joint, and inventory combinations for a given demand.

Three everyday problems in Section 4 dealt with health issues. These are a person's weight distribution, precision of a weight scale, and determining percentage of patients without high glucose and cholesterol levels.

Main focus of the Section 5 is to solve six everyday problems using inputs, calculations, and outputs in a spreadsheet format. In a spreadsheet format, what if scenarios can be applied with ease by changing the inputs and seeing effects on the outputs.

Four of the everyday problems that are tackled in Section 5 are in economic arena. These are determining the required drinks for a party, home energy usage, house remodeling, and foreign currency exchange.

Two of the everyday problems in Section 5 fall into health category. These are carbon footprint and nutrition facts.

About the Author

M. Kemal Atesmen completed his high school studies at Robert Academy in Istanbul Turkey in 1961. He received his B. Sc. degree from Case Western Reserve University, his M. Sc. degree from Stanford University, and his Ph. D. degree from Colorado State University, all in mechanical engineering. He is a life member of ASME. He initially pursued an academic and an industrial career in parallel and became an associate professor in mechanical engineering before dedicating his professional life to international engineering project management and engineering management for thirty-three years. He helped many young engineers in the international arena to bridge the gap between college and professional life in automotive, computer component, data communication, and offshore oil industries.

He published two books, sixteen technical papers, and has four patents. His books are "Global Engineering Project Management", CRC Press, 2008 and "Everyday Heat Transfer Problems – Sensitivities to Governing Variables", ASME Press, 2009.

Section 1 – Everyday Problems Solved Using Basic Equations

Main focus of Section 1 is to be able to set up the basic linear and non-linear equations from the given information for a specific problem. Once the equations are setup, the unknowns are determined and all the conditions are understood, solutions to an everyday problem falls out with ease.

Section 1 provides solutions to fourteen everyday problems that use linear equations. Linear equations with two, three, and four unknowns are solved by elimination method. Chapter 1-9 uses determinants to solve three equations and three unknowns to determine required fertilizer nutrients. Chapter 1-10 solves an optimization problem for a set of linear equations for minimizing costs for fertilizer mixing. In Problem 1-13 two linear equations and three unknowns are treated to show the resulting multiple solutions for possible combination of a movie theater customers. Problem 1-14 treats a pair of linear equations with no solution with the information provided for the ages of two children.

Also in Section 1, twenty-three everyday problems are treated to cover a wide range of non-linear functions. These functions are logarithmic functions to base 10, natural logarithmic functions to base e, fractional functions, square root functions, exponential functions, other power functions, and trigonometric functions. In Chapter 1-1 limit of functions, i.e. L'Hopital's Rule, is used to determine continuous compounding for credit card debt. In Chapter 1-24 optimization is done to a combination of linear and non-linear equations to determine optimum house wall insulation and heating costs. In Chapter 1-32 solutions to a quadratic equation are determined for buying a common lounge TV.

Chapter 1-1 Credit Card Interest

When you owe money to a credit card company, the amount you owe increases every interest period at a specified certain interest rate for that period. If the amount you owe is called P_o and the interest rate for an interest period is called i, then at the end of the first interest period the amount you owe increases to

$$P_1 = P_o + iP_o = P_0(1 + i) \quad \text{(1-1-1)}$$

At the end of the second interest period the amount you owe increases to

$$P_2 = P_1 + iP_1 = P_o(1 + i)^2 \quad \text{(1-1-2)}$$

At the end of the n interest periods the amount you owe increases to

$$P_n = P_0(1 + i)^n \quad \text{(1-1-3)}$$

Eq. (1-1-3) is the compound interest formula. Interest can be compounded annually, semiannually, quarterly, monthly, daily, and sometimes continuously.

The converse problem of Eq. (1-1-3) is to determine the present value or present amount of an investment or debt that would amount to P_n at the end of n interest periods at an interest rate for an interest period of i.

$$P_o = P_n(1 + i)^{-n} \quad \text{(1-1-4)}$$

Most interest rates are quoted as annual interest rates. So the annual interest rates have to be converted to the interest period that is needed for the calculations in Eqs. (1-1-3) and (1-1-4).

For example, if you owe \$20,000 to a credit card company that applies 20% annual interest rate monthly to your debt, how high your debt will be in a year? What is your effective annual percentage rate (APR) without considering any fees that are tacked on to your debt? Using Eq. (1-1-3) and the given information

$$P_o = \$20{,}000\,,\ i = \frac{20}{100x12} = 0.0167\,, \text{ and } n = 1x12 = 12\ months$$

$$P_{12\ interest\ periods} = \$24{,}387.82$$

$$APR = \frac{\$24{,}387.82 - \$20{,}000}{\$20{,}000} \times 100 = 21.94\%$$

Another example is for a long term deposit in a bank for investment. If \$100,000 is invested at a fixed 6% annual interest rate compounding monthly, what will your investment amount to in 20 years? Using Eq. (1-1-3) and the given information

$$P_o = \$100{,}000\,,\ i = \frac{6}{100x12} = 0.005\,, \text{ and } n = 20x12 = 240$$

$P_{240\ interest\ periods} = \$331,020.45$

Another example, if \$200,000 is needed in 18 years for college education of a child, what amount should be invested now at a 5% annual interest rate compounding quarterly? Using Eq. (1-1-4) and the given information

$P_n = \$200,000\ ,\ i = \frac{5}{100x4} = 0.0125\ ,$ and $n = 18x4 = 72$

$P_o = \$81,768.81$

For continuous compounding Eq. (1-1-3) can be reduced to a limit where the interest period goes to zero or the number of interest periods go to infinity. If $i = a/x$ where a is the annual interest rate and x is the number of interest periods in a year and if $n = yx$ where y is the number of years of investment, then Eq. (1-1-3) can be rewritten as

$$P_{yx} = P_0(1 + a/x)^{yx} \tag{1-1-5}$$

For continuous compounding, the limit of Eq. (1-1-5) has to be determined as $x \to \infty$. Eq. (1-1-5) can be rewritten in natural log form as

$$Ln\left(\frac{P_{yx}}{P_o}\right) = yx\ Ln\left(1 + \frac{a}{x}\right) = \frac{Ln(1+\frac{a}{x})}{1/yx} \tag{1-1-6}$$

Limit of Eq. (1-1-6) for the numerator and for the denominator, as $x \to \infty$, gives $\frac{0}{0}$. Then L'Hopital's rule, see Reference [1], can be applied to determine the limit of Eq.(1-1-6). The numerator and the denominator in Eq. (1-1-6) are differentiated independently with respect to x to determine the limit as $x \to \infty$.

$$\lim_{x\to\infty} Ln(\frac{P_{yx}}{P_o}) = \lim_{x\to\infty} \frac{\frac{1}{(1+\frac{a}{x})}(-\frac{a}{x^2})}{\left(-\frac{1}{yx^2}\right)} = ay \tag{1-1-7}$$

Since $Ln(\frac{P_{yx}}{P_o})$ is a continuous function of $\frac{P_{yx}}{P_o}$,

$$\lim_{x\to\infty} Ln(\frac{P_{yx}}{P_o}) = Ln(\lim_{x\to\infty} \frac{P_{yx}}{P_0}) = ay$$

and

$$\lim_{x\to\infty} \frac{P_{yx}}{P_0} = e^{ay}$$

or

$$P_{y\infty} = P_o e^{ay} \tag{1-1-8}$$

As expected, continuous compounding increases your credit card debt the most. For example, if \$10,000 is owed at a 30% annual interest rate, the amounts shown in Table 1-1-1 are expected after a year for different applied interest periods.

Interest Period	**Amount After A Year**	**Effective APR**
Annual	13000.00	30.00
Semiannual	13225.00	32.25
Monthly	13448.89	34.49
Weekly	13486.96	34.87
Daily	13496.92	34.97
Continuous (Eq. 1-1-8)	13498.59	34.99

Table 1-1-1 Debt amounts after a year for different applied interest periods

Eq. (1-1-8) for continuous compounding can also be obtained by a different method using small incremental increases in the amount invested for a annual interest rate a, namely a small incremental change dP in the amount invested during a very short interest period dy is

$$dP = aPdy \qquad (1\text{-}1\text{-}9)$$

or

$$\int \frac{dP}{P} = \int ady \qquad (1\text{-}1\text{-}10)$$

Eq. (1-1-10) can be integrated to give

$$Ln(P) = ay + C \qquad (1\text{-}1\text{-}11)$$

and by setting the constant $C_1 = e^C$

$$P = C_1 e^{ay} \qquad (1\text{-}1\text{-}12)$$

Since at $y = 0$, the principal amount is P_o, then $C_1 = P_o$. Same continuous compounding equation as in Eq. (1-1-8) is obtained.

Chapter 1-2 World Oil Production

World oil production has been steadily increasing since 1900's. World's annual oil production for the last hundred years is shown in billions of barrels per year in Figure 1-2-1 (data source from US Department of Energy). One barrel equals 42 US gallons or 159 liters.

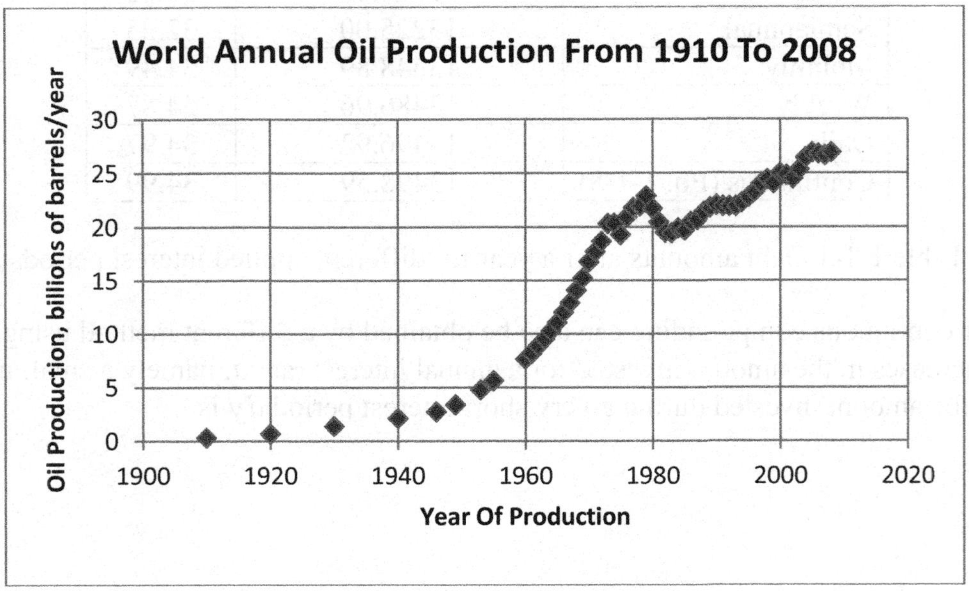

Figure 1-2-1 World annual oil production from 1910 to 2008

Demand for oil increased exponentially until 1980's. 1973 OPEC (Organization of the Petroleum Exporting Countries) nations' embargo on US and 1980 Middle East crises put dips into demand growth. After 1980's the demand for oil decreased and the rate of increase for demand is reduced due to more efficient vehicles on the roads and high price of fuels.

Orders of magnitude increase in world annual oil production is best represented in a semi-logarithmic graph. Logarithm to base 10 for world annual oil production versus year of production is shown in Figure 1-2-2.

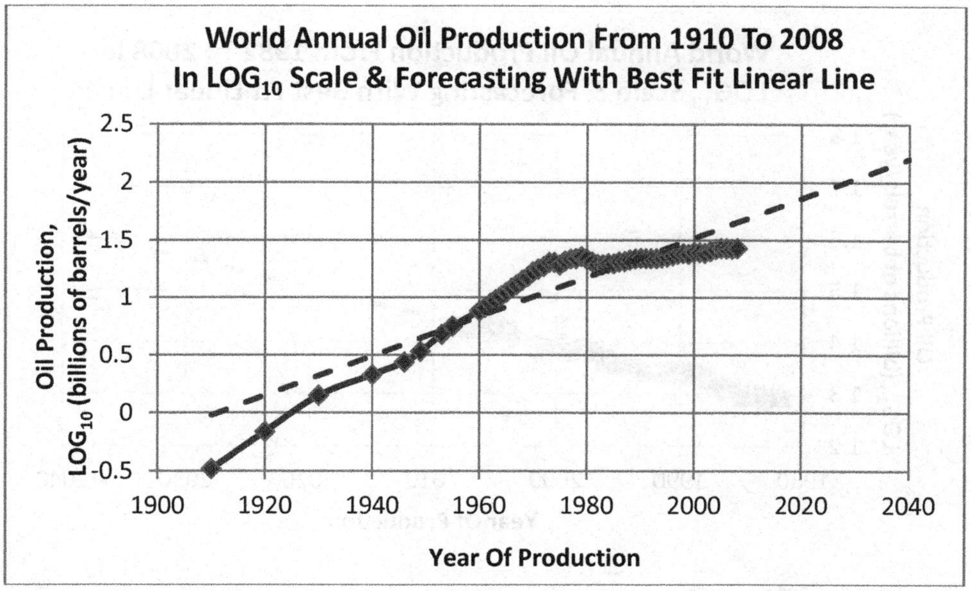

Figure 1-2-2 World annual oil production from 1910 to 2008 in LOG_{10} scale and forecasting with best fit linear line

For determining best fit curves, see Chapter 2-3. In Figure 1-2-2, the best fit linear line is

$$\log_{10} OIL\ PRODUCTION = 0.0171 \times YEAR - 32.681 \qquad (1\text{-}2\text{-}1)$$

with a correlation coefficient value of $r = 0.918$. See Chapter 2-3 for definition of correlation coefficient r . Extending the best fit line in Eq. (1-2-1) to year 2040 gives a LOG_{10} oil production value of 2.203 which is $10^{2.203} = 159.6\ billions\ of\ barrels\ per\ year$. Since the demand curve has a much shallower slope from 1982 on, it will be better to forecast the 2040 oil production by fitting the best fit linear line to data from 1982 to 2008. The analysis of data from 1982 to 2008 is shown in Figure 1-2-3.

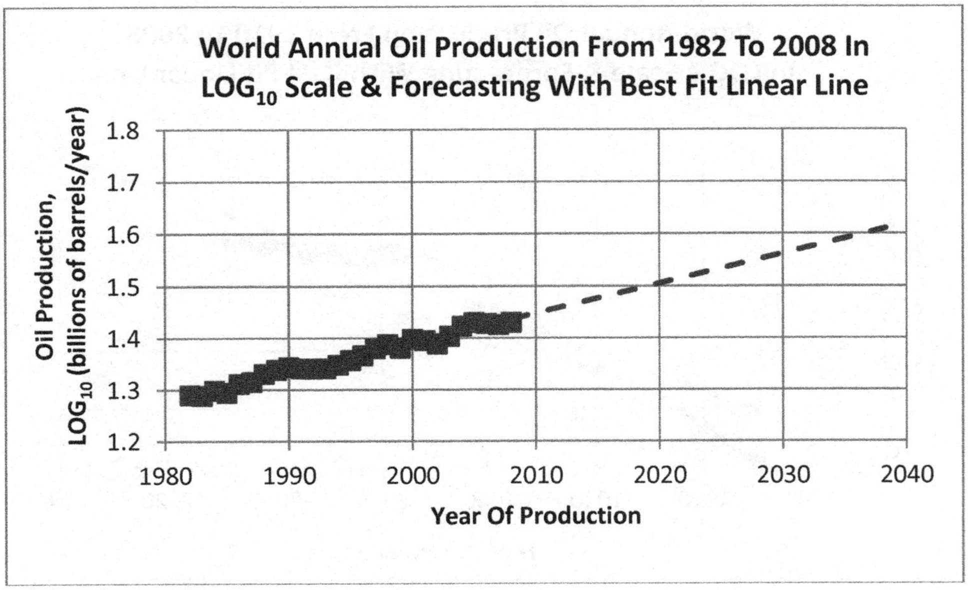

Figure 1-2-3 World annual oil production from 1982 to 2008 in LOG_{10} scale and forecasting with best fit linear line

In Figure 1-2-3, the best fit linear line is

$$\log_{10} OIL\ PRODUCTION = 0.05715 \times YEAR - 10.0386 \qquad (1\text{-}2\text{-}2)$$

with a correlation coefficient value of $r = 0.988$. Extending the best fit line in Eq. (1-2-2) to year 2040 gives a LOG_{10} oil production value of 1.62 which is $10^{1.62} = 41.7\ billions\ of\ barrels\ per\ year$. This is a much more reasonable forecast of oil production in the year 2040. However, this forecast might still be too high due to constantly increasing usage of alternate energy sources and due to more efficient vehicles on the roads.

Chapter 1-3 Basal Metabolic Rate

Basal metabolic rate (BMR) is the energy that a human or an animal's body produces at rest. This energy that is produced at rest is just enough energy to maintain vital functions such as respiration, blood circulation, nervous system, etc. Basal metabolic rate is generally given in kilo calories per day or kcal/day.

In thermodynamics, one calorie is defined as the energy required to raise the temperature of one gram of water from 14.5 C to 15.5 C at one atmospheric pressure. In our daily lives, humans talk about food Calories (with capital C) and 1 food Calorie, 1 C, equals to 1,000 thermodynamic calories. When a man at a certain weight, height, and age says he need 1,700 Calories a day in order to maintain his weight, he actually means 1,700,000 thermodynamic calories.

BMR has been studied by many scientists from microbes to elephants. BMR varies many orders of magnitudes depending upon the weight of the living being. Most famous of these studies is the Kleiber's Law which correlates BMR to an animal's weight as follows.

$$BMR = 70 \times W^{0.75} \quad (1\text{-}3\text{-}1)$$

where BMR is in kcal/day and weight W is in kilograms. Figure 1-3-1 shows BMR of several animals (data from Nature, Volume 403, 10 February 2000) from a 0.02 kg mouse to a 3,500 kg elephant and also compares Kleiber's Law, Eq. (1-3-1).

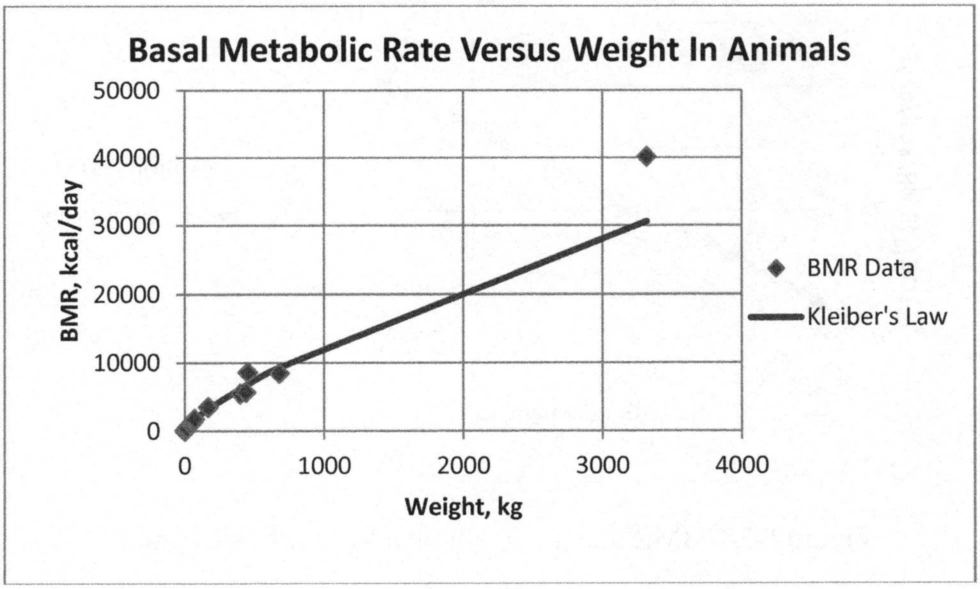

Figure 1-3-1 BMR versus weight in animals

Both the BMR range and the weight range in Figure 1-3-1 cover more than five orders of magnitude. Even if the orders of magnitude for weight is shrunk to show only small animals, shown in Figure 1-3-2, it is hard to see what is going on in the low end of the weight scale.

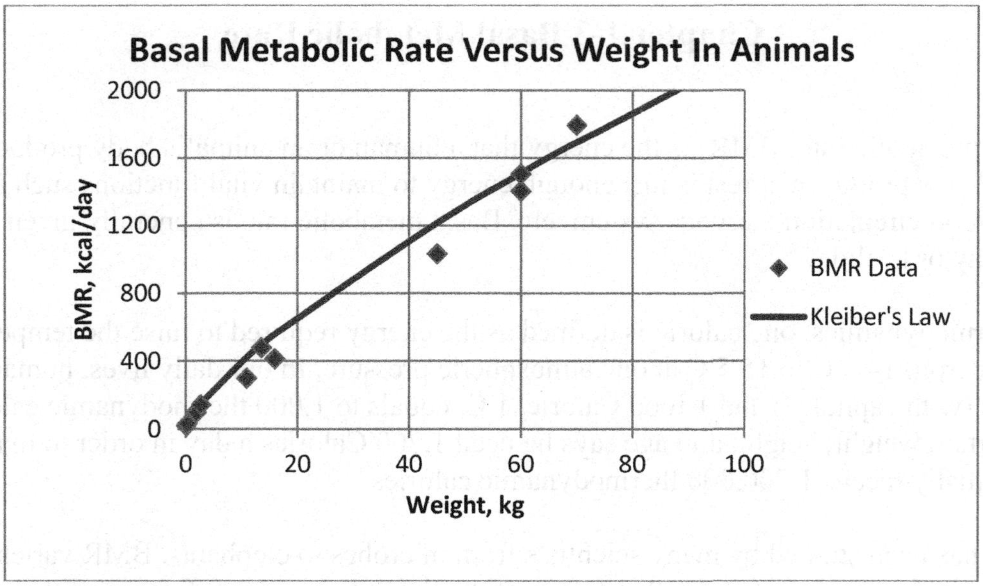

Figure 1-3-2 BMR versus weight in small animals

In this case where both axes are spanning many orders of magnitudes, it is best to represent them on a LOG_{10}-LOG_{10} graph as shown in Figure 1-3-3.

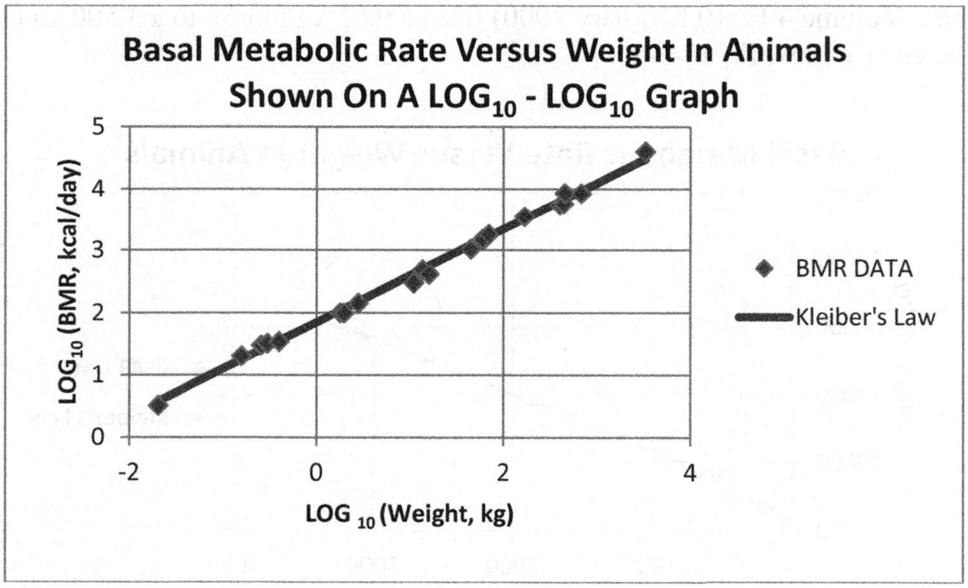

Figure 1-3-3 BMR versus weight on a LOG_{10}-LOG_{10} graph

As seen in Figure 1-3-3, Kleiber's Law, Eq. (1-3-1), represents BMR of animals from tiny ones to large ones very well. Whole range of animal weights and their BMRs can be seen with more ease on a LOG_{10}-LOG_{10} graph. As the animals get larger, they get more efficient in their energy usage and therefore they need less BMR. For example, an elephant can weigh 162,000 times more than a mouse, but it needs only 12,400 times more energy at rest than a mouse. BMR per weight is shown in Figure 1-3-4 using Kleiber's Law.

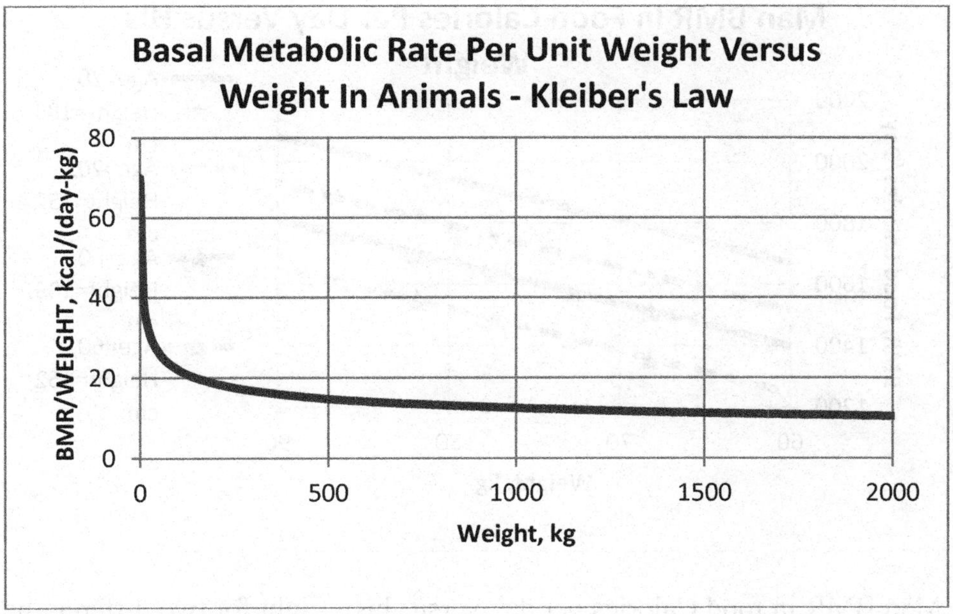

Figure 1-3-4 BMR per unit weight versus weight in animals by Kleiber's Law $\frac{BMR}{W} = 70 \times W^{-0.25}$

For humans, BMR predictions are mostly given in food Calories. Most popular human BMR prediction equations are Harris-Benedict linear equations (source: Cornell University Pediatric Critical Care Medicine, 9 March 2000).

For man:

$$BMR_{man} = 66.5 + 13.75 \times W + 5.003 \times H - 6.775 \times AGE \qquad (1\text{-}3\text{-}2)$$

For woman:

$$BMR_{woman} = 655.1 + 9.5663 \times W + 1.85 \times H - 4.676 \times AGE \qquad (1\text{-}3\text{-}3)$$

where BMR is in food Calories, weight W in kilograms, height H in centimeters, and AGE in years. BMR for man and for woman are shown in Figure 1-3-5 and Figure 1-3-6 respectively. BMR is plotted as a function of weight for two different heights and two different ages. BMR for woman is about 15% less than man. Man has to take in about 20% less Calories when he is 60 than when he is 20 years old. Woman has to take in about 15% less Calories when she is 60 than when she is 20 years old.

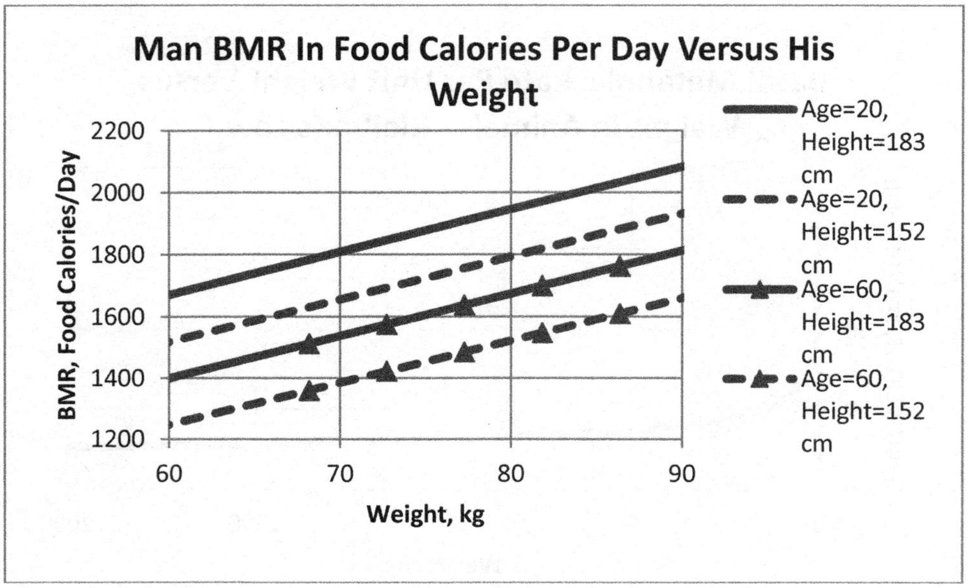

Figure 1-3-5 Man BMR in food Calories per day versus his weight for two different heights and for two different ages

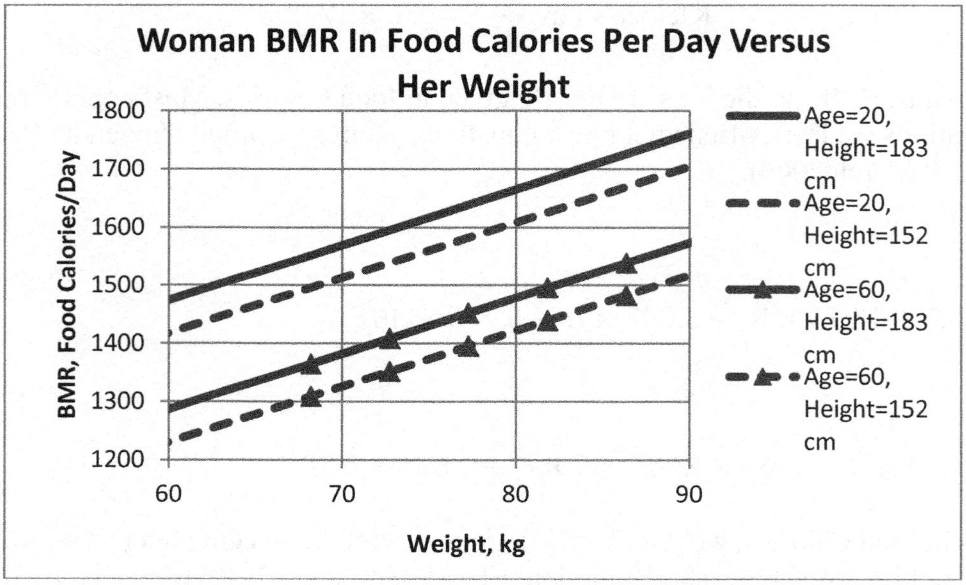

Figure 1-3-6 Woman BMR in food Calories per day versus her weight for two different heights and for two different ages

Chapter 1-4 Wind Chill Factor

Exposed human skin temperature can drop fast as the speed of the wind increases. Drops in skin temperature can cause frostbites and hypothermia. There have been many experiments that monitor the heat loss from human faces and the temperature of human faces under varying wind speeds and ambient air temperatures. The empirical wind chill temperature formula used by the US National Weather Service, Reference [2], is as follows.

$$T_{wc} = 35.74 + 0.6215T_a - 35.75V^{0.16} + 0.4275T_aV^{0.16} \quad (1\text{-}4\text{-}1)$$

where T_{wc} is the wind chill temperature in °F, T_a is the ambient (below calm wind speed of 3 mph) air temperature in °F, and V is the wind speed in miles per hour at a height of five feet. Wind chill temperature versus ambient air temperature at different wind speeds is shown in Figure 1-4-1.

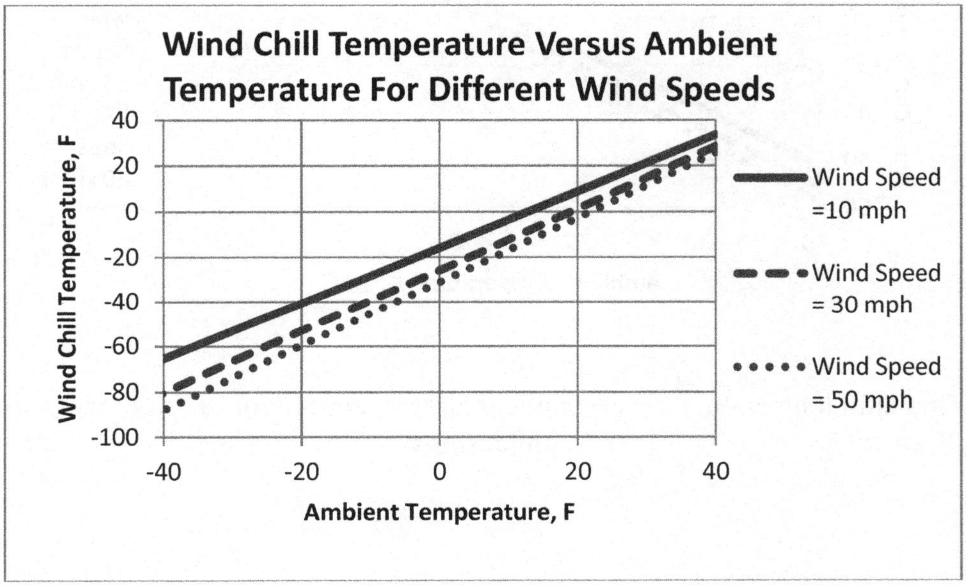

Figure 1-4-1 Wind chill temperature versus ambient air temperature for different wind speeds

For example, a -20 °F ambient air temperature goes down to -40 °F at a wind speed of 10 mph and down to -60 °F at a wind speed of 50 mph.

Eq. (1-4-1) can also be presented in metric form where the temperatures are in degrees centigrade and the wind speed is in kilometers per hour. Noting the dimensional conversions, $1.8 \times T_{°C} + 32 = T_{°F}$ and $V_{mph} = V_{kmph}/1.609$, Eq. (1-4-1) can be rewritten as

$$1.8T_{wc} + 32 = 35.74 + 0.6215(1.8T_a + 32) - 35.75\left(\frac{V_{kmph}}{1.609}\right)^{0.16} +$$

$$0.4275(1.8T_a + 32)\left(\frac{V_{kmph}}{1.609}\right)^{0.16} \quad (1\text{-}4\text{-}2)$$

where T_{wc} is the wind chill temperature in °C , T_a is the ambient (below calm wind speed of 5 kilometers per hour) air temperature in °C , and V is the wind speed in kilometers per hour at a height of 1.5 meters. Eq. (1-4-2) can be simplified to take the following form.

$$T_{wc} = 13.12 + 0.6215T_a - 11.37V^{0.16} + 0.3965T_aV^{0.16} \quad (1\text{-}4\text{-}3)$$

Wind chill temperature versus ambient air temperature at different wind speeds in metric dimensions, Eq. (1-4-3), is shown in Figure 1-4-2.

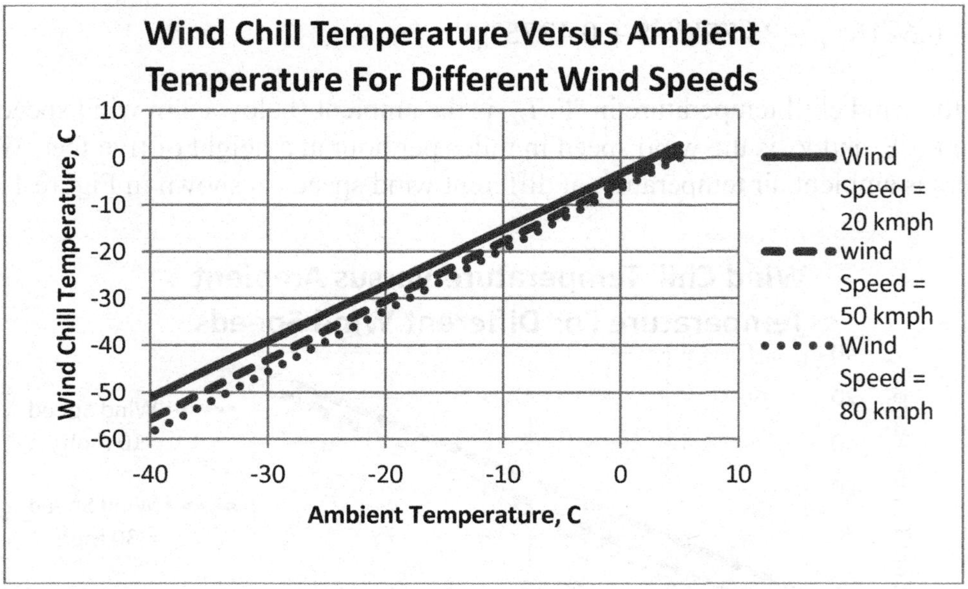

Figure 1-4-2 Wind chill temperature versus ambient air temperature for different wind speeds in metric dimensions

Chapter 1-5 Purity and Mass of a Solid Gold Bar

Precious metal such as gold has a completely unique metal purity (fineness) system and mass system. For many centuries in many countries gold's purity was quantified by carat. A carat is defined as

$$Carat = 24 \times \frac{Mass\ of\ Gold}{Total\ Mass} \qquad (1\text{-}5\text{-}1)$$

A 24-carat gold has a minimum of 99% purity. An 18-carat gold has a minimum of 75% gold and 25% other metals mixed with it such as copper or silver. Since gold is a very precious metal, about $35 per gram at the time of this write up on February 2010, a more precise gold purity system has been adopted. It is called a millesimal gold purity system in which the amount of gold in a mixture is defined to tenth of a percent. The millesimal gold purity system is shown in Table 1-5-1.

CARAT	MILLESIMAL	% GOLD
24	999	99.9
22	916	91.6
20	833	83.3
18	750	75.0
15	625	62.5
14	585	58.5
10	417	41.7
9	375	37.5
8	333	33.3
1	42	0.42

Table 1-5-1 Millesimal gold purity (fineness) system

In addition to its different purity system, the mass system for gold is also different than the standard metric mass system of kilogram. Let us calculate the mass of a 24-carat gold bar (a rectangular parallelepiped) from its length L, width W, and height H dimensions. The mass of the gold bar is its density ρ in grams per cm^3 times its volume V in cm^3.

$$Mass = \rho V = \rho \times L \times W \times H \qquad (1\text{-}5\text{-}2)$$

Consider a gold bar whose dimensions and density are as follows:

$$L = 7\ inches, W = 3.625\ inches, H = 1.75\ inches, and\ \rho = 19.32\ grams/cm^3$$

Inserting above inputs into Eq. (1-5-2) gives

$$Mass = 19.32\ \frac{g}{cm^3}(7 \times 2.54)(3.625 \times 2.54)(1.75 \times 2.54)cm^3 = 14{,}059\ g$$

or

$$Mass = 14.059\ kg = 14.059\ kg \times \frac{1\ pound}{0.45359237\ kg} = 31.0\ pounds$$

Since there are 16 ounces in a pound, the mass of this gold bar is 496 ounces. However, these masses are in standard avoirdupois pounds and in standard avoirdupois ounces, namely 1 avoirdupois pound = 0.45359237 kilograms and 1 avoirdupois ounce = 28.34952313 grams.
The gold mass system uses troy ounces as a standard which comes from the medieval Roman times.

$$1\ troy\ ounce = 31.1034768\ grams$$

This gold bar has a mass of (14058/31.1) = 452 troy ounces.

If you are buying this gold bar, the sellers usually quote you in $ per troy ounce prices, namely $35 per gram x 31.1034768 grams per troy ounce = $1,088.62 troy ounce. If the seller does not mention the $ per troy ounce price and mentions only $ per ounce price and by mistake multiplies $ per ounce price by the mass of 496 avoirdupois ounces, you would pay an extra $47,800.37 for this gold bar. It would be a 9.7% overcharge by the seller.

Gold bar price using troy ounces:

$$Price = 1{,}088.62 \frac{\$}{ounce} \times 452\ troy\ ounces = \$492{,}062.67$$

Gold bar price using avoirdupois ounces:

$$Price = 1{,}088.62 \frac{\$}{ounce} \times 496\ avoirdupois\ ounces = \$539{,}863.04$$

Chapter 1-6 Head and Tail Wind Effects to Airplane Travel Time

As the airplane travel times get longer and the flying altitudes get higher, head winds and tail winds affect the airplane travel times tremendously. As an example, let us consider a direct flight from Los Angeles to London and back to Los Angeles.

Our captain informs us that in still air, the jet airplane's average speed for this trip is 490 air miles per hour. The captain is going to use the same air travel route both ways. Travel time from Los Angeles to London is 10 hours and 12 minutes and from London to Los Angeles 12 hours and 21 minutes. What is the average head and tail wind speed? What is the air travel distance between Los Angeles and London?

Since the trip from Los Angeles to London takes a shorter time, the airplane should experience a tail wind during this portion of the trip. The airplane should experience a head wind during the return trip. The air travel distance between Los Angeles to London is total average speed times travel time.

$$S = (V + W) \times T_{Los\ Angeles\ to\ London} \qquad (1\text{-}6\text{-}1)$$

where S is the air travel distance between Los Angeles and London in miles, V is the average airplane speed in still air in miles per hour, W is the average wind speed in miles per hour, and T is the travel time in hours between Los Angeles and London.

For the return trip with a head wind, assuming the same air travel distance, Eq. (1-6-1) takes the following form.

$$S = (V - W) \times T_{London\ to\ Los\ Angeles} \qquad (1\text{-}6\text{-}2)$$

Eqs. (1-6-1) and (1-6-2) can be equated and solved for the average wind speed.

$$W = V \times \frac{(T_{London\ to\ Los\ Angeles} - T_{Los\ Angeles\ to\ London})}{(T_{London\ to\ Los\ Angeles} + T_{Los\ Angeles\ to\ London})} \qquad (1\text{-}6\text{-}3)$$

In Eq. (1-6-3) times can be expressed in minutes, since minutes dimensions cancel out. The average wind speed is calculated to be

$$W = 490\ mph \times \frac{(741-612)minutes}{(741+612)minutes} = 46.72\ mph \qquad (1\text{-}6\text{-}4)$$

Either Eq. (1-6-1) or Eq. (1-6-2) can be used to find the air travel distance between Los Angeles and London. However, travel time has to be expressed in hours.

$$S = (490 + 46.72)mph \times \left(\frac{612}{60}\right) hours = 5{,}475\ miles \qquad (1\text{-}6\text{-}5)$$

Chapter 1-7 Cost Benefit Of Cooking At Home Or Buying Cooked Food

Cost of cooking at home or buying cooked food outside for one person is analyzed. Same analysis is extended to five persons eating the same food. Costs for one person are shown in Table 1-7-1. Buying cooked food costs are modeled from a moderately priced takeout restaurant. Costs for five persons are shown in Table 1-7-2.

COST OF COOKING AT HOME FOR ONE PERSON			
ITEM	**UNIT COST**	**USAGE QTY**	**TOTAL COST**
Chicken Breast	$3.00	1	$3.00
Rice	$0.50	1	$0.50
Green Salad	$0.50	1	$0.50
Fruit Plate	$0.75	1	$0.75
Natural Gas	$1.00/therm	0.50 therms	$0.50
Water	$0.04/feet^3	5.00 feet^3	$0.20
Electricity	$0.14/kWh	1.00 kWh	$0.14
Preparation Effort	$30.00/hr	1.00 hour	$30.00
	GRAND TOTAL COST		**$35.59**
COST OF BUYING COOKED FOOD OUTSIDE FOR ONE PERSON			
Chicken Breast	$6.00	1	$6.00
Rice	$1.50	1	$1.50
Green Salad	$2.00	1	$2.00
Fruit Plate	$2.00	1	$2.00
Car Costs	$1.00/mile	4 miles	$4.00
Driving & Pickup Effort	$30.00/hr	0.50 hours	$15.00
	GRAND TOTAL COST		**$30.50**

Table 1-7-1 Cost of cooking at home or buying cooked food outside for one person

COST OF COOKING AT HOME FOR FIVE PERSONS			
ITEM	UNIT COST	USAGE QTY	TOTAL COST
Chicken Breast	$3.00	5	$15.00
Rice	$0.50	5	$2.50
Green Salad	$0.50	5	$2.50
Fruit Plate	$0.75	5	$3.75
Natural Gas	$1.00/therm	0.50 therms	$0.50
Water	$0.04/feet^3	10.00 feet^3	$0.40
Electricity	$0.14/kWh	2.00 kWh	$0.28
Preparation Effort	$30.00/hr	1.50 hour	$45.00
	GRAND TOTAL COST		**$69.93**
COST OF BUYING COOKED FOOD OUTSIDE FOR FIVE PERSONS			
Chicken Breast	$6.00	5	$30.00
Rice	$1.50	5	$7.50
Green Salad	$2.00	5	$10.00
Fruit Plate	$2.00	5	$10.00
Car Costs	$1.00/mile	4 miles	$4.00
Driving & Pickup Effort	$30.00/hr	0.60 hours	$18.00
	GRAND TOTAL COST		**$79.50**

Table 1-7-2 Cost of cooking at home or buying cooked food outside for five persons

Unit costs and food consumption quantities in Tables 1-7-1 and 1-7-2 are typical. These values can be adjusted to one's personal preferences. Home cooking costs and buying cooked food outside costs versus number of persons are plotted in Figure 1-7-1. Food costs of one and five persons are assumed to be connected linearly, namely by a straight line. Buying cooked food costs starts lower than the home cooking costs and crosses over the home cooking costs around 2.4 persons. So in this model, home cooking is more feasible, for three or more persons.

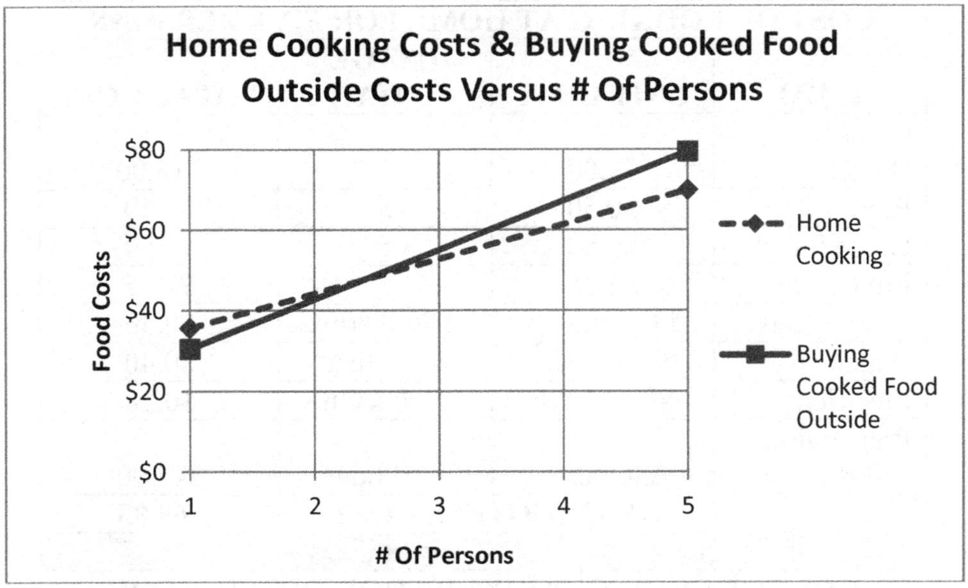

Figure 1-7-1 Home cooking costs and buying cooked food outside costs versus number of persons

Assuming that food costs increase linearly, linear equations for home cooking and for buying cooked food outside can be determined and then these linear equations can be extrapolated to higher number of persons.

Let us designate number of persons, the independent variable, with X. Let us also designate grand total cost, the dependent variable, with C. Linear equation for grand total cost versus number of persons is

$$C = aX + b \qquad (1\text{-}7\text{-}1)$$

For home cooking, the coefficients $a\ and\ b$ in Eq. (1-7-1) are determined from the following known values in Tables 1-7-1 and 1-7-2.

$$\$35.59 = a \times 1 + b \qquad (1\text{-}7\text{-}2)$$

$$\$69.93 = a \times 5 + b \qquad (1\text{-}7\text{-}3)$$

Solving Eqns. (1-7-2) and (1-7-3) for $a\ and\ b$ gives,

$$a = 8.585\ \$/person\ and \qquad b = \$27.01$$

or for home cooking, grand total cost versus number of persons is estimated by the following linear equation.

$$C = 8.585X + 27.01 \qquad (1\text{-}7\text{-}4)$$

For buying cooked food outside, the coefficients $a\ and\ b$ in Eq. (1-7-1) are determined from the following known values in Tables 1-7-1 and 1-7-2.

$$\$30.50 = a \times 1 + b \qquad (1\text{-}7\text{-}5)$$

$$\$79.50 = a \times 5 + b \tag{1-7-6}$$

Solving Eqns. (1-7-5) and (1-7-6) for a and b gives,

$a = 12.25\ \$/person\ and \quad b = \$18.25$

or for buying cooked food outside, grand total cost versus number of persons is estimated by the following linear equation.

$$C = 12.25X + 18.15 \tag{1-7-7}$$

Home cooking costs and buying cooked food outside costs can be extrapolated to higher number of persons by assuming that these linear relationships in Eqs. (1-7-4) and (1-7-7) hold for higher number of persons. Extrapolated results to higher number of persons are shown in Figure 1-7-2. For example, for a dozen persons, it costs \$130.03 to cook at home and \$165.25 to buy cooked food outside.

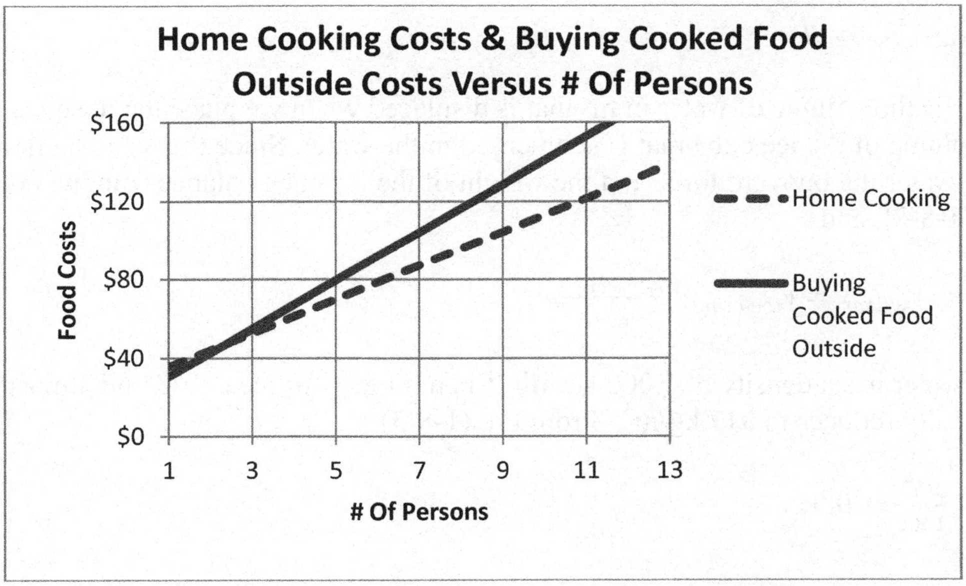

Figure 1-7-2 Home cooking costs and buying cooked food outside costs versus number of persons extrapolated to higher number of persons

Chapter 1-8 Volume of Ice below Sea Surface

Water molecules form ice crystals which are less dense and therefore the volume of water expands when it freezes at zero degrees centigrade at atmospheric pressure. When we put a cube of ice into our glass of water, what volume of ice is exposed?

The ice cube has a weight, $W_{ice\ cube}$, which forces it down the water in the glass which is jointly proportional to the density of the ice cube times the volume of the ice cube.

$$W_{ice\ cube} = \rho_{ice\ cube} V_{ice\ cube}\, g \qquad (1\text{-}8\text{-}1)$$

where $\rho_{ice\ cube}$ is the density of ice cube in kg/m^3, $V_{ice\ cube}$ is the volume of ice cube in m^3, and g is the gravitational acceleration 9.8 m/s^2. The ice cube is buoyed up by a buoyant force, $F_{buoyant}$, which has a magnitude that is equal to the weight of the water displaced by the ice cube. The buoyant force is also proportional to the density of the water times the volume of water displaced by the ice cube.

$$F_{buoyant} = \rho_{water} V_{water}\, g \qquad (1\text{-}8\text{-}2)$$

where V_{water} is the volume of water in m^3 that is displaced when we place the ice cube in water which is the volume of the ice cube that is submerged in the water. Since the ice cube floats in equilibrium in water the buoyant force and the weight of the ice cube balances out, namely Eq. (1-8-1) equals to Eq. (1-8-2), and

$$V_{ice\ submerged} = V_{water} = V_{ice\ cube} \times \frac{\rho_{ice\ cube}}{\rho_{water}} \qquad (1\text{-}8\text{-}3)$$

Drinkable tap water has a density of 1,000 kg/m^3. When it becomes ice at 0°C and atmospheric pressure its density reduces to 917 kg/m^3. From Eq. (1-8-3)

$$\frac{V_{ice\ submerged}}{V_{ice\ cube}} = \frac{917}{1000} = 0.917$$

Therefore 91.7% of the ice cube volume is submerged in water and 8.3% of the ice cube volume is above the surface of the water.

Similar analysis, Eq. (1-8-3), can be applied to icebergs, but this time the densities are different due to salt content of the seas. Typical sea water has a density of 1,030 kg/m^3, namely 30 kg of salt is added to one m^3 of fresh water to make up the salty sea water. Sea water density varies from ocean to ocean and from sea to sea depending upon how much fresh water mixes up with it from rivers, ice melts, etc. The iceberg density is very close to frozen fresh water density, about 920 kg/m^3, since most icebergs split out of glaciers.

$$\frac{V_{iceberg\ submerged}}{V_{iceberg}} = \frac{\rho_{iceberg}}{\rho_{sea\ water}} = \frac{920}{1030} = 0.893$$

Therefore 89.3% of the iceberg volume is submerged in sea water and 10.7% of the iceberg volume is above the surface of the sea water.

Chapter 1-9 Required Fertilizer Nutrients

A farmer wants to fertilize his lawns and trees. The lawns need 3-1-2 combination of nitrogen, phosphorus, and potassium by weight, i.e. in a six pound bag the farmer should have 3 lbs of nitrogen, 1 lb of phosphorus, and 2 lbs of potassium. Trees need 2-1-1 combination of nitrogen, phosphorus, and potassium by weight. He needs 50 pounds of fertilizes for the lawns and 100 pounds of fertilizer for the trees.

He has three types of fertilizer options. Option X is a 15-5-5, percentage of nitrogen, phosphorus, and potassium by weight in a bag, fertilizer which comes in a 40 pound bag at $20 each bag. Option Y is a 15-15-5, percentage of nitrogen, phosphorus, and potassium by weight in a bag, fertilizer which comes in a 20 pound bag at $15 each bag. Option Z is a 10-0-10, percentage of nitrogen, phosphorus, and potassium by weight in a bag, fertilizer which comes in a 20 pounds bag at $10 each bag.

How many bags of each type should he buy to fertilize his lawns and trees?

For the lawns, each amount of nutrient needed is as follows:

$$Nitrogen_{lawn} = \left(\frac{3}{6}\right) \times 50 = 25\ pounds \tag{1-9-1}$$

$$Phosphorus_{lawn} = \left(\frac{1}{6}\right) \times 50 = 8.3\ pounds \tag{1-9-2}$$

$$Potassium_{lawn} = \left(\frac{2}{6}\right) \times 50 = 16.7\ pounds \tag{1-9-3}$$

For the trees, each amount of nutrient needed is as follows:

$$Nitrogen_{tree} = \left(\frac{2}{4}\right) \times 100 = 50\ pounds \tag{1-9-4}$$

$$Phosphorus_{tree} = \left(\frac{1}{4}\right) \times 100 = 25\ pounds \tag{1-9-5}$$

$$Potassium_{tree} = \left(\frac{1}{4}\right) \times 100 = 25\ pounds \tag{1-9-6}$$

Total amount of each nutrient needed for fertilizing the lawns and the trees are obtained by summing the appropriate Eqns. (1-9-1) through (1-9-6).

$$Nitrogen_{total} = Nitrogen_{lawn} + Nitrogen_{tree} = 75\ pounds \tag{1-9-7}$$

$$Phosphorus_{total} = Phosphorus_{lawn} + Phosphorus_{tree} = 33.3\ pounds \tag{1-9-8}$$

$$Potassium_{total} = Potassium_{lawn} + Potassium_{tree} = 41.7\ pounds \tag{1-9-9}$$

Next let us analyze to see what nutrients are available in each $X, Y, and\ Z$ fertilizer bags by weight.

$$Nitrogen_X = \left(\frac{15}{100}\right) \times 40 = 6\ pounds \qquad (1\text{-}9\text{-}10)$$

$$Phosphorus_X = \left(\frac{5}{100}\right) \times 40 = 2\ pounds \qquad (1\text{-}9\text{-}11)$$

$$Potassium_X = \left(\frac{5}{100}\right) \times 40 = 2\ pounds \qquad (1\text{-}9\text{-}12)$$

$$Nitrogen_Y = \left(\frac{15}{100}\right) \times 20 = 3\ pounds \qquad (1\text{-}9\text{-}13)$$

$$Phosphorus_Y = \left(\frac{15}{100}\right) \times 20 = 3\ pounds \qquad (1\text{-}9\text{-}14)$$

$$Potassium_Y = \left(\frac{5}{100}\right) \times 20 = 1\ pound \qquad (1\text{-}9\text{-}15)$$

$$Nitrogen_Z = \left(\frac{10}{100}\right) \times 20 = 2\ pounds \qquad (1\text{-}9\text{-}16)$$

$$Phosphorus_Z = \left(\frac{0}{100}\right) \times 20 = 0\ pounds \qquad (1\text{-}9\text{-}17)$$

$$Potassium_Z = \left(\frac{10}{100}\right) \times 20 = 2\ pounds \qquad (1\text{-}9\text{-}18)$$

We now have all the coefficients needed in Eqs. (1-9-7) through (1-9-18) to set up the following three linear equations.
For nitrogen requirements

$$6X + 3Y + 2Z = 75 \qquad (1\text{-}9\text{-}19)$$

For phosphorus requirements

$$2X + 3Y = 33.3 \qquad (1\text{-}9\text{-}20)$$

For potassium requirements

$$2X + Y + 2Z = 41.7 \qquad (1\text{-}9\text{-}21)$$

Eqs. (1-9-19) through (1-9-21) is a system of three equations and three unknowns. Solving these equations will provide a single solution for $X, Y, and\ Z$, if the determinant of the coefficients of the unknowns is different than zero. This is known in literature as the Cramer's rule, see Reference [3].

For a single solution for $X, Y, and\ Z$, the following has to be true

$$\begin{vmatrix} 6 & 3 & 2 \\ 2 & 3 & 0 \\ 2 & 1 & 2 \end{vmatrix} \neq 0 \qquad (1\text{-}9\text{-}22)$$

Let us find the value of the 3x3 determinant in Eq. (1-9-22). See Reference [3] for evaluating determinants.

$$6\begin{vmatrix}3 & 0\\1 & 2\end{vmatrix} - 2\begin{vmatrix}3 & 2\\1 & 2\end{vmatrix} + 2\begin{vmatrix}3 & 2\\3 & 0\end{vmatrix} = 6(3x2 - 0x1) - 2(3x2 - 2x1) + 2(3x0 - 2x3) = 16 \qquad (1\text{-}9\text{-}23)$$

Since the determinant of the coefficients of the unknowns in Eq. (1-9-22) is different than zero, there is single solution for $X, Y, and\ Z$. If the value of the determinant in Eq. (1-9-22) comes out to zero, then there can be many solutions or no solution to the system of equations (1-9-19), (1-9-20), and (1-9-21).

X is determined by replacing the X coefficients of the determinant in Eq. (1-9-22) by the right hand side values of Eqs. (1-9-19), (1-9-20), and (1-9-21), and dividing it by the determinant value obtained in Eq. (1-9-23).

X is

$$\frac{\begin{vmatrix}75 & 3 & 2\\33.3 & 3 & 0\\41.7 & 1 & 2\end{vmatrix}}{16} = \frac{\left[75\begin{vmatrix}3 & 0\\1 & 2\end{vmatrix} - 33.3\begin{vmatrix}3 & 2\\1 & 2\end{vmatrix} + 41.7\begin{vmatrix}3 & 2\\3 & 0\end{vmatrix}\right]}{16} = \frac{66.6}{16} = 4.2 \qquad (1\text{-}9\text{-}24)$$

Y is determined by replacing the Y coefficients of the determinant in Eq. (1-9-20) by the right hand side values of Eqs. (1-9-19), (1-9-20), and (1-9-21), and dividing the value of the determinant by the value obtained in Eq. (1-9-23).

Y is

$$\frac{\begin{vmatrix}6 & 75.0 & 2\\2 & 33.3 & 0\\2 & 41.7 & 2\end{vmatrix}}{16} = \frac{\left[6\begin{vmatrix}33.3 & 0\\41.7 & 2\end{vmatrix} - 2\begin{vmatrix}75.0 & 2\\41.7 & 2\end{vmatrix} + 2\begin{vmatrix}75.0 & 2\\33.3 & 0\end{vmatrix}\right]}{16} = \frac{133.3}{16} = 8.3 \qquad (1\text{-}9\text{-}25)$$

Z is determined by replacing the Z coefficients of the determinant in Eq. (1-9-20) by the right hand side values of Eqs. (1-9-19), (1-9-20), and (1-9-21), and dividing the value of the determinant by the value obtained in Eq. (1-9-23).

Z is

$$\frac{\begin{vmatrix}6 & 3 & 75.0\\2 & 3 & 33.3\\2 & 1 & 41.7\end{vmatrix}}{16} = \frac{\left[6\begin{vmatrix}3 & 33.3\\1 & 41.7\end{vmatrix} - 2\begin{vmatrix}3 & 75.0\\1 & 41.7\end{vmatrix} + 2\begin{vmatrix}3 & 75.0\\3 & 33.3\end{vmatrix}\right]}{16} = \frac{200}{16} = 12.5 \qquad (1\text{-}9\text{-}26)$$

Since the farmer cannot buy partial bags, he has to round up the values in Eqs. (1-9-24), (1-9-25), and (1-9-26). He has to buy 5 bags of fertilizer X , 9 bags of fertilizer Y , and 13 bags of fertilizer Z at a cost of $365. Total available nutrients to him are given in Table 1-9-1.

Fertilizer Type	# of bags	Nitrogen, lb	Phosphorus, lb	Potassium, lb
X	5	30	10	10
Y	9	27	27	9
Z	13	26	0	26
Totals In Bags	27	83	37	45
Total Required Nutrients		75	33.3	41.7

Table 1-9-1 Fertilizer bags $X, Y, and\ Z$ to be purchased to meet total nutrient requirements

Next let us determine the number of $X, Y, and\ Z$ fertilizer bags needed to satisfy the lawn nutrient requirements.

X fertilizer bags needed for lawn nutrient requirements can be determined by replacing the X coefficients of the determinant in Eq. (1-9-22) by the lawn nutrient values in Eqs. (1-9-1), (1-9-2), and (1-9-3), and dividing it by the determinant value obtained in Eq. (1-9-23).

$$X_{lawn} = \frac{\begin{vmatrix} 25.0 & 3 & 2 \\ 8.3 & 3 & 0 \\ 16.7 & 1 & 2 \end{vmatrix}}{16} = 1.0 \quad (1\text{-}9\text{-}27)$$

Y fertilizer bags needed for lawn nutrient requirements can be determined by replacing the Y coefficients of the determinant in Eq. (1-9-22) by the lawn nutrient values in Eqs. (1-9-1), (1-9-2), and (1-9-3), and dividing it by the determinant value obtained in Eq. (1-9-23).

$$Y_{lawn} = \frac{\begin{vmatrix} 6 & 25.0 & 2 \\ 2 & 8.3 & 0 \\ 2 & 16.7 & 2 \end{vmatrix}}{16} = 2.1 \quad (1\text{-}9\text{-}28)$$

Z fertilizer bags needed for lawn nutrient requirements can be determined by replacing the Z coefficients of the determinant in Eq. (1-9-22) by the lawn nutrient values in Eqs. (1-9-1), (1-9-2), and (1-9-3), and dividing it by the determinant value obtained in Eq. (1-9-23).

$$Z_{lawn} = \frac{\begin{vmatrix} 6 & 3 & 25.0 \\ 2 & 3 & 8.3 \\ 2 & 1 & 16.7 \end{vmatrix}}{16} = 6.3 \quad (1\text{-}9\text{-}29)$$

Next let us determine the number of $X, Y, and\ Z$ fertilizer bags needed to satisfy the tree nutrient requirements.

X fertilizer bags needed for tree nutrient requirements can be determined by replacing the X coefficients of the determinant in Eq. (1-9-22) by the tree nutrient values in Eqs. (1-9-4), (1-9-5), and (1-9-6), and dividing it by the determinant value obtained in Eq. (1-9-23).

$$X_{tree} = \frac{\begin{vmatrix} 50 & 3 & 2 \\ 25 & 3 & 0 \\ 25 & 1 & 2 \end{vmatrix}}{16} = 3.1 \quad (1\text{-}9\text{-}30)$$

Y fertilizer bags needed for tree nutrient requirements can be determined by replacing the Y coefficients of the determinant in Eq. (1-9-22) by the tree nutrient values in Eqs. (1-9-4), (1-9-5), and (1-9-6), and dividing it by the determinant value obtained in Eq. (1-9-23).

$$Y_{tree} = \frac{\begin{vmatrix} 6 & 50 & 2 \\ 2 & 25 & 0 \\ 2 & 25 & 2 \end{vmatrix}}{16} = 6.3 \qquad (1\text{-}9\text{-}31)$$

Z fertilizer bags needed for tree nutrient requirements can be determined by replacing the Z coefficients of the determinant in Eq. (1-9-22) by the tree nutrient values in Eqs. (1-9-4), (1-9-5), and (1-9-6), and dividing it by the determinant value obtained in Eq. (1-9-23).

$$Z_{tree} = \frac{\begin{vmatrix} 6 & 3 & 50 \\ 2 & 3 & 25 \\ 2 & 1 & 25 \end{vmatrix}}{16} = 6.3 \qquad (1\text{-}9\text{-}32)$$

Chapter 1-10 Minimizing Costs for Fertilizer Mixing

Minimum cost for a fertilizer mix requirement can be determined by writing down the linear constraints for necessary nutrient mix and the cost function that has to be minimized. These linear equations are graphed and the optimum cost point that satisfies all the constraints is found. This procedure is called linear programming in literature, see Reference [3].

For example, a farmer wants to fertilize his trees by using a 2-1-1 weight proportions for nitrogen, phosphorus, and for potassium. With this weight proportions he needs 200 pounds of nitrogen, 100 pounds of phosphorus, and 100 pounds of potassium.

Two types of fertilizers bags are available to him.

Type X fertilizer bag weighs 40 pounds and it contains 30% nitrogen, 8% phosphorus, and 2% potassium and it costs $20 per bag.

Type Y fertilizer bag weighs 20 pounds and it contains 7% nitrogen, 4% phosphorus, and 9% potassium and it costs $10 per bag.

Type X fertilizer bag contains $40 \times 0.3 = 12$ pounds of nitrogen, $40 \times 0.08 = 3.2$ pounds of phosphorus, and $40 \times 0.02 = 0.8$ pounds of potassium.

Type Y fertilizer bag contains $20 \times 0.07 = 1.4$ pounds of nitrogen, $20 \times 0.04 = 0.8$ pounds of phosphorus, and $20 \times 0.09 = 1.8$ pounds of potassium.

Constraint equations for the necessary nutrient mix are

$$X \geq 0 \; for \; positive \; X \; fertilizer \; bags, \tag{1-10-1a}$$

$$Y \geq 0 \; for \; positive \; Y \; fertilizer \; bags, \tag{1-10-1b}$$

$$12X + 1.4Y \geq 200 \; for \; required \; nitrogen, \tag{1-10-1c}$$

$$3.2X + 0.8Y \geq 100 \; for \; required \; phosphorus, and \tag{1-10-1d}$$

$$0.8X + 1.8Y \geq 100 \; for \; required \; potassium \tag{1-10-1e}$$

Cost function that has to be minimized is

$$COST = 20X + 10Y \tag{1-10-2}$$

Linear equations (1-10-1a) through (1-10-1e) can be plotted and the feasible region that satisfies all these equations can be found as illustrated in Figure 1-10-1.

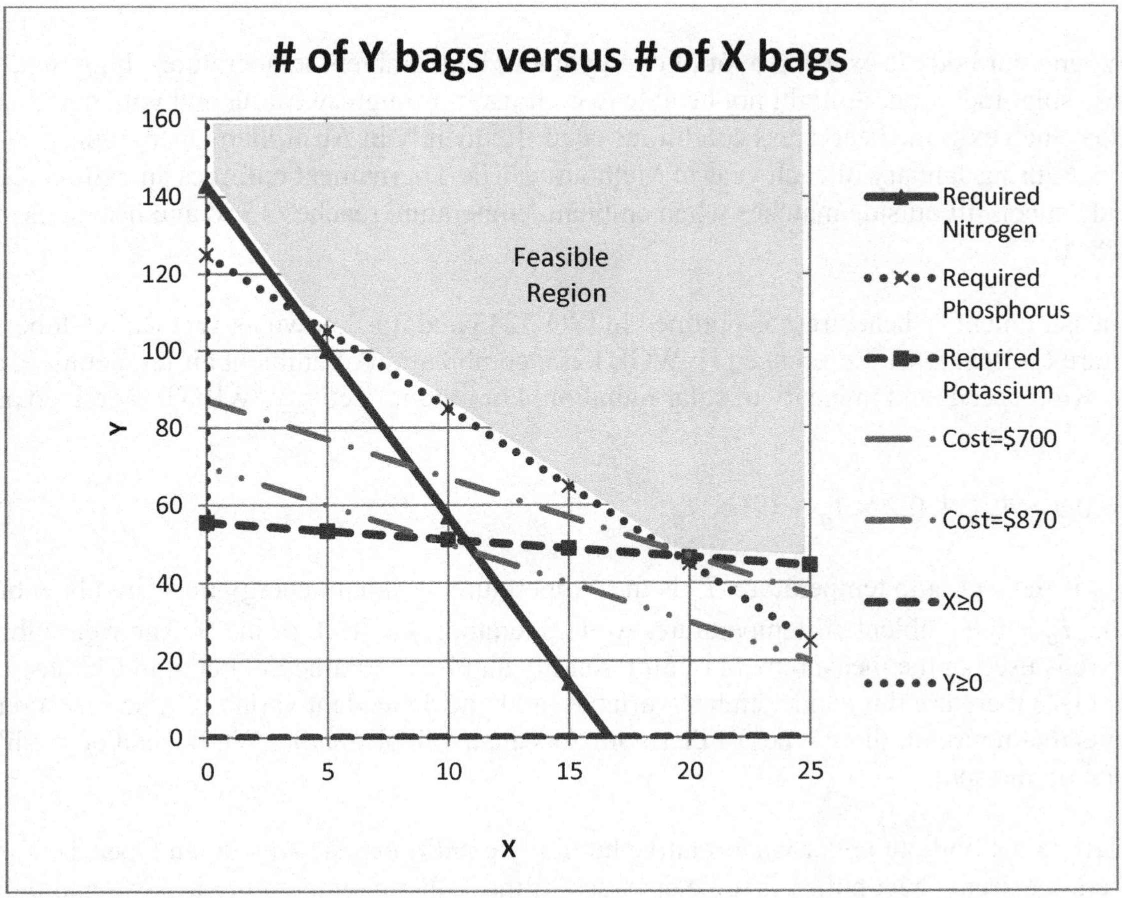

Figure 1-10-1 Number of Y bags versus number of X bags and the feasible region

Figure 1-10-1 also gives plots of Eq. (1-10-2) for two different cost values. When the cost value increases from \$700 to \$870, the line that represents the cost function starts to go into the feasible region of required fertilizer nutrients. The cost function always has the same slope. The cost function, Eq. (1-10-2), enters the feasible region at the intersection of required phosphorus constraint line, Eq. (1-10-1d), and required potassium constraint line, Eq. (1-10-1e).

The intersection of required phosphorus constraint line, Eq. (1-10-1d), and required potassium constraint line, Eq. (1-10-1e) is at $X = 19.5$ *and* $Y = 46.9$. Therefore, by rounding up these values, 20 bags of fertilizer *X* are required and 47 bags of fertilizer *Y* are required to meet all the constraints in Eqs. (1-10-1a) through (1-10-1e) at a minimum cost of \$870.

Chapter 1-11 Heat Stress

When your body is exerting a lot of energy under high ambient temperatures, high humidity, and intense solar radiation, it might not be able to cool itself through sweating and you might have a heat stroke. Such extreme heat stress conditions occur frequently in Australian Open Tennis Tournament during January of each year in Melbourne. The Tournament enforces an extreme heat policy and cancels all outside matches when ambient temperature reaches 35 °C and heat stress level reaches 28 °C.

Measurement of heat stress is outlined in ISO-7243 and it is known as Wet Bulb Globe Temperature (WBGT), see Reference [4]. WGBT is a combination of ambient air temperature, air humidity, wind speed, and intensity of solar radiation. For outdoor activity, WBGT is defined as follows.

$$WGBT = 0.7 \times T_{wb} + 0.2 \times T_g + 0.1 \times T_a \qquad (1\text{-}11\text{-}1)$$

where T_{wb} is the wet bulb temperature, T_g is the temperature of radiant energy measured by a black bulb globe, T_a is the ambient air temperature. All temperatures are in °C or in °F. The wet bulb temperature is used in the measurement of air humidity and it is explained in detail in Chapter 1-35. In Eq. (1-11-1) there are three independent variables and one dependent variable. Also measurement instruments that house all three types of thermometers are available so that WGBT can be readily determined on the spot.

Here let us find out under what relative humidity conditions, the Australian Open Tennis Tournament's extreme heat policy is to be enforced, if the ambient temperature is 35 °C and radiant energy's black bulb globe temperature, T_g, is 45 °C.

From Eq. (1-11-1), maximum wet bulb temperature can be obtained.

$$28\ °\text{C} = 0.7 \times T_{wb} + 0.2 \times 45\ °\text{C} + 0.1 \times 35\ °\text{C}$$

or

$$T_{wb} \le 22.1\ °\text{C}$$

Using Eq. (1-35-1) and assuming barometric pressure to be 1 atmosphere, maximum relative humidity before enforcing the extreme heat policy is determined to be 32%.

Chapter 1-12 Temperature Scales

There are four different temperature scales that are used commonly around the World. In International System of Units, SI, the absolute temperature scale, a scale that starts with zero temperature, is in degrees Kelvin (symbol K). In our daily lives, the degrees Celcius or the degrees centigrade (symbol °C) temperature scale is used which is also called the metric temperature scale.

In some parts of the World such as the US, Liberia, and Myanmar, English System of Units is used. The absolute temperature scale for the English System of Units, the scale that starts with zero temperature, is in degrees Rankine (symbol $°R$). In US, Liberia, and Myanmar, the daily temperature scale is degrees Fahrenheit (symbol °F).

In practice and in experiments, it is very difficult to reach absolute zero temperatures. As absolute zero temperature is approached atoms start to flow freely without any resistance and they are called super fluids. Absolute temperature scales are mostly used in engineering and in scientific experiments and calculations. One has to utilize absolute temperature scales when any non-linear power of temperature is calculated, i.e. T^4 for radiation heat transfer, $T^{0.5}$ for speed of sound, etc.

The difference between Kelvin, T_K, and degrees centigrade, T_C, temperature scales is a constant of 273.16.

$$T_K = T_C + 273.16 \qquad \text{(1-12-1)}$$

The most familiar metric temperatures are the freezing and boiling points of water at sea level. $T_C = 0\ °C, T_K = 273.16\ K$ for freezing temperature of water and $T_C = 100\ °C, T_K = 373.16\ K$ for boiling temperature of water.

The degrees Rankine absolute temperature scale, T_R, is related to the Kelvin absolute temperature scale by a 1.8 constant multiplier.

$$T_R = 1.8 \times T_K \qquad \text{(1-12-2)}$$

The difference between degrees Rankine, T_R, and degrees Fahrenheit, T_F, temperature scales is a constant of 459.69.

$$T_R = T_F + 459.69 \qquad \text{(1-12-3)}$$

By combining Eqs. (1-12-1), (1-12-2), and (1-12-3), the following linear relationship between the degrees Fahrenheit and degrees Centigrade temperatures are obtained.

$$T_F = 1.8 \times T_C + 32 \qquad \text{(1-12-4)}$$

Again, considering the most familiar metric temperatures at the freezing and boiling points of water at sea level,

when $T_C = 0\ °C, T_K = 273.16\ K, T_F = 32\ °F, T_R = 491.69\ °R$

and when $T_C = 100\ °C, T_K = 373.16\ K, T_F = 212\ °F, T_R = 671.69\ °R$.

The linear relationship between the degrees Fahrenheit and degrees centigrade, Eq. (1-12-4), is plotted in Figure 1-12-1. The slope of the linear relationship in Figure 1-12-1 is 1.8. Also the degrees Fahrenheit and degrees centigrade are the same at -40 degrees.

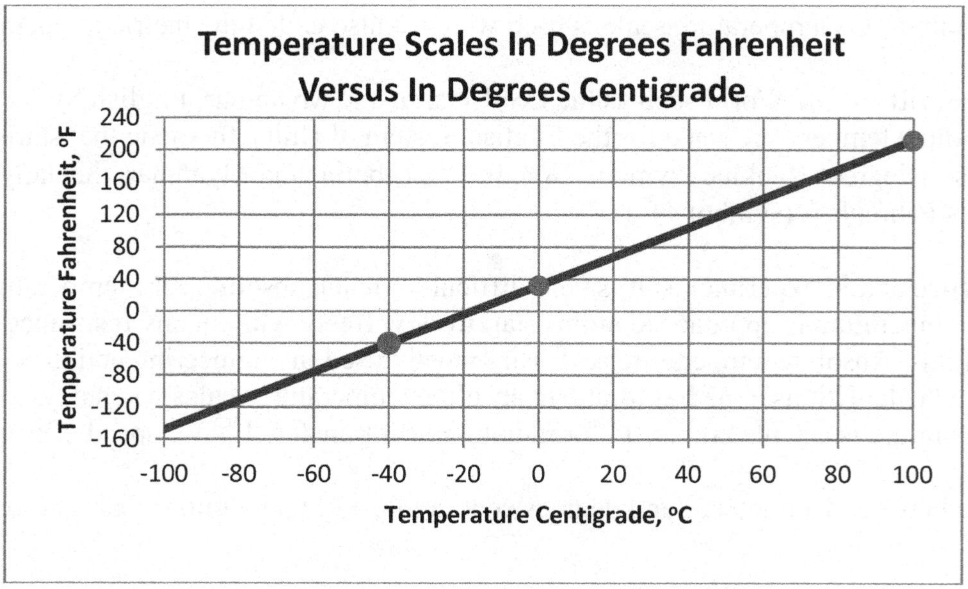

Figure 1-12-1 Temperature scales in degrees Fahrenheit versus in degrees centigrade

Since the slope of Eq. (1-12-4) is 1.8, it is easy to convert from degrees centigrade to degrees Fahrenheit. Every 10 degrees increase or decrease in degrees centigrade corresponds to an 18 degrees increase or decrease in degrees Fahrenheit. A 10 degrees centigrade corresponds to 32+18=50 degrees Fahrenheit. A -10 degrees centigrade corresponds to 32-18=14 degrees Fahrenheit and so on.

Chapter 1-13 Movie Theater Audience Mix

In some situations, we confront with less information than the number of unknowns in a problem. For example, a movie theater sells 750 tickets for a show. Total proceeding from ticket sales is $6,150. Ticket prices are $12 for adults, $8 for seniors, and $5 for children. How many adults, seniors, and children are attending this show?

There are three unknowns in this problem. They are the number of adults, A, the number of seniors, S, and the number of children, C, attending this show. First information is that the sum of all the tickets should be 750.

$$A + S + C = 750 \quad (1\text{-}13\text{-}1)$$

Second information is that the total proceeding from ticket sales is $6,150.

$$12A + 8S + 5C = 6{,}150 \quad (1\text{-}13\text{-}2)$$

Eqs. (1-13-1) and (1-13-2) provide two equations and three unknowns. These two equations can have multiple solutions or no solution. Since Eq. (1-13-2) is not a constant multiple of Eq. (1-13-1), no solution case does not apply. In order to obtain the multiple solutions, multiply Eq. (1-13-1) by 5 and subtract it from Eq. (1-13-2) to obtain the following.

$$7A + 3S = 2{,}400 \quad (1\text{-}13\text{-}3)$$

In Eq. (1-13-3), the variable C is eliminated. There are two conditions that have to be satisfied while finding multiple solutions to Eq. (1-13-3). First, the variables $A\ and\ S$ have to be positive integers and second $A + S \leq 750$. In order to get positive integer values for $A\ and\ S$, A has to be a multiples of 3 or

$$S = \frac{2{,}400 - 7(3N)}{3} = 800 - 7N\ where\ 13 \leq N \leq 114 \quad (1\text{-}13\text{-}4)$$

$$A = 3N \quad (1\text{-}13\text{-}5)$$

where $N = 13, 14, 15, \ldots, 114$, to satisfy $A + S \leq 750$ and Eq. (1-13-4) does not go negative. Possible number of seniors versus number of adults, from Eqs. (1-13-4) and (1-13-5), are shown in Figure 1-13-1. Then for each possible case, Eq. (1-13-1) is used to obtain number of children. Possible number of children versus number of adults and the maximum number of people at the show, 750, are also shown in Figure 1-13-1.

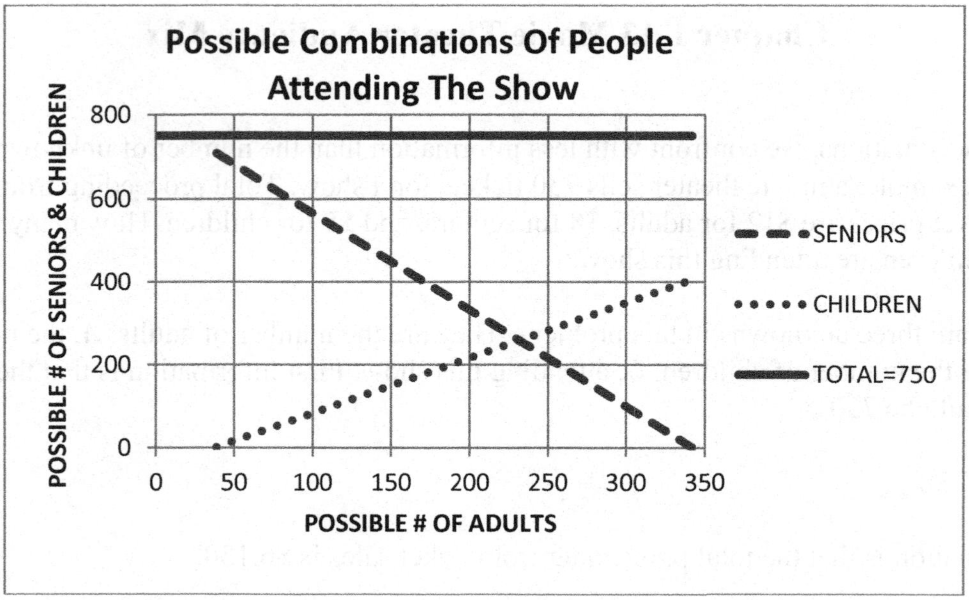

Figure 1-13-1 Possible combinations of people attending the show

Some of the possible combinations of people attending the show are shown in Table 1-13-1. There are a total of 102 possible combinations that comply with the given information.

N	Adults	Seniors	Children	Total	Combination #
13	39	709	2	750	1
14	42	702	6	750	2
15	45	695	10	750	3
16	48	688	14	750	4
…	…	…	…	…	…
111	330	23	394	750	99
112	333	16	398	750	100
113	339	9	402	750	101
114	342	2	406	750	102

Table 1-13-1 Possible combinations of people attending the show

Chapter 1-14 Children's Ages

A proud mom explains that she has two children named Lisa and Tom. She also adds that the sum of their ages is 14 years. Then the dad cuts in to add more information. He says that 3 times the age of Lisa added to 3 times the age of Tom is 36 years. Then they quiz you as to what Lisa's and Tom's ages are?

Mom's information gives the following relationship for Lisa's age, let us call her age L, and Tom's age, let us call his age T.

$$L + T = 14 \qquad (1\text{-}14\text{-}1)$$

Dad's information gives the following relationship for Lisa's age, , and Tom's age, T.

$$3L + 3T = 36 \qquad (1\text{-}14\text{-}2)$$

If Eq. (1-14-2) is divided into 3 and then plotted along with Eq. (1-14-1) and two parallel lines are obtained as shown in Figure 1-14-1. The lines are plotted between the ages of 0 and 14, since these are the minimum and maximum ages that Lisa and Tom can have.

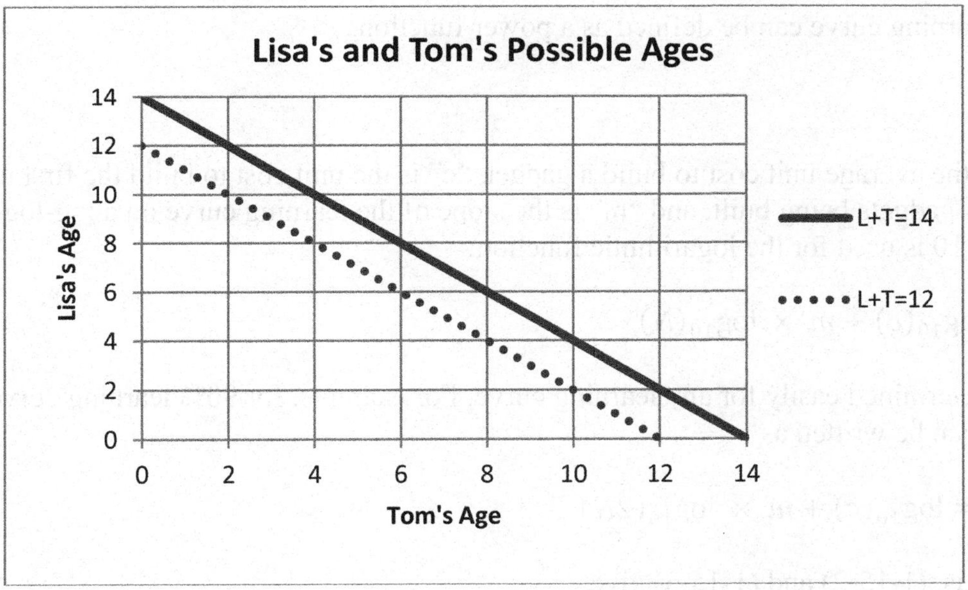

Figure 1-14-1 No solution for Lisa's and Tom's ages

The two lines shown in Figure 1-14-1 are parallel, namely they have no point in common. Therefore there is no solution to Lisa's and Tom's ages with the information given.

Chapter 1-15 Learning Curve

Learning curve has different meanings in different contexts. For example "this subject has a steep learning curve" means that the subject is difficult to absorb and master in a short time. "The new soccer team will go through a learning curve as the season progresses" means that the new team will make mistakes and will fail in collective performance initially, but as the season matures the team will gain stability towards good team performance. However, in industrial engineering, "learning curve" was originated to describe the improvements in production efficiency and in cost reduction as the produced quantities increased.

The learning curves will vary for different types of production lines. A labor intensive production line might have an 80% learning curve, namely if the average unit cost of building 1,000 gadgets is $20, then the average unit cost of building 2,000 gadgets should be $16 or 0.8x$20=$16. The average unit cost of building 4,000 gadgets should be 0.8x$16=$12.80. An automated high volume production line might have a 95% learning curve, namely if the average unit cost of building one million toys is $2, then the average unit cost of building two million toys should be $2 or 0.95x$2=$1.9. This type of learning curve is only applicable at the beginning of a production program because production efficiency and cost reduction start to plateau out and a constant rate of unit cost reduction as the production quantity doubles becomes very difficult.

Production learning curve can be defined as a power function.

$$T = cN^m \qquad (1\text{-}15\text{-}1)$$

where "T" is the average unit cost to build a gadget, "c" is the unit cost to build the first unit, "N" is the number of gadgets being built, and "m" is the slope of the learning curve on a log-log graph. In this exercise base 10 is used for the logarithmic function.

$$\log_{10}(T) = \log_{10}(c) + m \times \log_{10}(N) \qquad (1\text{-}15\text{-}2)$$

"m" can be determined easily for any learning curve. For example, for 90% learning curve, Eq. (1-15-2) can be written as

$$\log_{10}(0.9T) = \log_{10}(c) + m \times \log_{10}(2N) \qquad (1\text{-}15\text{-}3)$$

Combining Eqs. (1-15-2) and (1-15-3) gives

$$\log_{10} T - \log_{10}(0.9T) = m[\log_{10} N - \log_{10} 2N]$$

or

$$m = \frac{\log_{10}(0.9)}{\log_{10}(2)} = -0.152 \qquad (1\text{-}15\text{-}4)$$

If the average unit cost is $200 to build the initial 10 gadgets and if the learning curve is running at 90%, then "c" can be obtained from Eqs. (1-15-1) and (1-15-4) to be $283.81. This is the cost to build the first gadget. Similar analysis can be done for 80% and for 70% learning curves for this example. The results are shown in Figure 1-15-1.

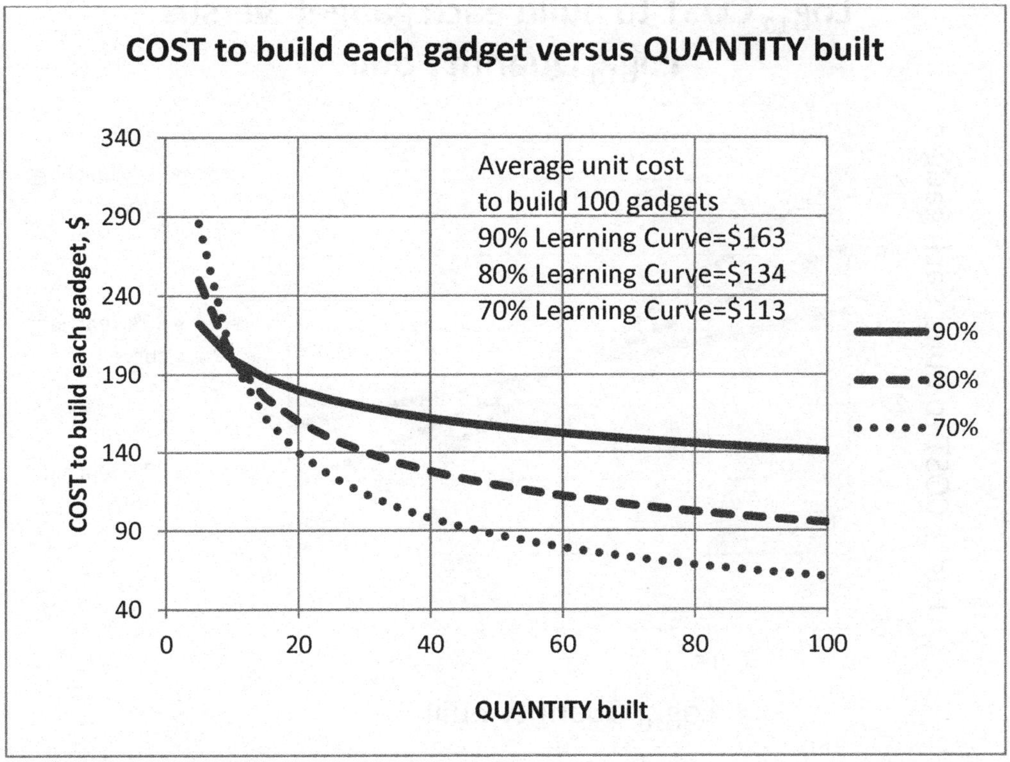

Figure 1-15-1 Cost to build each gadget versus quantity built

Figure 1-15-1 shows that as the learning curve percentage decreases, improvements in production efficiency and in cost reduction increases as more gadgets are produced. Figure 1-15-1 can be re-plotted on a log-log graph in Figure 1-15-2 to be able to observe the slopes “m”.

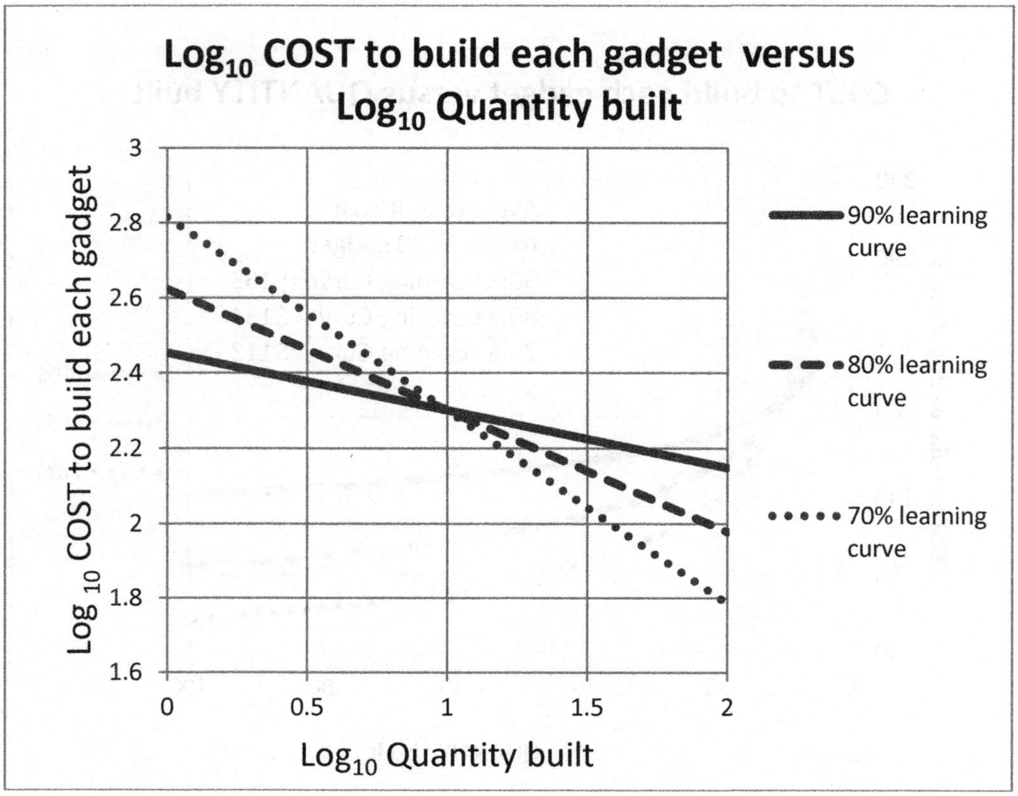

Figure 1-15-2 Cost to build each gadget versus quantity built
shown on a log-log graph

In the production environment context, “steeper learning curve” means faster improvements in production efficiency and in cost reduction.

Chapter 1-16 Row, Row, Row Your Boat

A boy roving a boat upstream in a river passes a girl who is sun tanning on a raft that is drifting downstream. He says "Hello" when his boat crosses her raft. He rows upstream a while and decides to go back and catch up with her. He catches up with her 1 mile downstream from the point where he said "Hello" to her. If the river is flowing at 3 mph and the boy's speed at still water is 6 mph, how long did it take him to decide to turn back and catch up with her after he said "Hello" to her? How many miles did he row after he turned back to catch up with her?

There are three locations of importance in this problem. First location is where he said "Hello" to her while rowing upstream. Let us call the first location P_{Hello} . Second location is where he decides to turn back and catch up with her. Let us call the second location $P_{Decision}$. Third location is where he catches up to her raft. Let us call the third location $P_{Catch\ up}$. An illustration of the locations as shown in Figure 1-16-1 can be helpful.

Figure 1-16-1 Illustration of three locations

Next speed, time, and distance relationships are determined. While the boy is rowing upstream from the location P_{Hello} to $P_{Decision}$, the following constant linear speed, time, and distance relationship holds.

$$(6-3)\times t_{Hello\ to\ Decision} = D \qquad (1\text{-}16\text{-}1)$$

where D is the distance in miles between location P_{Hello} and location $P_{Decision}$ and t denotes time in hours between designated locations . While he is rowing downstream from the location $P_{Decision}$ to $P_{Catch\ up}$, the following constant linear speed, time, and distance relationship holds.

$$(6+3)\times t_{Decision\ to\ Catch\ Up} = D+1 \qquad (1\text{-}16\text{-}2)$$

Girl's raft goes downstream from the location P_{Hello} to location $P_{Catch\ up}$ that complies with the following constant linear speed, time, and distance relationship.

$$3\times (t_{Hello\ to\ Decision} + t_{Decision\ to\ Catch\ Up}) = 1 \qquad (1\text{-}16\text{-}3)$$

In the above three equations, there are three unknowns, namely $D, t_{Hello\ to\ Decision}, t_{Decision\ to\ Catch\ Up}$. Let us see if there is a unique solution to these unknowns. Subtracting Eq. (1-16-2) from Eq. (1-16-1), eliminates the unknown D and the following relationship is obtained.

$$3\times t_{Hello\ to\ Decision} - 9\times t_{Decision\ to\ Catch\ Up} = -1 \qquad (1\text{-}16\text{-}4)$$

Subtracting Eq. (1-16-3) from Eq. (1-16-4) eliminates $t_{Hello\ to\ Decision}$ and provides the results for

$$t_{Decision\ to\ Catch\ Up} = \frac{1}{6}\ hours\ or\ 10\ minutes$$

Plugging the above result to either Eq. (1-16-3) or Eq. (1-16-4) provides the time it takes from "Hello" to "Decision".

$$t_{Hello\ to\ Decision} = \frac{1}{6}\ hours\ or\ 10\ minutes$$

It is observed that $t_{Hello\ to\ Decision}$ and $t_{Decision\ to\ Catch\ Up}$ are equal to each other. It can be shown that this equality in time always holds for this problem. Inserting the known times into either Eq. (1-16-1) or Eq. (1-16-2) gives the distance *D* from "Hello" to "Decision".

$$D = 0.5\ miles$$

He rows $D + 1$ miles after he turns his boat back to catch up with her. This distance is 1.5 miles.

Chapter 1-17 Office Cleaners

Consider three office cleaners Jim, Joe, and John. Jim and Joe clean an office in 4 hours. Jim and John clean the same office in 5 hours. Joe and John clean the same office in 6 hours. If all three work together, how long will it take them to clean this office?

Let us write down the given three conditions for cleaning the office in percentages per hour for each person while the total cleanup adds up to 100%.

$$4 \times Jim + 4 \times Joe = 100 \qquad (1\text{-}17\text{-}1)$$

$$5 \times Jim + 5 \times John = 100 \qquad (1\text{-}17\text{-}2)$$

$$6 \times Joe + 6 \times John = 100 \qquad (1\text{-}17\text{-}3)$$

Unknowns $Jim, Joe, and\ John$ in Eqs. (1-17-1), (1-17-2), and (1-17-3) are percentages of office cleaned per hour by each person. There are several different methods of solving these three equations and three unknowns. Here we will use elimination of unknowns. If Eq. (1-17-3) is subtracted from 1.5 times Eq. (1-17-1), the following relationship is obtained.

$$6 \times Jim - 6 \times John = 50 \qquad (1\text{-}17\text{-}4)$$

Then taking 6/5 th of Eq. (1-17-2) gives

$$6 \times Jim + 6 \times John = 120 \qquad (1\text{-}17\text{-}5)$$

Adding Eqs. (1-17-4) and (1-17-5) gives percentage of office cleaned per hour by Jim.

$$Jim = \frac{170}{12} = 14.17\%\ of\ office\ cleaned\ per\ hour \qquad (1\text{-}17\text{-}6)$$

Now from either Eq. (1-17-2) or Eq. (1-17-4) or Eq. (1-17-5) percentage of office cleaned per hour by $John$ can be determined.

$$John = \frac{35}{6} = 5.83\%\ of\ office\ cleaned\ per\ hour \qquad (1\text{-}17\text{-}7)$$

Now from either Eq. (1-17-1) or Eq. (1-17-3) percentage of office cleaned per hour by Joe is determined.

$$Joe = \frac{65}{6} = 10.83\%\ of\ office\ cleaned\ per\ hour \qquad (1\text{-}17\text{-}8)$$

It seems that $John$ is the slowest worker and Jim is the fastest while Joe is hovering in the middle pace. If three of them worked together without interfering each other, it will take X amount hours to clean the office.

$$(Jim + Joe + John)X = \left(\frac{85}{6} + \frac{65}{6} + \frac{35}{6}\right)X = 100 \qquad (1\text{-}17\text{-}9)$$

or $X = \frac{600}{185} = \frac{120}{37} = 3.24\ hours$.

Chapter 1-18 Ratio of Boys to Girls

A school has a total of 15,120 students. 4/7 th of the total students are foreign. 3/13 th is the ratio of foreign student girls to foreign student boys. Also 1/5 th is the ratio of the total student girls to total student boys. Let us find out ratios, percentages, and numbers of foreign students that are girls, foreign students that are boys, domestic students that are girls, and domestic students that are boys in the whole student body.

First condition is that the sum of four ratios in the whole student body should be unity.

$$G_F + B_F + G_D + B_D = 1 \tag{1-18-1}$$

where G_F is the ratio of foreign girls, B_F is the ratio of foreign boys, G_D is the ratio of domestic girls, and B_D is the ratio of domestic boys to the whole student body.

Second condition is the ratio of foreign students to the whole student body.

$$G_F + B_F = \frac{4}{7} \tag{1-18-2}$$

Third condition is the ratio of foreign student girls to foreign student boys.

$$\frac{G_F}{B_F} = \frac{3}{13} \tag{1-18-3}$$

Fourth condition is the ratio of total student girls to total student boys.

$$\frac{G_F+G_D}{B_F+B_D} = \frac{1}{5} \tag{1-18-4}$$

In this case there are four unknowns and four equations. It is possible that there is a unique solution to this problem. First Eqs. (1-18-2) and (1-18-3) can be combined to obtain G_F and B_F.

$$\frac{13}{3}G_F + G_F = \frac{4}{7} \qquad \text{or} \qquad G_F = \frac{3}{28} \quad \text{and} \quad B_F = \frac{13}{28}$$

Next Eqs. (1-18-1) and (1-18-4) can be combined to get the ratio of the total girls to the whole student body.

$$G_F + G_D + \frac{5}{1}(G_F + G_D) = 1 \qquad \text{or} \qquad G_F + G_D = \frac{1}{6}$$

Total girls make up 1/6 th of the whole student body. We already know, G_F , the ratio of foreign student girls to the whole student body. Then the ratio of domestic girls to the whole student body becomes

$$G_D = \frac{1}{6} - \frac{3}{28} = \frac{5}{84}$$

Since three of the unknowns are determined, the fourth one, B_D , can be determined from Eq. (1-18-1).

$$B_D = 1 - \frac{5}{84} - \frac{3}{28} - \frac{13}{28} = \frac{31}{84}$$

Domestic students that are boys have a ratio of 31/84 to the whole student body. They make up 36.9% of the student body and there are 5,580 of them.

Domestic students that are girls have a ratio of 5/84 to the whole student body. They make up 6.0% of the student body and there are 900 of them.

Foreign students that are boys have a ratio of 13/28 to the whole student body. They make up 46.4% of the student body and there are 7,020 of them.

Foreign students that are girls have a ratio of 3/28 to the whole student body. They make up 10.7% of the student body and there are 1,620 of them.

Chapter 1-19 World Track Records

World track running distances cover quite a range from 100 meters to marathon which is 40,000 meters. A runner can run 100 meters in 9.58 seconds in record time. The record running time extends to 2 hours: 3 minutes: 59 seconds (7439 seconds) for the marathon distance. World record running times versus distances are shown in Figure 1-19-1 for the whole range, data source International Association of Athletics Federations. From Figure 1-19-1, it is hard to visualize the relationship between running time and distance for short distances because data is all bunched up there.

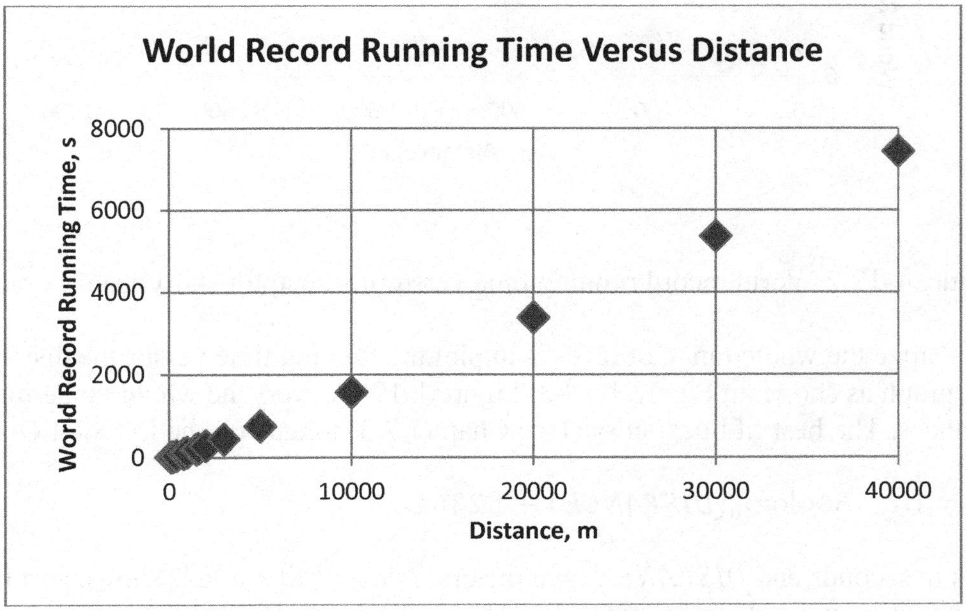

Figure 1-19-1 World record running time versus distance

The scale of the coordinates can be altered and running distances can go up to 1000 meters as shown in Figure 1-19-2. Figure 1-19-2 gives a better view of short distance races.

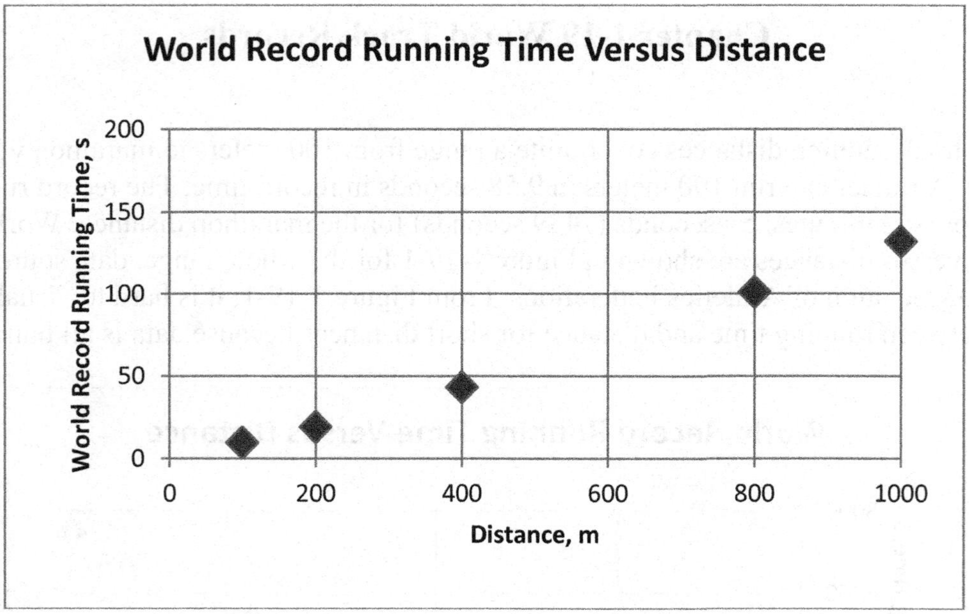

Figure 1-19-2 World record running time versus distance for short distance races

Best way to visualize the whole range of races is to plot the running time versus the distance on a LOG_{10}-LOG_{10} graph as shown in Figure 1-19-3. Figure 1-19-3 covers the whole range for running times and distances. The best fit linear curve (see Chapter 2-3) to data on the LOG_{10}-LOG_{10} graph is

$$\log_{10}(TIME) = 1.1116 \times \log_{10}(DISTANCE) - 1.2361 \quad (1\text{-}19\text{-}1)$$

where $TIME$ is in seconds and $DISTANCE$ is in meters. The best fit curve is shown as a line in Figure 1-19-3 and it is an excellent fit to data with a correlation coefficient value of $r = 0.99975$.

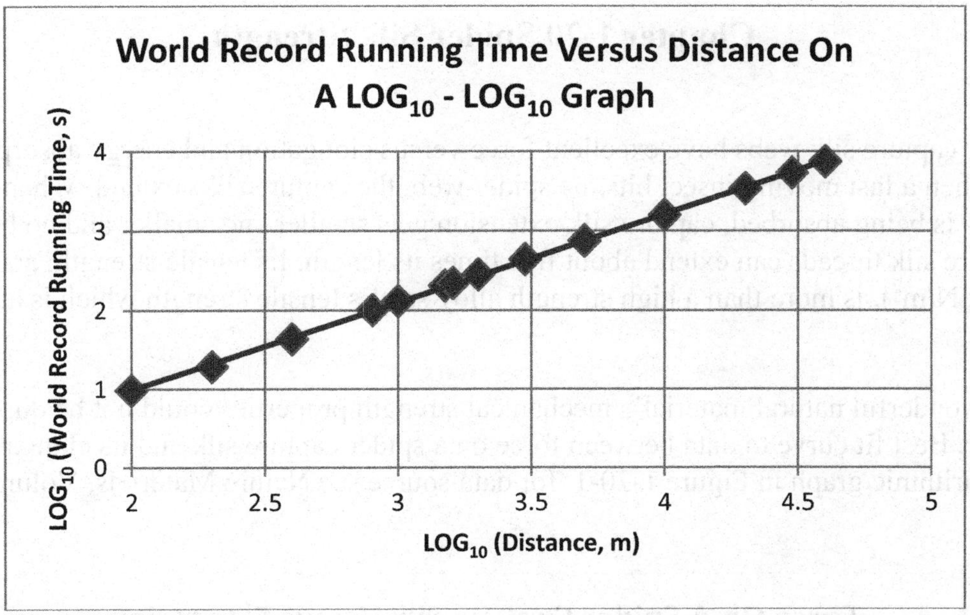

Figure 1-19-3 World record running time versus distance on a LOG_{10}-LOG_{10} graph

Chapter 1-20 Spider Silk Strength

Spider capture silk webs have excellent force versus elongation and energy absorption properties. When a fast moving insect hits the spider web, the capture silks extend. When the insect's kinetic energy is being absorbed, capture silk extension gets smaller and smaller and no bounce back occurs. Capture silk threads can extend about five times its length. Its tensile strength, about 1 Giga Pascal (10^{12} x N/m^2), is more than a high strength alloy steel's tensile strength which is about 0.76 Giga Pascal.

This wonderful natural material's mechanical strength properties could not be duplicated by humans so far. Best fit curve to data between force on a spider capture silk and its elongation is shown in a semi-logarithmic graph in Figure 1-20-1, for data source see Nature Materials, Volume 2, April 2003.

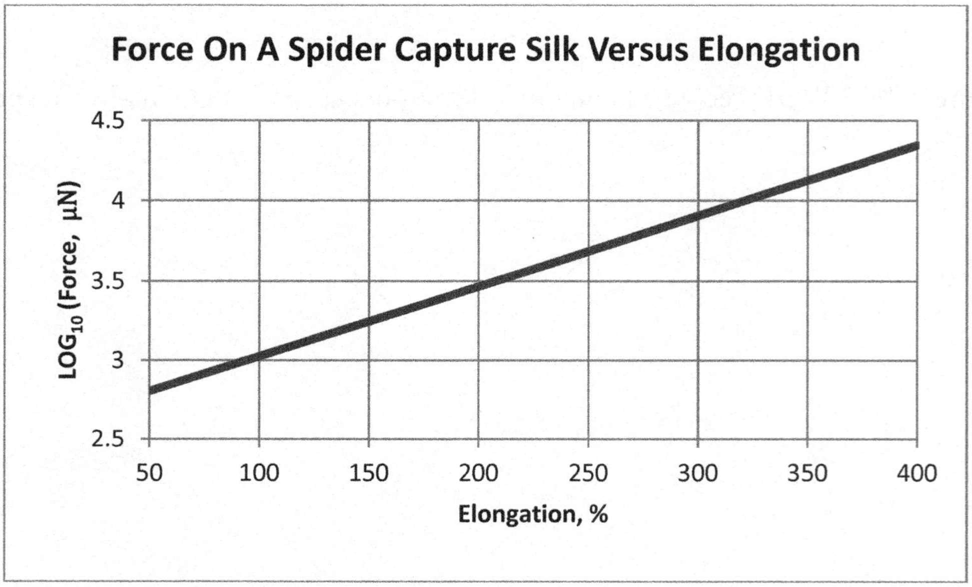

Figure 1-20-1 Force on s spider capture silk versus elongation

Figure 1-20-1 shows that force versus elongation for a spider capture silk thread behaves logarithmically. Force in μN (micro Newtons) versus % elongation equation is given below.

$$F = 385 \times 10^{0.0044 \times \%\, Elongation} \quad (1\text{-}20\text{-}1)$$

or

$$\log_{10} F = 2.585 + 0.0044 \times \%\, Elongation \quad (1\text{-}20\text{-}2)$$

A capture silk thread has a diameter which is approximately one tenth the diameter of a human hair. If a typical capture silk diameter, D, of 5 μm is considered, then the stressed area, $0.25\pi D^2$, of the capture silk thread is 2×10^{-11} m^2. Tensile stress on the capture silk thread at 400% (4 times the elongation of its original length) is about 1 Giga Pascals. The tensile force on the capture silk thread is 22,154 μN.

Chapter 1-21 – Depreciation in Value of a Car

Value of a car starts go down the moment you drive it off the dealer's lot. Here let us consider a car with normal wear and tear that depreciates at an annual fixed percentage. There is also an initial lot drive off depreciation percentage, IDP , of the car's value, V, after it is driven off the dealer's lot. So after a day of owning the car, your car's value is

$$V_{after\ 1\ day} = \left(\frac{100-IDP}{100}\right) \times V \tag{1-21-1}$$

After one, two, three, and N years, with annual fixed depreciation percentage, $AFDP$, the value of the car becomes

$$V_{after\ 1\ year} = \left(\frac{100-AFDP}{100}\right) \times V_{after\ 1\ day}$$
$$V_{after\ 2\ years} = \left(\frac{100-AFDP}{100}\right) \times V_{after\ 1\ year} = \left(\frac{100-AFDP}{100}\right)^2 \times V_{after\ 1\ day}$$
$$V_{after\ 3\ years} = \left(\frac{100-AFDP}{100}\right) \times V_{after\ 2\ years} = \left(\frac{100-AFDP}{100}\right)^3 \times V_{after\ 1\ day}$$
$$V_{after\ N\ years} = \left(\frac{100-AFDP}{100}\right) \times V_{after\ (N-1)\ years} = \left(\frac{100-AFDP}{100}\right)^N \times V_{after\ 1\ day}$$

After N years, the value of the car can have a ratio to $V_{after\ 1\ day}$ as follows.

$$\frac{V_{after\ N\ years}}{V_{after\ 1\ day}} = \left(\frac{100-AFDP}{100}\right)^N \tag{1-21-2}$$

If a value of a $40,000 car depreciates to $8,000 after 7 years of normal usage and if the initial lot drive off depreciation percentage is 10%, what is its annual fixed percentage depreciation?
From Eq. (1-21-1), the car's value after one day is $36,000. Using Eq. (1-21-2), taking logarithms of both sides to base ten, and rearranging gives the relationship for $AFDP$.

$$AFDP = 100 \times \left[1 - 10^{\left(\frac{1}{N}\right)\log_{10}\left(\frac{V_{after\ N\ years}}{V_{after\ 1\ day}}\right)}\right] \tag{1-21-3}$$

or

$$AFDP = 100 \times \left[1 - 10^{\left(\frac{1}{7}\right)\log_{10}\left(\frac{8000}{36000}\right)}\right] = 19.34\%$$

For this car, its depreciated value versus the car's age is plotted in Figure 1-21-1. The plot also shows the initial value drop of 10%.=

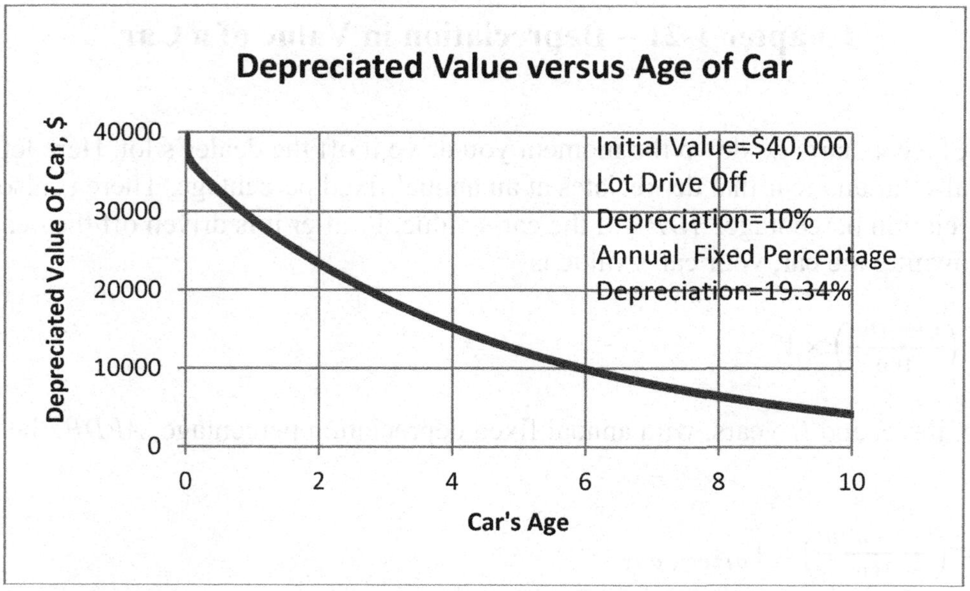

Figure 1-21-1 Depreciated value versus age of car

Using Eq. (1-21-2), depreciated value of the car divided by its value after one day of purchase versus car's age are shown in Figure 1-21-2 for different annual fixed depreciation percentages.

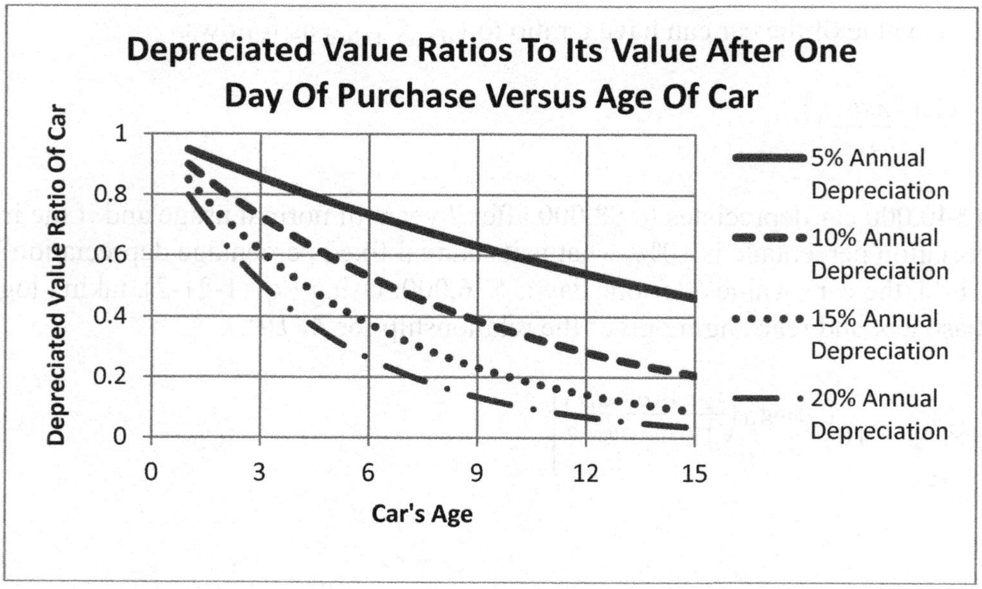

Figure 1-21-2 Depreciated value ratios to its value after one day of purchase versus car's age

Chapter 1-22 Earthquake Magnitude - Richter Scale

Charles Richter formulated his earthquake magnitude scale called the "Richter Scale" in 1935. The Richter scale is in a logarithmic form, namely a tenfold increase in the recorded earthquake wave amplitude increases the Richter scale by one, see US Geological Survey web site. Also the distance of the recording seismometer with respect to the epicenter of the earthquake has to be corrected for. The seismometer location correction factor with respect to the epicenter of the earthquake is formulated from differing speeds of two type waves an earthquake generates. One type is compression waves called the P-waves and the other type is shear waves called the S-waves. P-waves travel faster than the S-waves and they travel perpendicular to each other. P-waves arrive first to a seismometer and then the S-waves arrive. The difference in P-wave and S-wave arrival times determines the distance correction factor between the seismometer and the epicenter of the earthquake.

Earthquake magnitude, M, is expressed as a logarithm to base 10 as follows.

$$M = \log_{10} A + 3 \times \log_{10}(8T) - 2.92 \qquad (1\text{-}22\text{-}1)$$

where A is the maximum amplitude measured from a seismograph in millimeters and T is the time difference in the arrival of S-waves and P-waves to the seismograph in seconds. So the second and the third terms in Eq. (1-22-1) are the epicenter distance correction factors. Earthquake magnitudes on Richter scale for different amplitudes and for different distance corrections are shown in Figure 1-22-1.

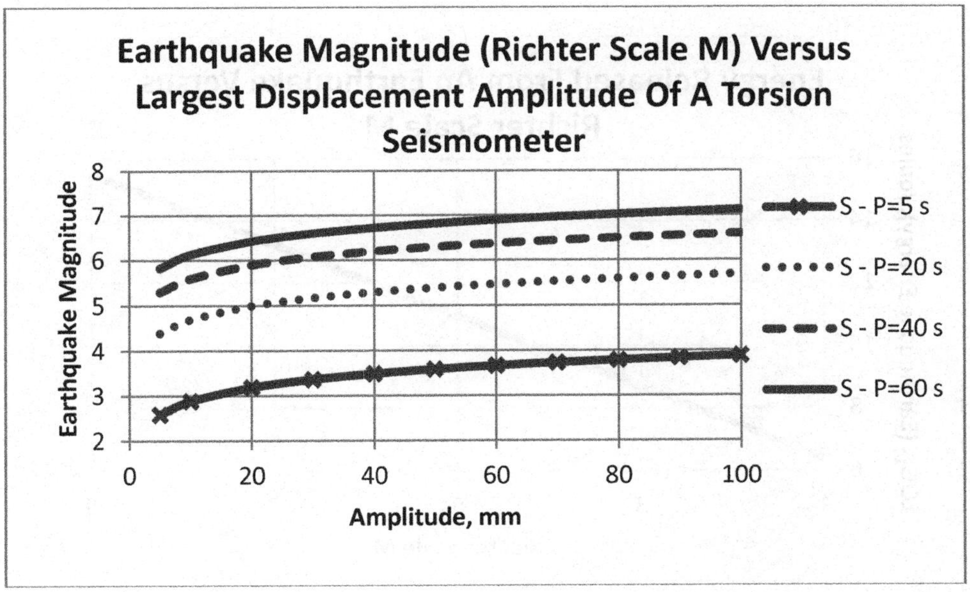

Figure 1-22-1 Earthquake magnitude (Richter scale M) versus largest displacement amplitude of a torsion seismometer for different earthquake epicenter distances with respect to the seismometer

It can be seen from Figure 1-22-1 that the earthquake magnitude increases by unity when the seismograph maximum amplitude increases by tenfold for the same epicenter distance. Earthquake magnitudes in Richter scale are classified as in Table 1-22-1.

Richter Scale	Earthquake Classification
Below 2.0	micro
2.0 to 3.9	minor
4.0 to 4.9	light
5.0 to 5.9	moderate
6.0 to 6.9	strong
7.0 to 7.9	major
8.0 to 9.9	great

Table 1-22-1 Earthquake classifications in Richter scale

The energy released by an earthquake increases by an order of magnitude each time the Richter scale increases by one. Energy released by an earthquake in Joules can be approximated by an empirical logarithmic expression as follows.

$$\log_{10}\left(\frac{E}{1.74}\right) = 5 + 1.14 \times M \qquad \text{(1-22-2)}$$

Energy released from an earthquake versus Richter scale M is shown in Figure 1-22-2. The slope of the curve is 13.8, namely energy released from an earthquake increases 13.8 times for a unit increase in Richter scale.

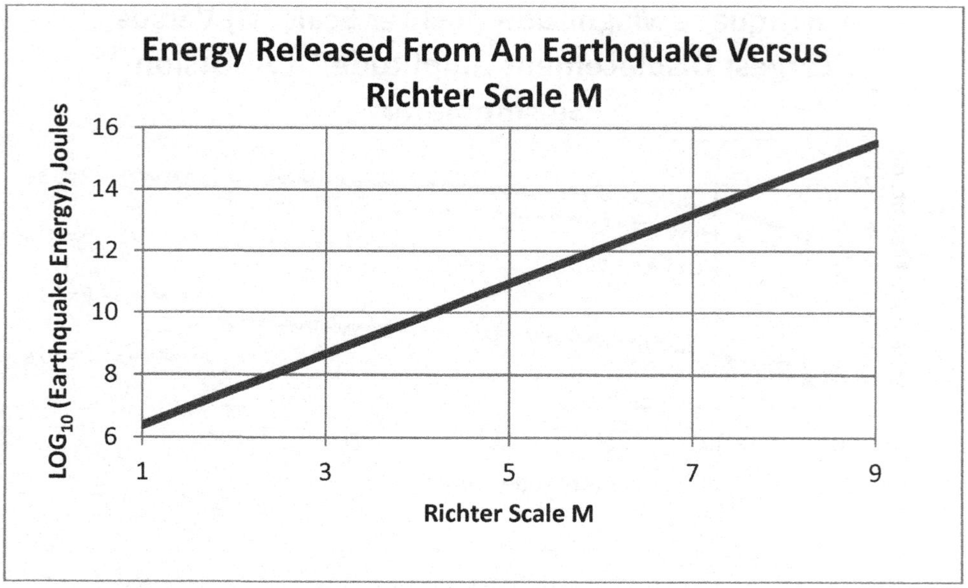

Figure 1-22-2 Logarithm to base 10 of Energy released from an earthquake versus Richter scale M

The slope of the curve in Figure 1-22-2 can be determined from Eq. (1-22-2) by setting $E = E_1$ for $M = M_1$ and $E = E_2$ for $M = M_1 + 1$.

$$\frac{E_2}{E_1} = 10^{1.14} = 13.8$$

Chapter 1-23 Disk Drive Areal Density Increase

Areal density of disk drives has been increasing at a phenomenal rate since 1970's. Areal density of a disk drive is the product of its linear bit density and its track density. Linear bit density in bits per inch is the maximum amount of magnetic flux changes that can be squeezed on the circumference of a track. Magnetic flux changes per inch can be converted to user bits per inch once the drive channel coding type is known. In 2003, linear bit densities were up to 1,000,000 bits per inch, track densities were up to 100,000 tracks per inch which comes out to an areal density of $10^6 x 10^5 = 10^{11}$ bits per square inch or 100 gigabits per square inch. In 1970 the areal density of the most advanced disk drive was 1 megabits per square inch.

In three decades, there is an advancement of 100,000 fold in disk drive areal density. This kind of improvement can easily be shown on a logarithmic scale.

$$\log_{10} AD_i = m \times (YEAR_i - 1970) \qquad (1\text{-}23\text{-}1)$$

where AD_i is the disk drive areal density in megabits per square inch (Mb/in^2), m is the slope of the areal density growth curve, called the areal density compound growth rate, and $YEAR_i$ is the disk drive introduction year after 1970. Notice that Eq. (1-23-1) satisfies the initial condition of 1 Mb/in^2 areal density in 1970, i.e. $AD_{1970} = 10^0 = 1$.

The areal density compound growth rate has three different values between 1970 and 2010. From 1970 through 1990, the compound growth rate was 26%. In 1991, magnetoresistive heads were introduced into disk drives which pushed the compound growth rate to 51% and this compound growth rate lasted through 1996. After 1996, giant magnetoresistive heads were introduced which accelerated the compound growth rate to 100%. m for 1970 through 1990 is calculated from Eq. (1-23-1) as follows.

$$\log_{10} AD_{i+1} = \log_{10}(1.26 \times AD_i) = m \times (YEAR_i + 1 - 1970)$$

$$\log_{10} AD_{i+1} = \log_{10} 1.26 + \log_{10} AD_i = m \times (YEAR_i - 1970) + m$$

or

$$m = \log_{10} 1.26 = 0.10$$

For years 1991 through 1996 m is

$$m = \log_{10} 1.51 = 0.18$$

For years 1997 on m is

$$m = \log_{10} 2 = 0.30$$

The areal density increase for three different regions in time is shown in Figure 1-23-1. Figure 1-23-1 is also extrapolated to year 2020 assuming continuation of the 100% compound growth rate.

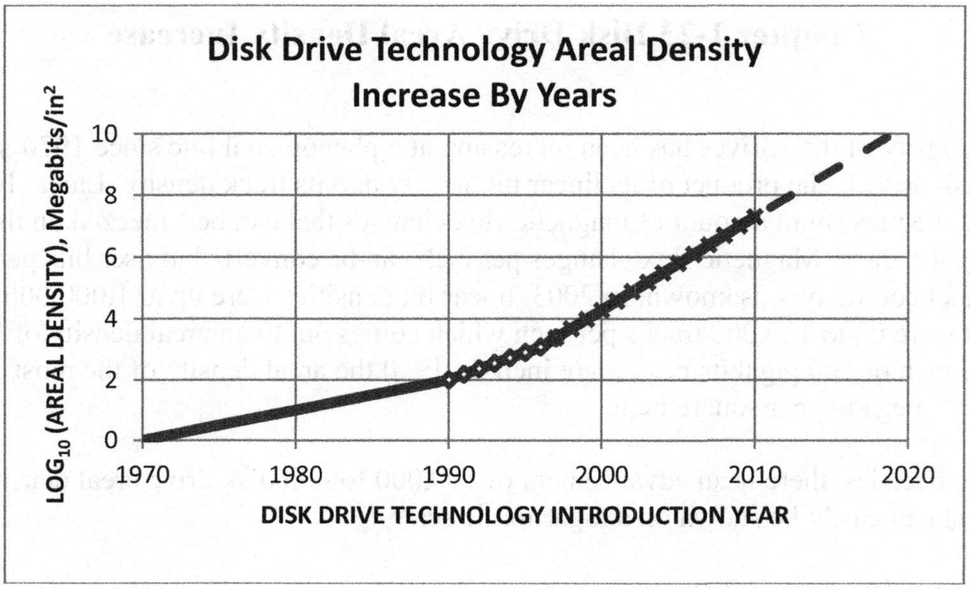

Figure 1-23-1 Disk drive technology areal density increase by years

Chapter 1-24 Optimize House Wall Insulation and Heating Costs

How thick insulation material should we put into vertical walls of a house so that the cost of heating plus cost of insulation can be minimized?

As the insulation thickness increases, the heat loss will decrease and therefore the heating costs will decrease. However, as the insulation thickness increases, the cost of insulation will increase. The object is to determine the total cost, namely the cost of heating plus cost of insulation, for a period of time. Consider this period to be 25 years or the life of the building.

First step is to determine heat loss from vertical walls of a house for incrementally increasing thicknesses of insulation. Heat transfer from vertical walls of a house can be calculated by considering one-dimensional and steady-state convection and conduction heat transfer mechanisms and by neglecting the radiation heat transfer mechanism as shown in Figure 1-24-1.

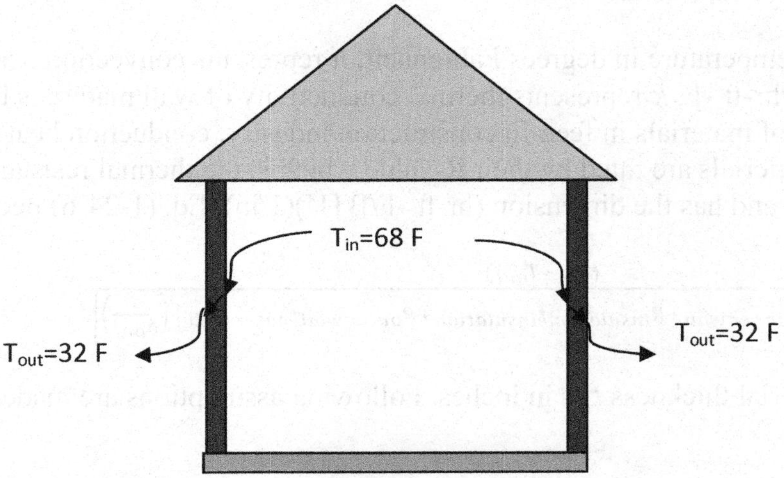

Figure 1-24-1 Heat loss from vertical walls of a house by convection and conduction heat transfer mechanisms

Assume that there is no air leakage through the vertical walls. The walls are to be constructed by three layers of materials, i.e. inner wall board, insulation layer, and outer wall board. Assuming winter conditions, the temperature inside the house is higher than the temperature outside the house.

The convection heat transferred per unit area from inside of the house to the inner surface of inner wall board is:

$$\frac{Q}{A} = h_{in}(T_{in} - T_{inside\ surface\ inner\ wall}) \qquad (1\text{-}24\text{-}1)$$

The heat transferred per unit area from the three wall layers is by conduction and they are given by the following rate equations:

$$\frac{Q}{A} = \left(\frac{k_{inner\ wall}}{t_{inner\ wall}}\right)(T_{inside\ surface\ inner\ wall} - T_{outside\ surface\ inner\ wall}) \qquad (1\text{-}24\text{-}2)$$

$$\frac{Q}{A} = \left(\frac{k_{insulation}}{t_{insulation}}\right)(T_{outside\ surface\ inner\ wall} - T_{inside\ surface\ outer\ wall}) \tag{1-24-3}$$

$$\frac{Q}{A} = \left(\frac{k_{outer\ wall}}{t_{outer\ wall}}\right)(T_{inside\ surface\ outer\ wall} - T_{outside\ surface\ outer\ wall}) \tag{1-24-4}$$

The heat transferred from the outer surface of the outer wall board to the environment is by convection and can be expressed by the following rate equation:

$$\frac{Q}{A} = h_{out}(T_{outside\ surface\ outer\ wall} - T_{out}) \tag{1-24-5}$$

After eliminating all wall temperatures from Eqs. (1-24-1) through (1-24-5), the heat loss per unit area from the vertical walls of a house can be rewritten as:

$$\frac{Q}{A} = \frac{(T_{in} - T_{out})}{\left[\left(\frac{1}{h_{in}}\right)+\left(\frac{t_{inner\ wall}}{k_{inner\ wall}}\right)+\left(\frac{t_{insulation}}{k_{insulation}}\right)+\left(\frac{t_{outer\ wall}}{k_{outer\ wall}}\right)+\left(\frac{1}{h_{out}}\right)\right]} \tag{1-24-6}$$

where T represents temperature in degrees Fahrenheit, h represents convection heat transfer coefficients in BTU/hr-ft^2-F, k represents thermal conductivity of wall materials in BTU/hr-ft-F, and t represents thickness of materials in feet. In construction industry, conduction heat transfer (insulation goodness) of wall materials are rated by their R-value which is the thermal resistance, (1/k), per inch thickness of material and has the dimension (hr-ft^2-F/BTU)(1/in). Eq. (1-24-6) becomes

$$\frac{Q}{A} = \frac{(T_{in} - T_{out})}{\left[\left(\frac{1}{h_{in}}\right)+R_{inner\ wall}t_{inner\ wall}+R_{insulation}t_{insulation}+R_{outer\ wall}t_{outer\ wall}+\left(\frac{1}{h_{out}}\right)\right]} \tag{1-24-7}$$

In Eq. (1-24-7), material thickness t is in inches. Following assumptions are made for the variables that govern Eq. (1-24-7).

$T_{in} = 68\ °\text{F}$, $T_{out} = 32\ °\text{F}$, $h_{in} = 5\ \frac{BTU}{hr-ft^2-°F}$, $R_{inner\ wall} = 0.85\ \frac{hr-ft^2-°F}{BTU-in}$, $t_{inner\ wall} = 1\ in$, $R_{insulation} = 3.5\ \frac{hr-ft^2-°F}{BTU-in}$, $t_{insulation} = varies\ 0\ to\ 8\ in$, $R_{outer\ wall} = 5\ \frac{hr-ft^2-°F}{BTU-in}$, $t_{outer\ wall} = 1\ in$, $h_{out} = 10\ \frac{BTU}{hr-ft^2-°F}$

Let us consider a house with 30 feet by 100 feet base and walls being 10 feet high. Total area of vertical walls is

$$A = (30\ ft + 30\ ft + 100\ ft + 100\ ft) \times 10\ ft = 2600\ ft^2$$

Also let us assume that the house is being heated 6 months out of a year. In 25 years, the heater will work the following total number of hours.

$$Heating\ Time = 24\frac{hours}{day} \times 30\frac{days}{month} \times 6\frac{months}{year} \times 25 years = 108{,}000\ hours$$

Total energy that is needed to heat the house due to the heat loss from its vertical walls is shown in Figure 1-24-2 as a function of insulation thickness.

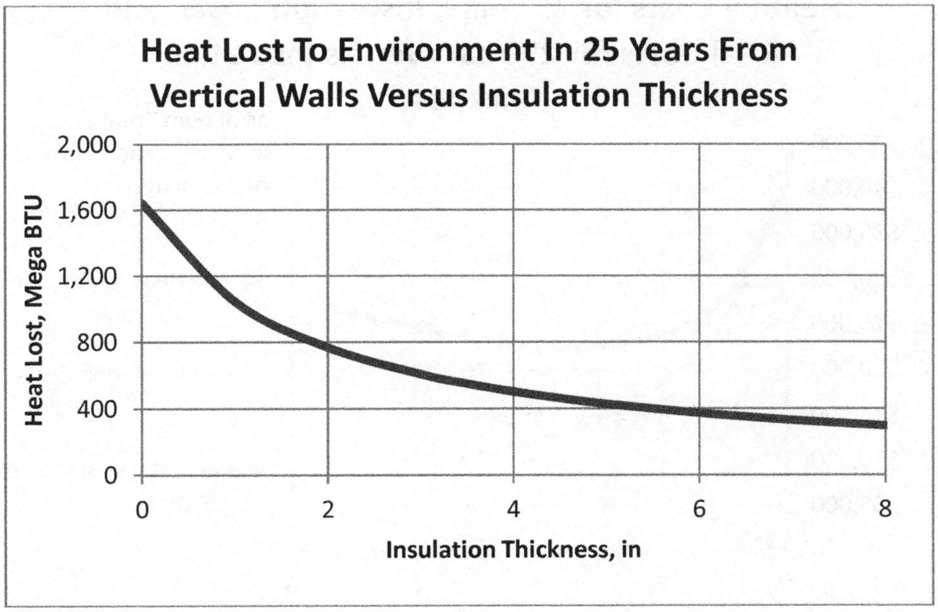

Figure 1-24-2 Heat lost to environment in 25 years in Mega BTU's from vertical walls of the house versus wall insulation thickness in inches

Next the cost of heating has to be considered. Assume that the house is heated with a natural gas heater with an 80% thermal efficiency. Natural gas has a heating value of 1,028 BTU/ft^3 and assuming that the cost of natural gas is \$20/1000 ft^3. Also assume that the natural gas price will increase as the treasury bills interest rates in 25 years. Therefore, the present value of natural gas prices will be a constant at \$20/1000 ft^3 for every year. The heating cost per Mega BTU is

$$Heating\ Cost\ per\ Mega\ BTU = \frac{\frac{\$20}{1{,}000 ft^3}}{\frac{1{,}028\ BTU}{ft^3}\times 0.8} \times \frac{1{,}000{,}000\ BTU}{Mega\ BTU} = \$24.32$$

Insulation cost is assumed to be \$1.00 per ft^2 per inch thick material, namely 1" thick insulation costs \$2,600, 2" thick insulation costs \$5,200, and so on. With all the above assumptions, present value of heating costs, insulation costs, and sum of these two costs can be calculated. All costs are shown as a function of insulation thickness in Figure 1-24-3.

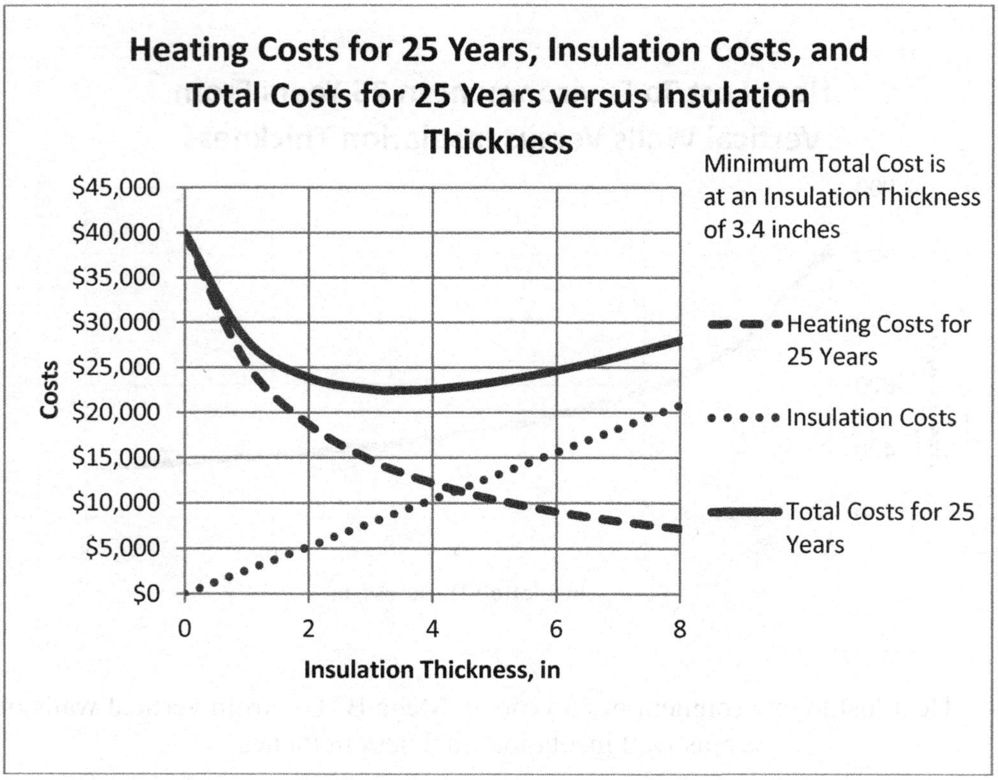

Figure 1-24-3 Present value of heating costs for 25 years, insulation costs, and total costs for 25 years versus insulation thickness

Total costs for 25 years starts high, $39,973 at no insulation material at the vertical walls and starts to decrease as the insulation material thickness increases. Total costs for 25 years go through a minimum at an insulation thickness of 3.4 inches and the total cost value at this point is $22,460. Total costs for 25 years starts increasing after this optimum insulation thickness. If more insulation thickness is added during construction, heating costs will keep going down, but the total cost will increase due to high insulation material costs.

Chapter 1-25 Different Raffle Payment Choices

A raffle gives payments to winners in six different ways. These six payment options are given in Table 1-25-1.

Raffle Payment Options	Time Of Payment	Payment Amount
1	After 10 Years	$25,000
2	After 5 Years	$20,000
3	After 3 Years	$17,500
4	After 1 Year	$16,000
5	When Winners Are Announced (0 Years)	$15,000
6	Distributed Payments After 1 Year After 2 Years After 3 Years After 4 Years	 $5,000 $4,000 $4,000 $4,000

Table 1-25-1 Raffle payment options

By looking at the payment options, it is hard to decide which one is the best. One has to do present value calculations on all the payments options to find the most favorable one. Present value, P_o , of a future payment, P_n , after n interest periods and for an interest rate i for an interest period is given as follows.

$$P_o = P_n \frac{1}{(1+i)^n} \quad (1\text{-}25\text{-}1)$$

Assume that a financial institution provides tax free interest on investment at a 5% annual interest rate compounded monthly.

For raffle payment option 1 the present value is:

$$\frac{\$25{,}000}{\left(1+\frac{0.05}{12}\right)^{10\times12}} = \$15{,}179.03 \quad (1\text{-}25\text{-}2)$$

For raffle payment option 2 the present value is:

$$\frac{\$20{,}000}{\left(1+\frac{0.05}{12}\right)^{5\times12}} = \$15{,}584.11 \quad (1\text{-}25\text{-}3)$$

For raffle payment option 3 the present value is:

$$\frac{\$17{,}500}{\left(1+\frac{0.05}{12}\right)^{3\times12}} = \$15{,}067.08 \quad (1\text{-}25\text{-}4)$$

For raffle payment option 4 the present value is:

$$\frac{\$16{,}000}{\left(1+\frac{0.05}{12}\right)^{1\times12}} = \$15{,}221.25 \qquad (1\text{-}25\text{-}5)$$

For raffle payment option 5 the present value is:

$$\frac{\$15{,}000}{\left(1+\frac{0.05}{12}\right)^{0\times12}} = \$15{,}000.00 \qquad (1\text{-}25\text{-}6)$$

For raffle payment option 6 the present value is:

$$\frac{\$5{,}000}{\left(1+\frac{0.05}{12}\right)^{1\times12}} + \frac{\$4{,}000}{\left(1+\frac{0.05}{12}\right)^{2\times12}} + \frac{\$4{,}000}{\left(1+\frac{0.05}{12}\right)^{3\times12}} + \frac{\$4{,}000}{\left(1+\frac{0.05}{12}\right)^{4\times12}} = \$15{,}096.93 \qquad (1\text{-}25\text{-}7)$$

Best raffle payment option is option 2 where you get paid $20,000 after 5 years and it has a present value of $15,584.11. Worst raffle payment option is option 5 where you get paid $15,000 when the winners are announced.

Best raffle option is annual interest rate dependent. For low annual interest rates, long term payment options are the best. As the annual interest rate increases, shorter term payment options become favorable. If you win the raffle and the annual interest rate is 7% or higher, you should take the $15,000 and enjoy it.

Chapter 1-26 Intensity of Sound

Intensity of sound I in Watts per meter squared is defined as average sound wave power per unit area. If spherical sound waves are emitted from a speaker at a power rating of P Watts, then the intensity of sound gets weaker as we get away from the speaker by the inverse square of the distance R in meters from the speaker.

$$I = \frac{P}{4\pi R^2} \tag{1-26-1}$$

Since the sound intensity varies many orders of magnitude from a whisper to a rock band, a logarithmic scale of base 10 is used to define it in units of decibels or dB.

$$dB = 10 \times \log\left(\frac{\mathrm{I}}{\mathrm{I_r}}\right) \tag{1-26-2}$$

where I_r is a reference sound intensity which is usually taken as the threshold of human hearing at 1×10^{-12} Watts/m^2. Eq. (1-26-2) can be rewritten as

$$dB = 10 \times \log\left(\frac{\mathrm{I}}{1\times10^{-12}}\right) \tag{1-26-3}$$

Typical sound sources and their sound intensity levels from low to high and their corresponding dB values are shown in Table 1-26-1.

Sound Source	**Sound Intensity Watts/m^2**	**Decibel, dB, Value Eq. (38-3)**
I_r reference sound, threshold of hearing	1x10^{-12}	0
Rustling leaf	1x10^{-11}	10
Library	1x10^{-8}	40
Vacuum cleaner at a distance of 1 m	1x10^{-5}	70
At a disco at a distance of 1 m from speaker	1x10^{-2}	100
Threshold of human ear pain	10	130
Jet engine at a distance of 50 m	100	140

Table 1-26-1 Typical sound sources and their sound intensity levels and decibel values

If we want to listen to music at a rock concert or in our home theater without damaging our hearing system, how far should we sit from the speakers? The answer to this question can be estimated by combining Eqs. (1-26-1) and (1-26-3).

$$R = \sqrt{\left(\frac{P}{4\pi}\right)\frac{1}{1\times10^{-12}\times10^{\left(\frac{dB}{10}\right)}}} \tag{1-26-4}$$

For a speaker power that is in full blast and for a desired listening decibel level, the distance that we should be away from the sound source is calculated using Eq. (1-26-4) and it is presented in Figure 1-26-1. Listening distance is shown in meters as a function of sound source power in Watts for different listening sound intensity levels in decibels.

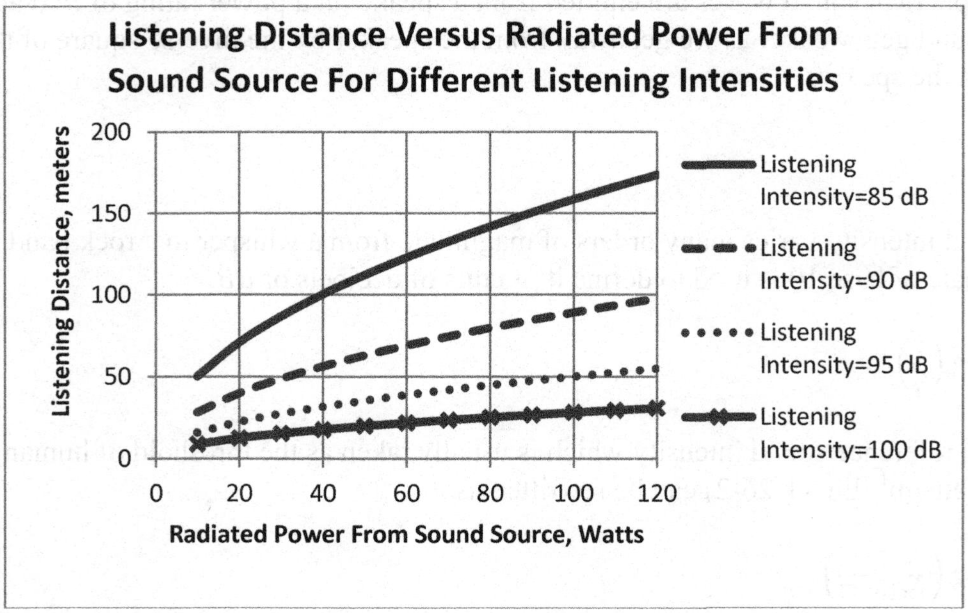

Figure 1-26-1 Listening distance versus radiated power from sound source for different listening intensities

Figure 1-26-1 can be interpreted best by an example. If we are going to listen to a music that is coming out of a 90 Watt speaker at full blast and if we want to keep the intensity of sound that is coming to our ears around 85 dB, than we should move away from the speaker about 150 meters.

Chapter 1-27 Brightness of Stars

The astronomers starting with the Greeks classified the stars according to their brightness. Human eye on Earth responds to luminous flux coming from the stars in base ten logarithmic scale. A star's apparent brightness magnitude, M, to a human eye can be written as follows.

$$M = C_1 \log_{10} F + C_2 \tag{1-27-1}$$

where F is the luminous flux coming from a star in lumens/m^2 and C_1 and C_2 are constants. If a bright star with an apparent brightness magnitude of M_1 is compared to a dim star with an apparent brightness magnitude of M_2, Eq. (1-27-1) takes the following form.

$$M_1 - M_2 = C_1 \log_{10}\left(\frac{F_1}{F_2}\right) \tag{1-27-2}$$

The astronomers categorized the brightest stars in an apparent brightness magnitude of 1 and the dimmest stars in an apparent brightness magnitude of 6. They also determined that the brightest stars have 100 times more luminous flux than the dimmest stars that the eye can see. With these conditions, the constant, C_1, in Eq. (1-27-2) can be determined.

$$1 - 6 = C_1 \log_{10} 100 = 2C_1 \quad \text{or} \quad C_1 = -2.5 \tag{1-27-3}$$

Eq. (1-27-2) can be rewritten as

$$\frac{F_1}{F_2} = 10^{\frac{M_1 - M_2}{-2.5}} \tag{1-27-4}$$

As telescopes and light detection instrumentation improved, the apparent brightness magnitude values were extended to negative values in order to be able to categorize very bright objects in the sky such as the Sun or the full Moon. The following Table 1-27-1 gives the range apparent brightness magnitudes detectable by humans at this time.

Objects In The Sky	**Apparent Brightness Magnitude**
Sun	-26.8
Full Moon	-12.6
Venus	-4.4
Sirius	-1.4
Dimmest Star To Naked Eye	+6
Dimmest Star To Largest Telescope	+30

Table 1-27-1 Range of apparent brightness magnitudes from objects in the sky

Ratio of luminous flux coming from the Sun to the human eye as compared to the full Moon can be calculated from Eq. (1-27-4) even if one should not look into the Sun with unprotected eyes.

$$\frac{F_{Sun}}{F_{Moon}} = 10^{\frac{-26.8-(-12.6)}{-2.5}} = 478{,}630 \quad (1\text{-}27\text{-}5)$$

Eq. (1-27-5) shows how powerful the rays we receive from the Sun are in the visible band. They are 478,630 times as powerful as the full Moon's rays in the visible band. Luminuous flux coming from the Sun as compared to the luminuous flux coming from other objects in the sky are shown in LOG_{10} scale in Figure 1-27-1. The Sun has 10^{23} times more luminuous flux coming to Earth than the faintest star in the sky.

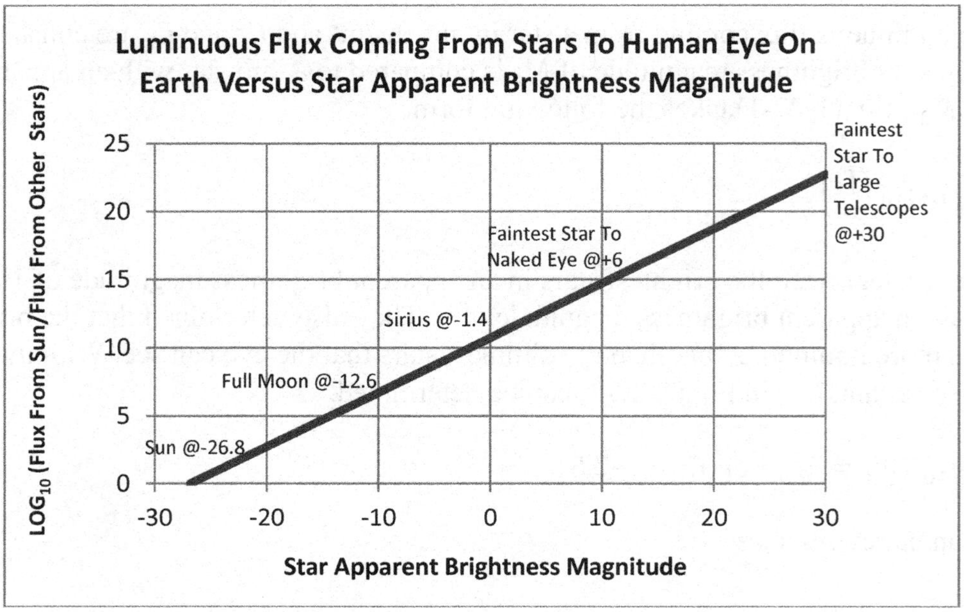

Figure 1-27-1 Luminuous flux coming from stars to human eye on Earth versus star apparent brightness magnitude

Chapter 1-28 Water Vapor in Saturated Air

Experimentally it was determined that water vapor in saturated air behaves exponentially with air temperature. Water vapor pressure versus dry bulb air temperature in saturated or 100% humid air is shown in Figure 1-28-1. The empirical exponential equation that defines the water vapor pressure curve is given in Eq. (1-28-1) , see World Meteorological Organization, Technical Regulations, No. 49.

$$P_{water\ vapor} = 611 \times exp\left(\frac{17.27 \times T}{237.3 + T}\right) \qquad (1\text{-}28\text{-}1)$$

where $P_{water\ vapor}$ is in Pascals (N/m^2) and T is the air temperature in C. At sea level, water vapor pressure approaches 1 atmosphere (101,325 N/m^2) at boiling temperature of 100 C.

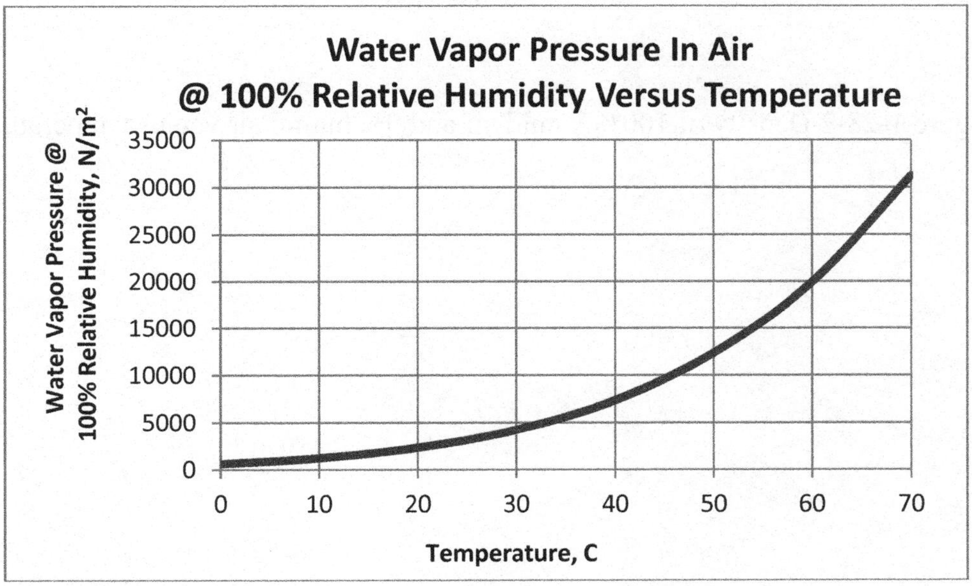

Figure 1-28-1 Water vapor pressure in saturated air versus air temperature

Humid air is heavier than dry air. This can be shown by calculating the densities of water vapor and dry air using the ideal gas law.

$$\rho_{humid\ air} = \rho_{dry\ air} + \rho_{water\ vapor} \qquad (1\text{-}28\text{-}2)$$

Equation (1-28-2) can be rewritten as follows.

$$\rho_{humid\ air} = \frac{P_{dry\ air}}{R_{dry\ air} \times T} + \frac{P_{water\ vapor}}{R_{water\ vapor} \times T} \qquad (1\text{-}28\text{-}3)$$

where the pressure of dry air at sea level is $P_{dry\ air} = 101{,}325 - P_{water\ vapor}$ in N/m^2, the air temperature T is in degrees Kelvin, the gas constant for dry air is $R_{dry\ air} = 287.05$ in N-m/kg-K, and the gas constant for water vapor is $R_{water\ vapor} = 461.495$ in N-m/kg-K. Densities of 100% humid air and 0% humid air are shown as a function of air temperature in Figure 1-28-2. As air temperature increases, 100% humid air weighs less and less.

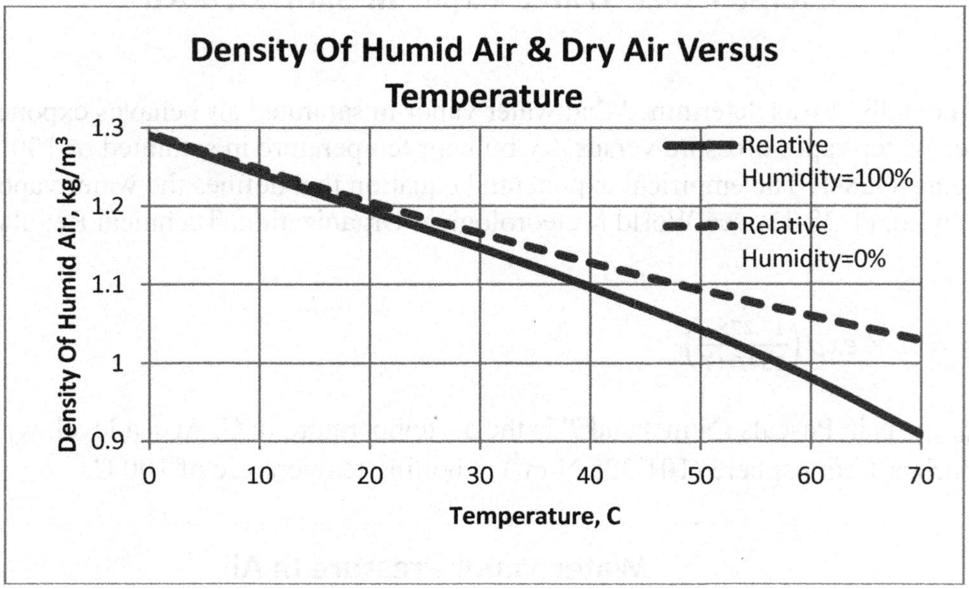

Figure 1-28-2 Density of 100% humid air and 0% humid air versus temperature

Chapter 1-29 Supply and Demand

Demand on a product increases as its price goes down. The opposite trend occurs for the producer side. As the as the price goes down, so does the supply to the market. A vehicle manufacturing companies' marketing department estimates that unit price of one of their products, P, decreases exponentially as the quantity per year demand, Q, increases.

$$P = A \times e^{(-B \times Q)} \tag{1-29-1}$$

The marketing group has two data points to find out constants A and B in Eq. (1-29-1).

When = $1{,}000\ units/year$, P = $\$200{,}000\ per\ unit$.
When = $10{,}000\ units/year$, P = $\$80{,}000\ per\ unit$.

From these two conditions, two equations are obtained by taking natural logarithm of both sides of Eq. (1-29-1).

$$\ln(200{,}000) = \ln(A) - 1{,}000 \times B \tag{1-29-2}$$

and

$$\ln(80{,}000) = \ln(A) - 10{,}000 \times B \tag{1-29-3}$$

From Eqs. (1-29-2) and (1-29-3), first constant B is determined. Then B is plugged into one of the equations to determine the constant A. Final form of the demand equation, Eq. (1-29-1), becomes

$$P = 221{,}435 \times e^{(-0.000102 \times Q)} \tag{1-29-4}$$

On the supply side, the production department estimates that vehicle unit price increases linearly with supply quantity per year.

$$P = M \times Q + N \tag{1-29-5}$$

The production department has the following conditions to determine constants M and N for the supply Eq. (1-29-5).

When = $1{,}000\ units/year$, P = $\$50{,}000\ per\ unit$.
When = $10{,}000\ units/year$, P = $\$120{,}000\ per\ unit$.

From these two conditions, two equations are obtained to solve for constants M and N.

$$50{,}000 = M \times 1{,}000 + N \tag{1-29-6}$$

and

$$120{,}000 = M \times 10{,}000 + N \tag{1-29-7}$$

From Eqs. (1-29-6) and (1-29-7), first constant M is determined. Then M is plugged into one of the equations to determine the constant N. Final form of the supply equation, Eq. (1-29-5), becomes

$$P = 7.78 \times Q + 42{,}222 \qquad (1\text{-}29\text{-}8)$$

Supply and demand curves for this particular vehicle are shown in Figure 1-29-1. The point where the two curves intersect is the ideal or equilibrium point for the manufacturer. Equilibrium unit price in this case is \$101,676 and the equilibrium production (supply) quantity per year is 7,644. At this equilibrium quantity there will be no left over inventories. Everything that is being built will be sold.

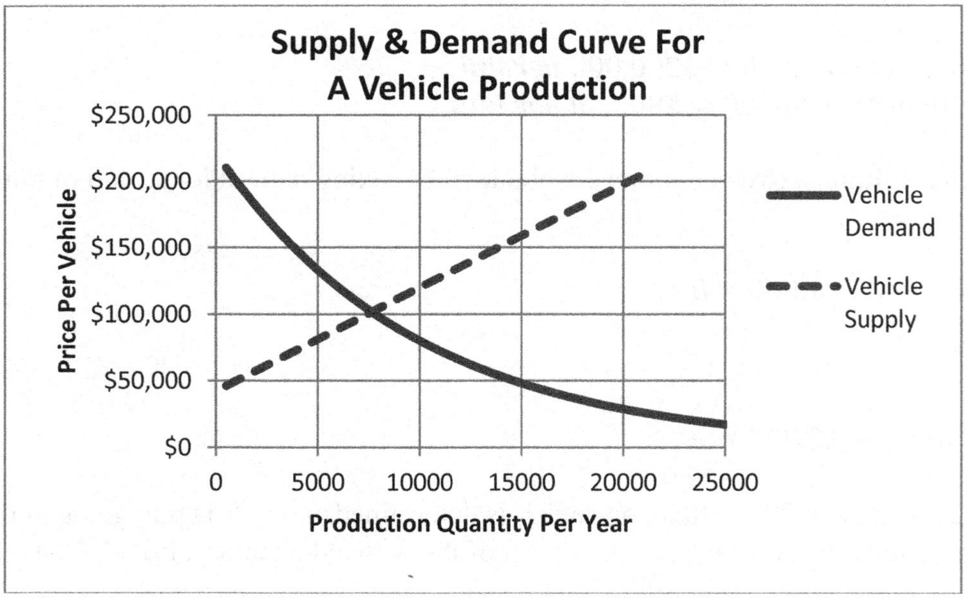

Figure 1-29-1 Supply and demand curve for a vehicle production

Chapter 1-30 Venturi Effect

Venturi effect causes the pressure to drop when a fluid goes through a throat in an internal incompressible flow as shown in Figure 1-30-1. There is a limit to how much fluid can go through a throat or how narrow the throat can be. At this limit the fluid flow is choked. If the fluid is a gas, like air, the choking limit is the speed of sound or approaching Mach number equals to one. If the fluid is water, choking occurs as the pressure in the throat approaches water vapor pressure which is close to zero pressure for water. As the pressure at the throat approaches zero for water flow, water vaporizes into small bubbles and causes cavitation.

Venturi effect is used in carburetors to feed fuel to air flow. Air stream is regulated from our accelerator pedal. As the air stream speed increases, the pressure at the carburetor throat decreases and it sucks in more fuel. Similar effect is utilized in a wine aerator. Wine flows through a throat and the pressure decreases and the aerator sucks in air into the wine to enhance flavor and give a smoother taste.

For steady, incompressible, frictionless, and constant level flows it can be shown that mechanical energy between points 1 and 2 in Figure 1-30-1 is conserved as follows, see Reference [5].

$$\frac{P_1}{\rho} + 0.5 \times V_1^2 = \frac{P_2}{\rho} + 0.5 \times V_2^2 \qquad (1\text{-}30\text{-}1)$$

The flow continuity for a steady and incompressible fluid can be written as

$$V_1 A_1 = V_2 A_2 \qquad (1\text{-}30\text{-}2)$$

Combining Eqs. (1-30-1) and (1-30-2), the velocity at point 1, V_1 , can be determined, if the areas A_1 and A_2, pressure difference $(P_1 - P_2)$, and the fluid density ρ are known.

$$V_1 = A_2 \sqrt{\frac{2(P_1 - P_2)}{\rho(A_1^2 - A_2^2)}} \qquad (1\text{-}30\text{-}3)$$

If a differential manometer with mercury which has a density ρ_{Hg} is inserted between points 1 and 2 as shown in Figure 1-30-1, the pressure difference $(P_1 - P_2)$ can be determined from the following static pressure relation.

$$P_1 + \rho g H = P_2 + \rho_{Hg} g H + \rho g (R_1 - R_2) \qquad (1\text{-}30\text{-}4)$$

where R_1 and R_2 are radii at points 1 and 2 for the fluid pipe. Eq. (1-30-3) can be rewritten as

$$V_1 = A_2 \sqrt{\frac{2(\rho_{Hg} - \rho) g H + \rho g (R_1 - R_2)}{\rho(A_1^2 - A_2^2)}} \qquad (1\text{-}30\text{-}5)$$

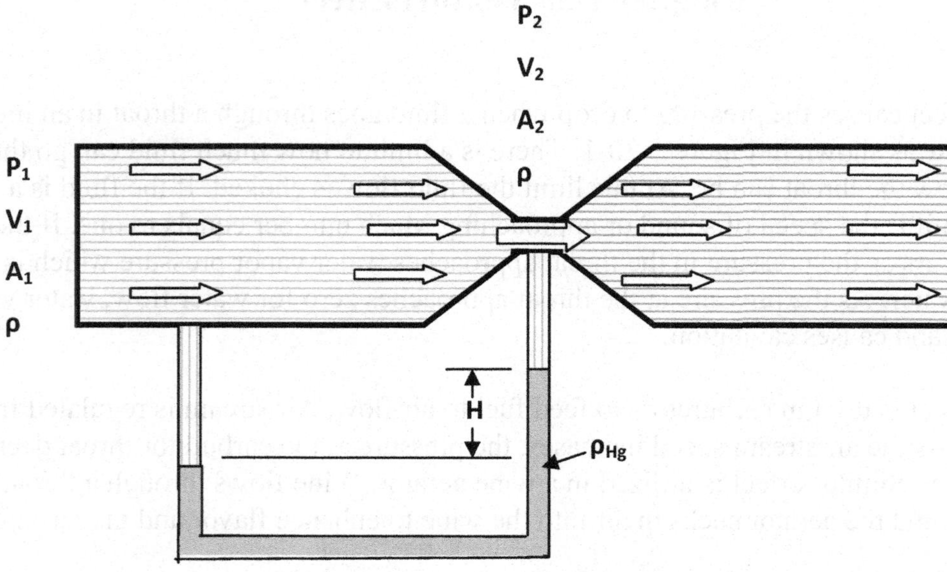

Figure 1-30-1 Venturi effect for a fluid going through a throat

For water fluid density $\rho = 1000\ kg/m^3$, manometer mercury density $\rho_{Hg} = 13500\ kg/m^3$, $A_1 = 0.1\ m^2$, and $A_2 = 0.02\ m^2$, Eq. (1-30-5) gives the results shown in Figure 1-30-2 for velocity V_1 at inlet of Venturi meter as a function of manometer fluid height difference, H. Velocity at point 1 increases as the square root of manometer fluid height difference.

Next step is to determine the choked flow limits for the velocity of water at point 1. At room temperature, say 20 C, the water vapor pressure is 0.023 atmospheres. If the pressure at point 1 is assumed to be 3 atmospheres (1 atmosphere = 101,300 N/m^2), the velocity at point 2 approaches choked flow when the velocity at point 1 approaches 5 m/s as shown in Figure 1-30-3. At choked flow, velocity at point 2 approaches 25 m/s.

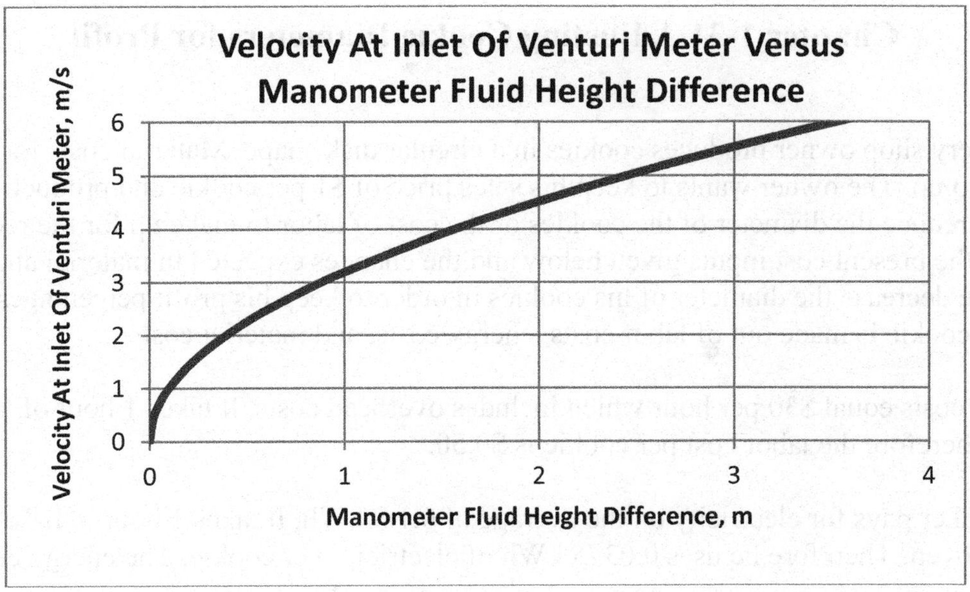

Figure 1-30-2 Velocity at inlet of Venturi meter versus manometer fluid height difference

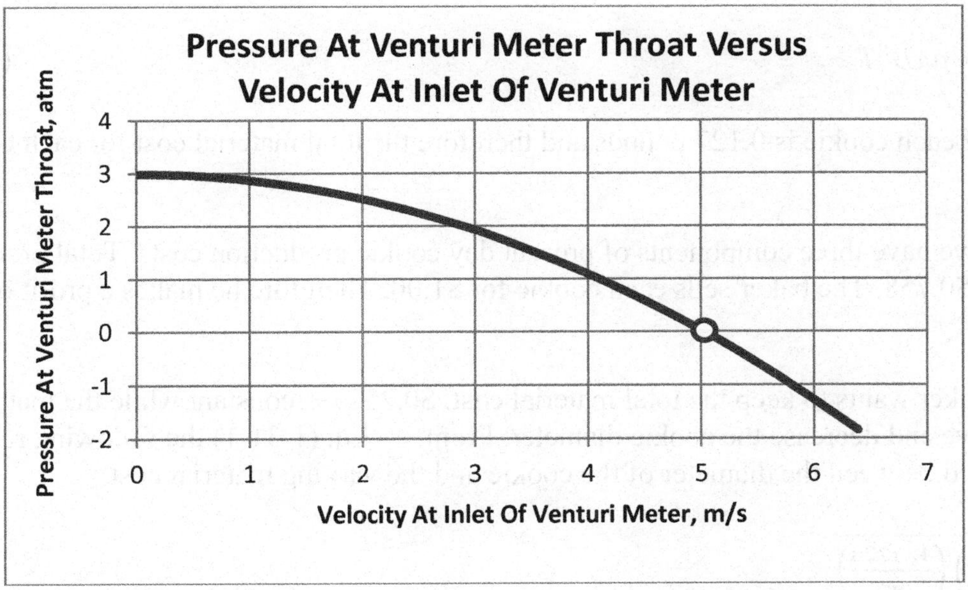

Figure 1-30-3 Pressure at Venturi meter throat versus velocity at inlet of Venturi meter

Chapter 1-31 Adjusting Cookie Diameters for Profit

A bakery shop owner produces cookies in a circular disk shape. Material costs for the cookies are going to go up. The owner wants to keep his sales price of \$1 per cookie and production volume the same, but reduce the diameter of the cookies or the cost of labor to make up for the reduction in his profits. With the present cost inputs given below and the changes expected in material and labor costs, how should he decrease the diameter of his cookies in order to keep his profit percentages the same? The cost of a cookie is made out of labor costs, energy costs, and material costs.

Labor costs equal \$30 per hour which includes overhead costs. It takes 1 hour of labor to make 60 cookies. Therefore the labor cost per cookie is \$0.50.

The baker pays for electricity on the average of 0.1 \$/kWh. It takes 1 hour to bake 40 cookies in a 1,500 W oven. Therefore he uses 0.0375 kWh of electricity per cookie. The energy cost per cookie is \$0.00375.

The material cost, C, is \$2.00 per pound at the present time. The diameter of every cookie, D, is 4 inches and its thickness, T, is 0.25 inches. The density of cookie material, ρ, is 70 pounds per cubic feet. The weight of a cookie, W, is given with the following formula.

$$W = \frac{1}{4\times1728} \times \rho\pi D^2 T \qquad (1\text{-}31\text{-}1)$$

The weight of each cookie is 0.127 pounds and therefore the total material cost for each cookie is \$0.2545.

Now we have three components of present day cookie production costs. Total production cost per cookie is \$0.758. The baker sells each cookie for \$1.00. Therefore he makes a profit of \$0.242 or 31.9%.

The baker wants to keep the total material cost, \$0.2545, a constant while the material is getting more expensive and decrease the cookie diameter. From Eq. (1-31-1) the following relationship can be obtained between the diameter of the cookie and the varying material cost.

$$D = \sqrt{\left(\frac{0.2545}{C}\right)\left(\frac{4\times1728}{\rho\pi T}\right)} \qquad (1\text{-}31\text{-}2)$$

Cookie diameter reduction versus material cost increase in order to be able to keep a constant 31.9% profit with no reduction in labor costs is shown in Figure 1-31-1. If the baker can reduce his labor costs by 5%, namely a reduction from \$0.50 per cookie to \$0.475 per cookie, he can increase his allowance, A, for material costs by \$0.025. If the baker can reduce his labor costs by 10%, namely a reduction from \$0.50 per cookie to \$0.45 per cookie, he can increase his allowance for material costs by \$0.05. Then Eq. (1-31-2) becomes

$$D = \sqrt{\left(\frac{0.2545+A}{C}\right)\left(\frac{4\times1728}{\rho\pi T}\right)} \qquad (1\text{-}31\text{-}3)$$

Results of reducing the labor costs by 5% and by 10% are also shown in Figure 1-31-1. If the baker can reduce his labor costs by 5%, he does not have to change the diameter of his cookies until the material costs go up by more than 9.8% or 2.196 \$/lb. If the baker can reduce his labor costs by 10%, he does not have to change the diameter of his cookies until the material costs go up by more than 19.65% or 2.393 \$/lb.

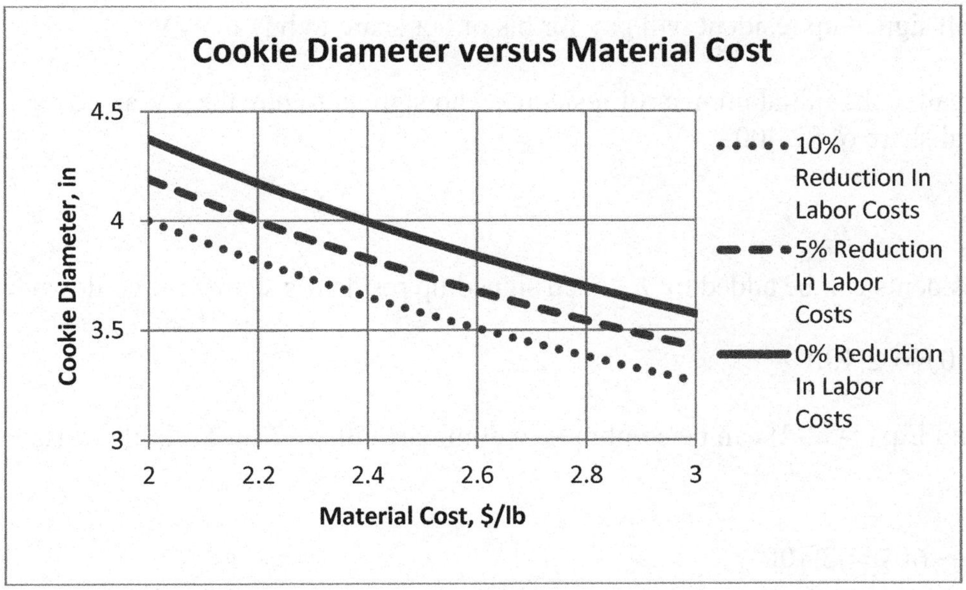

Figure 1-31-1 Cookie diameter versus material cost, \$/lb, for no reduction in labor costs, for 5% reduction in labor costs, and for 10% reduction in labor costs

Chapter 1-32 Buying a Common Lounge TV

A group of apartment residents want to buy a TV for their common lounge. TV costs $2,400. Initially 50% of the residents sign up for it. If they could sign up two more residents, each signed up resident will pay $60 less for their share. What is total number of residents in this apartment complex? How much each signed up resident will pay for his or her share to buy the TV?

Let us call X the initial number of residents who sign up to buy the TV and S each signed up resident's initial share of $2,400.

$$XS = 2{,}400 \tag{1-32-1}$$

If two more residents can be added to X, each signed up resident's share will go down by $60 or

$$(X + 2)(S - 60) = 2{,}400 \tag{1-32-2}$$

Eq. (1-32-1) and Eq. (1-32-2) can be combined to eliminate either X or S. In this case let us eliminate X.

$$\left(\frac{2{,}400}{S} + 2\right)(S - 60) = 2{,}400 \tag{1-32-3}$$

Multiplying both sides of Eq. (1-32-3) by S and rearranging gives a quadratic equation.

$$S^2 - 60S - 72{,}000 = 0 \tag{1-32-4}$$

Solutions to Eq. (1-32-4) by using the quadratic formula are

$$S = \frac{60 \pm \sqrt{60^2 + 4 \times 72{,}000}}{2} = 30 \pm \sqrt{72{,}900} = 30 \pm 270 \tag{1-32-5}$$

The negative solution to S is not applicable in this case. Therefore the initial share of each signed up resident is $300. From Eq. (1-32-1), the number of initially signed up residents is 8. If this is 50% of the total number of residents in the apartment complex, there are a total of 16 residents.

After adding 2 more residents to the TV purchase pool, there will be 10 residents who want to share the $2,400 cost. Each one of the 10 residents has to pay $240 which is $60 less than the initial $300 share.

Chapter 1-33 Typist Base and Bonus Pays

A publishing house pays the manuscript typists a base pay and a bonus pay. Base pay is directly proportional to the number of pages typed per hour. The bonus pay is added to the base pay, but it is inversely proportional to the average errors made per page.

Base pay is $5 per page. Each typist has a typing speed of W words minute. The manuscripts contain on the average 300 words per page. In this publishing house who earns more money, a typist with a speed of 30 words/min and makes on the average 1 error/page or a typist with a typing speed of 35 words/min and makes on the average 5 errors/page?

Base pay per hour of the typists can be expressed as follows.

$$Base\ Pay = \frac{(5)\frac{\$}{page}(W)\frac{words}{min}(60)\frac{min}{hr}}{(300)\frac{words}{page}} = W\frac{\$}{hr} \qquad (1\text{-}33\text{-}1)$$

Bonus pay is 3 \$ per page for zero errors per page, 1.5 \$ per page for 1 error per page, 1 \$ per page for 2 errors per page, etc. So the bonus pay goes inversely proportional to the number of errors per page as follows.

$$Bonus\ Pay = \frac{3}{(E+1)}\frac{\$}{page} \qquad (1\text{-}33\text{-}2)$$

where E is the number of errors per page and typists are allowed a maximum of 10 errors per page for bonus pay. Bonus pay can be plotted as a function of error per page as shown in Figure 1-33-1.

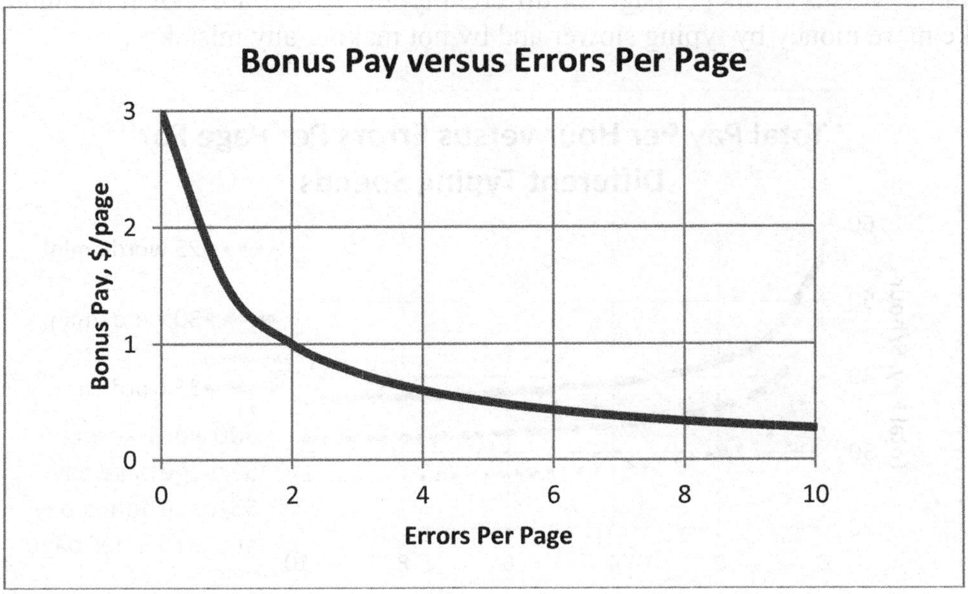

Figure 1-33-1 Bonus pay versus errors per page

Bonus pay that is given by Eq. (1-33-2) can be multiplied by the pages per hour to obtain the hourly bonus pay.

$$Bonus\ Pay = \left(\frac{3}{E+1}\right)\frac{\$}{page}\left[\frac{(W)\frac{words}{min}(60)\frac{min}{hr}}{(300)\frac{words}{page}}\right] = \left(\frac{3}{E+1}\right)\left(\frac{W}{5}\right)\frac{\$}{hr} \quad (1\text{-}33\text{-}3)$$

Total pay is the sum of the base pay and the bonus pay. For a typist with 30 words per minute speed, hourly base pay and total pay are shown in Figure 1-33-2.

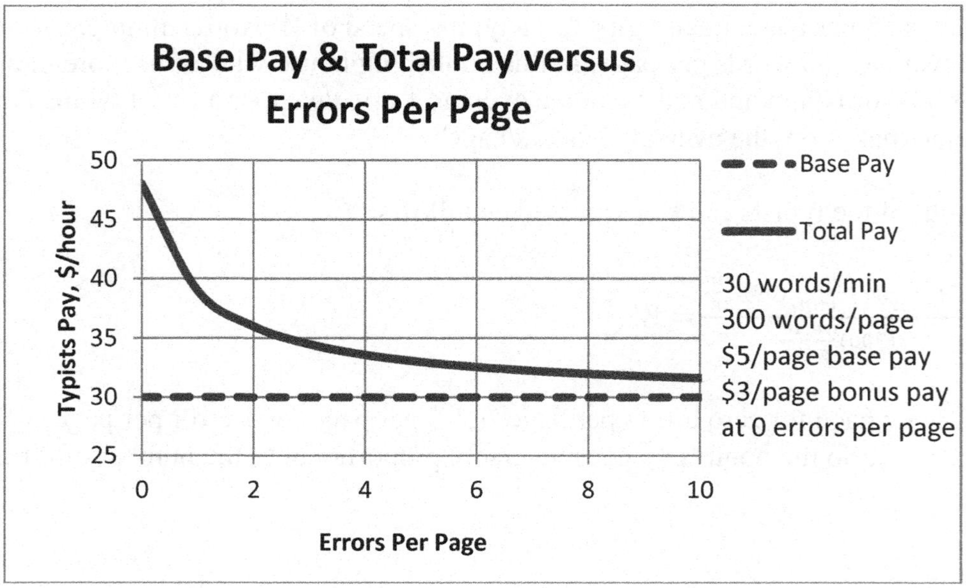

Figure 1-33-2 Base pay and total pay versus errors per page for a typist with a 30 words per minute speed

Total pays per hour versus errors per page for different typing speeds are shown in Figure 1-33-3. A typist can make more money by typing slower and by not making any mistakes.

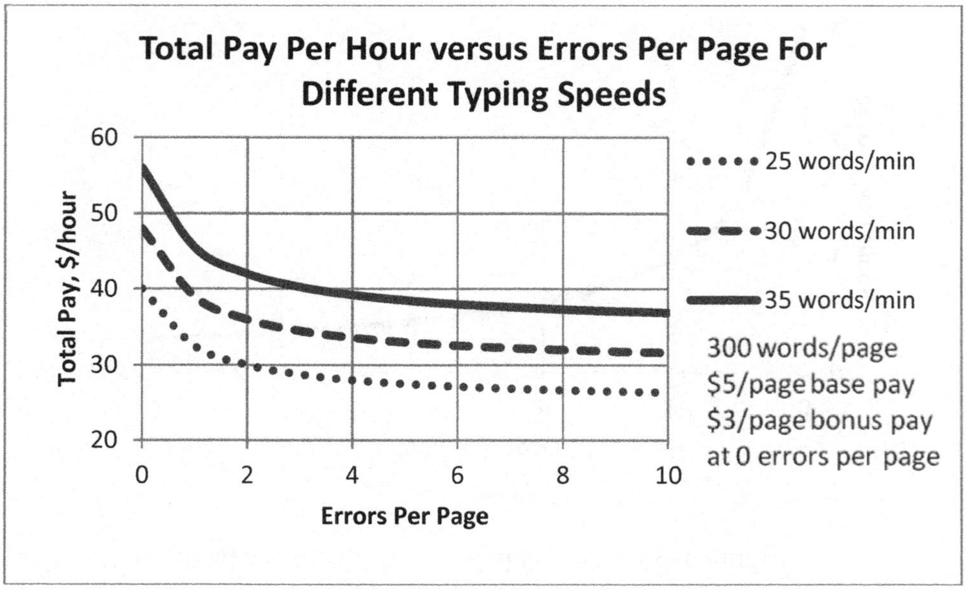

Figure 1-33-3 Total pay per hour versus errors per page for different typing speeds

From the results shown in Figure 1-33-3, a typist with a speed of 30 words/min and makes on the average 1 error/page earns $39 per hour and a typist with a typing speed of 35 words/min and makes on the average 5 errors/page earns $38.5 per hour.

Chapter 1-34 Distribution of Bonuses

A team of three people prepare for a TV contest. There is a team leader and two associates. The contest awards $242,000 to the winning team. If they win the contest, each team member has to pay 37% federal and 9% state taxes from their share of the prize. The team agrees among themselves before the contest that if they win the contest, the team leader will get a net prize of $25,000 more than the net equal prize of each associate.

They all start to dream that they are going to win the contest and start to calculate their net earnings from their upcoming success.

Let L denote the prize share for the leader and A denote the prize share for an associate. By considering the net prize amounts, the following relationship holds for the total net prize after taxes are taken out.

$$L_{net} + 2 \times A_{net} = (1 - 0.37 - 0.09) \times \$242{,}000 = \$128{,}260 \qquad (1\text{-}34\text{-}1)$$

The second condition is that the team leader will get a net prize of $25,000 more than the net prize of each associate.

$$L_{net} - A_{net} = \$25{,}000 \qquad (1\text{-}34\text{-}2)$$

There are two equations and two unknowns. If Eq. (1-34-2) is subtracted from Eq. (1-34-1), the following relationship is obtained.

$$3 \times A_{net} = \$103{,}260$$

or each associate will get a net prize of

$$A_{net} = \$34{,}420$$

Now the value of A_{net} can be substituted into either Eq. (1-34-1) or Eq. (1-34-2) to obtain L_{net}. The leader will get a net prize of

$$L_{net} = \$59{,}420$$

Since 47% of the gross prize goes to taxes, gross prize portion for each associate and the leader can be calculated from

$$(1 - 0.47) \times A_{gross} = A_{net} \quad \text{or} \quad A_{gross} = \$64{,}943.40$$

and

$$(1 - 0.47) \times L_{gross} = L_{net} \quad \text{or} \quad L_{gross} = \$112{,}113.30$$

Taxes to be paid by each associate and the leader can be calculated from

$$A_{taxes} = 0.47 \times A_{gross} = \$30{,}523.40$$

and

$$L_{taxes} = 0.47 \times L_{gross} = \$52{,}693.20$$

Double checking the results is always a good exercise.

$$(L_{net} + L_{taxes}) + 2 \times (A_{net} + A_{taxes}) = \$242{,}000$$

or

$$(\$59{,}420 + \$52{,}693.20) + 2 \times (\$34{,}420 + \$30{,}523.40) = \$242{,}000$$

Chapter 1-35 Relative Humidity

Relative humidity is related to how much water vapor is in the air. The ratio of how much water vapor (water vapor partial pressure) is in the air to the maximum amount water vapor (water vapor saturation pressure) that can be added to the air at a given ambient temperature and barometric pressure is the relative humidity that we hear from weather forecasters every day.

One way of measuring relative humidity is by using two different thermometers. One thermometer measures the ambient (dry bulb) temperature and the other measures the wet bulb temperature. The wet bulb thermometer is covered by a thin layer of water or ice. Evaporation from the wet bulb thermometer lowers the wet bulb temperature. The difference between the ambient and wet bulb temperature allows one to determine the relative humidity in the air. Once the ambient temperature, wet bulb temperature, and the barometric pressure are known, relative humidity can be determined from a psychrometric chart developed for that particular barometric pressure. However, there are also semi- empirical equations that were developed, see Reference [6], to determine the relative humidity by knowing the ambient temperature, wet bulb temperature, and the barometric pressure.

The semi-empirical equation for relative humidity that is considered here is for ambient temperatures -58 °F to +212 °F over water and for a wet bulb thermometer that is covered by a thin film of water.

$$RH = [(T + 459.4)^{4.9253}] \times \left[10^{\left(-23.2801+\frac{5287.32}{T+459.4}\right)}\right] \times \left\{[(TWB + 459.4)^{4.9253}] \times \left[10^{\left(23.2801-\frac{5287.32}{TWB+459.4}\right)}\right] - 10^{-4} \times (3.595 + 0.002336 \times TWB) \times P \times (T - TWB)\right\} \quad (1\text{-}35\text{-}1)$$

where T is the ambient temperature in °F, TWB is the wet bulb temperature in °F, and P is the barometric pressure in inches of mercury (in-Hg). Both partial pressure and saturation pressure of water vapor are anti-logarithmic to the base ten functions of ambient temperature, wet bulb temperature, and barometric pressure. Eq. (1-35-1) is a ratio of these anti-logarithmic functions. Eq. (1-35-1) is plotted to show relative humidity for different ambient and wet bulb temperatures at a barometric pressure of 29.7 in-Hg.

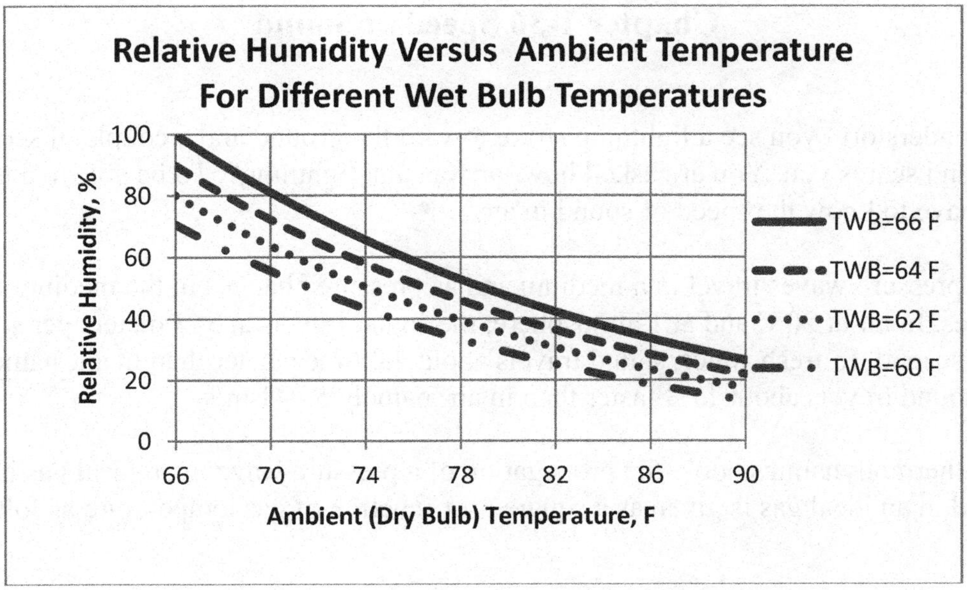

Figure 1-35-1 Relative humidity versus ambient temperature for different wet bulb temperatures at a barometric pressure of 29.7 in-Hg

Chapter 1-36 Speed of Sound

In a thunderstorm you see a lightning strike toward the ground and a couple of seconds later a thundering sound scares you. You are asked how far was that lightning? To be able to answer this question you have to know the speed of sound in air.

Sound pressure waves travel in a medium and as pressure changes in the medium the medium's volume changes. In air at 20°C and at 1 atmosphere, the sound travels at 343 meters per second or 1,125 feet per second. In fresh water, sound travels about 4.3 times faster than in air, namely 1,482 m/s. In steel, sound travels about 17.3 faster than in air, namely 5,941 m/s.

In any thermodynamics book, the propagation of a pressure wave in an ideal gas is treated. Speed of sound in an ideal gas is given as a square root function of gas temperature as follows, see Reference [7].

$$S = \sqrt{kRT} \qquad (1\text{-}36\text{-}1)$$

where S is the speed of sound in a gas medium is m/s, k is the ratio of specific heat at constant pressure to specific heat at constant volume (for air k = 1.4), R is the gas constant in m^2/s^2-K which is the same as J/kg-K (286 m^2/s^2-K for air), T is the temperature in degrees Kelvin which is 273.15+°C. For air the speed of sound, Eq. (1-36-1), can be rewritten as

$$S = 20\sqrt{T_c + 273.15} \quad m/s \qquad (1\text{-}36\text{-}2)$$

where T_C is the ambient air temperature in °C. Eq. (1-36-1) can also be expressed in the British system of units as follows.

$$S = \sqrt{kg_cRT} = \sqrt{1.4 \times 32.2 \frac{lbm{-}ft}{lbf{-}s^2} \times 53.3 \frac{ft{-}lbf}{lbm{-}°R} \times (T_F + 459.69)}$$

or

$$S = 49\sqrt{T_F + 459.69} \quad ft/s \qquad (1\text{-}36\text{-}3)$$

Speed of sound from Eq. (1-36-2) along with the time it takes for sound to travel one kilometer are shown as a function of ambient air temperature in Figure 1-36-1.

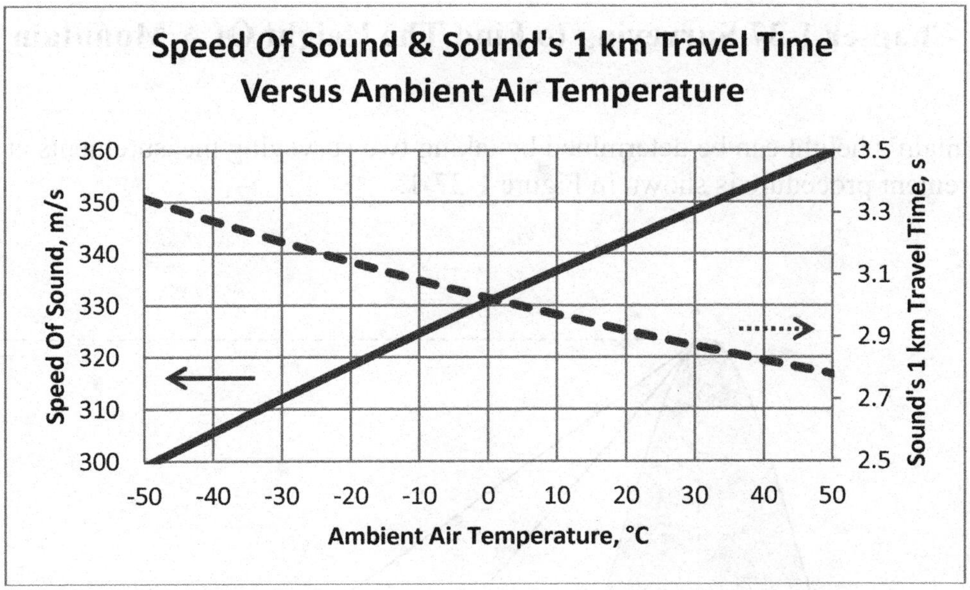

Figure 1-36-1 Speed of sound and sound's 1 km travel time versus ambient air temperature

Figure 1-36-1 shows that the speed of sound varies $\pm 10\%$ over the range of temperatures that are experienced on Earth's surface.

An approximation to the travel distance of sound in ambient air at 20°C can be remembered as sound travels 1 kilometer distance in 3 seconds (2.8% error) or 1 mile of in 5 seconds (6.1% error). During a thunderstorm, first you see a lightning because light travels 300,000 km/s or 186,000 mi/s in vacuum or very close to its vacuum value in ambient air (since the refractive index of air for visible light is almost unity, i.e. 1.0003). The speed of light is about 10^6 times faster than the speed of sound. For example, if the thunder sound is heard 9 seconds after than the lightning was seen, then you can estimate that the thundering clouds are about 9/3=3 kilometers or 9/5=1.8 miles away from where you are.

Chapter 1-37 Surveying to Find The Height Of A Mountain

A mountain's height can be determined by taking two surveying measurements at its base. A typical measurement procedure is shown in Figure 1-37-1.

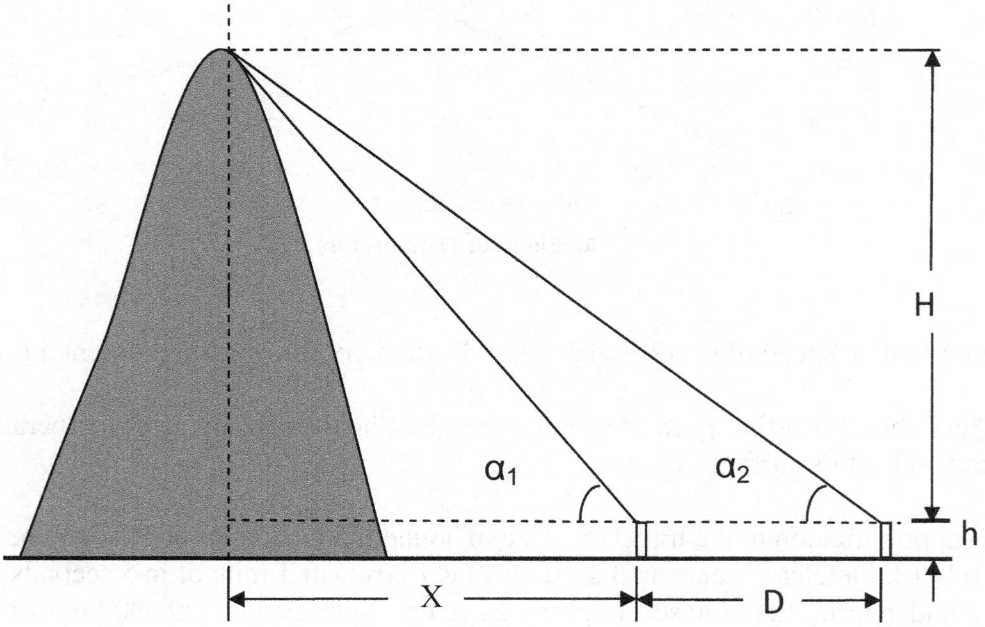

Figure 1-37-1 Surveying to find the height of a mountain

The surveyor measures two angles α_1 and α_2 to the peak of the mountain with his surveying equipment which is h feet high. Two angles are measured at a level surface at the base of the mountain at a distance of D feet apart. There are two equations, one from each triangle, and two unknowns, H and X, to solve this problem.

$$\frac{H-h}{X} = \tan\alpha_1 \qquad (1\text{-}37\text{-}1)$$

and

$$\frac{H-h}{X+D} = \tan\alpha_2 \qquad (1\text{-}37\text{-}2)$$

Combining Eqs. (1-37-1) and (1-37-2) provides the solution for the two unknowns.

$$X = \frac{D\tan\alpha_2}{\tan\alpha_1 - \tan\alpha_2} \qquad (1\text{-}37\text{-}3)$$

and

$$H = h + X\tan\alpha_1 = h + \frac{D\tan\alpha_1\tan\alpha_2}{\tan\alpha_1 - \tan\alpha_2} \qquad (1\text{-}37\text{-}4)$$

For a special case of α_1=40 degrees, α_2=35 degrees, D=1,000 feet, and h=6 feet, the height of the mountain, H, is 4,236.22 feet.

If the surveyor makes a $\pm$0.1 degree error in angle readings, maximum error comes when α_1=39.9 degrees, α_2=35.1 degrees. In this case, the height of the mountain, H, is 4,413.82 feet or 177.6 feet, 4.2%, over its true value.

Section 2 – Everyday Problems Solved Using Series Concepts

Main focus of the Section 2 is to be able to set up series solutions to everyday problems from the given information.

Finite series are utilized in solving seven everyday problems. Some of the series use arithmetic progression, in Problem 2-7, or geometric progression, in Problem 2-1, for determining payments for a fixed interest rate loan. In approximating functions, Taylor series expansion method is used in Problem 2-4 for tidal forces between two bodies. In Problem 2-2 for life cycle cost of owning a car, a double series is utilized. In Problem 2-3 best fit linear curve is derived by minimizing the sum of square of residuals. Best fit linear curve is used in maximizing profits from book sales. Best fit linear curve's application is also extended to power functions.

Chapter 2-1 Payments for a Fixed Interest Rate Loan

I want to buy a house or a car with a fixed interest rate loan. Let us call the principal loan amount "P". Loan term is the length of time over which all of the loan will be paid. Typical loan terms are 5, 15, or 30 years. Let us call the number of payment periods "N". For a 15-year loan, if the payments are monthly, the total number of payment periods will be 15x12=180. The payment period can be monthly, bi-monthly, weekly, etc. Also the fixed interest rate is always quoted as an annual interest rate. For example, if the annual fixed interest rate is 6%, the monthly fixed interest rate will be (6/12) = 0.5%. Fixed interest rate for the payment period will be called "i". The value of "i" for an annual fixed interest rate of 6% is (0.5/100) = 0.005. With all the information given above, the loan payment "M" for each payment period can be determined.

Let us look at the residual left after each payment period from the loan that was taken out. The residual of the loan at the time loan is given, payment period 0, is the principal "P". At the first payment period, the interest on the loan is "i x P". The residual after the first payment period is the sum of the residual from the previous payment period and the interest on that residual minus the payment "M". The residual after the second payment period is the sum of the residual from the previous payment period and the interest on that residual minus the payment "M". With this logic we can setup the Table 2-1-1 to show interest added and residual of loan for each payment period.

Payment Period	Interest Added	Residual of Loan
0	0	P
1	iP	P+iP-M=P(1+i)-M
2	i[P(1+i)-M]	P(1+i)-M+i[P(1+i)-M]-M= $P(1+i)^2-M(1+i)-M$
3	$i[P(1+i)^2-M(1+i)-M]$	$P(1+i)^3-M(1+i)^2-M(1+i)-M$
4	$i[P(1+i)^3-M(1+i)^2-M(1+i)-M]$	$P(1+i)^4-M(1+i)^3-M(1+i)^2-M(1+i)-M$
…	…	…
…	…	…
…	…	…
N	…	0

Table 2-1-1 Interest added and residual of loan for each payment period

It is obvious that after "N" payments, the residual should be zero which can be written as follows:

$$P(1+i)^N-[M(1+i)^{N-1}+M(1+i)^{N-2}+\ldots+M(1+i)^2+M(1+i)+M]=0 \quad (2\text{-}1\text{-}1)$$

Second term in Eq. (2-1-1) can be written in summation form as shown in Eq. (2-1-2).

$$P(1+i)^N - M\sum_{j=0}^{N-1}(1+i)^j = 0 \quad (2\text{-}1\text{-}2)$$

The series in Eq. (2-1-2) looks like a geometric progression with a common ratio of (1+i). The sum of this geometric progression series can be obtained by letting

$r = (1+i)$ and sum $= 1+r+r^2+r^3+\ldots+r^{N-1} = \sum_{j=0}^{N-1}(1+i)^j$.

"sum" can be obtained by finding "sum – r x sum" and solving for "sum" as follows:

r x sum = $r+r^2+r^3+\ldots+r^{N-1}+r^N$
sum – r x sum = $1 - r^N$
sum = $(1 - r^N)/(1 - r)$ if $r \neq 1$ (2-1-3)

"sum" in Eq. (2-1-3) is undefined for zero interest rate or r = 1. Eq. (2-1-2) can be rewritten by using Eq. (2-1-3).

$$P(1+i)^N - M\left\{\frac{[1-(1+i)^N]}{[1-(1+i)]}\right\} \quad (2\text{-}1\text{-}4)$$

Loan payment "M" for each payment period is

$$M = P\left\{\frac{i(1+i)^N}{(1+i)^N - 1}\right\} \quad (2\text{-}1\text{-}5)$$

Using Eq. (2-1-5), loan payments per payment period can be obtained for any combination of principal "P", fixed interest rate for the payment period "i", and number of payment periods "N". Typical loan payments "M" per payment period are given in Table 2-1-2.

Principal "P"	**Fixed Interest Rate Per Year**	**Loan Term In Years**	**Number Of Payment Periods**	**Loan Payments "M"**
\$100,000	5%	7.5	180 (bi-monthly)	\$666.79
\$100,000	5%	15	180 (monthly)	\$790.79
\$100,000	5%	15	360 (bi-monthly)	\$395.12
\$100,000	5%	15	780 (weekly)	\$182.29
\$100,000	5%	30	360 (monthly)	\$536.82
\$100,000	5%	30	720 (bi-monthly)	\$268.29
\$100,000	5%	30	1560 (weekly)	\$123.80

Table 2-1-2 Fixed interest rate loan payments per payment period

Let us calculate the interest paid at each payment period. The interest paid at the N'th payment period should be the fixed interest for the payment period times the residual after the (N-1)'st payment period or

Interest Paid At The Nth Payment = i x Residual After the $(N-1)^{th}$ Payment

For N=1, interest paid is $i \times P$
For N≥2, interest paid is $i \times \left[P(1+i)^{N-1} - M\sum_{j=0}^{N-2}(1+i)^j\right]$ (2-1-6)

The sum of this geometric progression series in Eq. (2-1-6) can be obtained similar to the one before to get to the final form of the interest equation for the N'th payment period.

For N=1, interest paid is $i \times P$
For N≥2, interest paid is $i \times \left[P(1+i)^{N-1} + M\left[\frac{1-(1+i)^{N-1}}{i}\right]\right]$ (2-1-7)

On the interest that is paid for home loans, there is always a federal tax break. This tax break depends on income and on deductibles, etc. In the present calculations, it will be assumed that a 25% federal tax break is achieved for the interest paid for a home loan.

Another factor that confounds the value of money that goes out of pocket is inflation. The best way to solve the inflation issue is to evaluate all the money that is going to go out of pocket in the future by bringing its value to present. This is called present value analysis. For present value analysis, a common method is to use the interest rate on treasury bills. In the present calculations, a 3% treasury bill interest rate is used for present value analysis. The present value of money is defined as

$$\text{Present Value} = \text{Future Value} / (1 + k)^{Y} \tag{2-1-8}$$

where "k" is the inflation rate and "Y" is the number of years in the future.

By using Eqs. (2-1-5) and (2-1-7) for each payment period and summing up the results for each loan term year give the annual out of pocket dollars in Table 2-1-3 for a 15-year and 5% fixed annual interest rate $100,000 home loan with monthly payments.

Year	Annual Payments	Annual Interest	25% Tax Break On Interest	Annual Out Of Pocket $	Annual Out Of Pocket $ Present Value At 3%
1	9489.52	4895.67	1223.92	8265.61	8024.86
2	9489.52	4660.64	1165.16	8324.36	7846.51
3	9489.52	4413.59	1103.40	8386.13	7674.49
4	9489.52	4153.89	1038.47	8451.05	7508.65
5	9489.52	3880.91	970.23	8519.30	7348.82
6	9489.52	3593.97	898.49	8591.03	7194.85
7	9489.52	3292.34	823.08	8666.44	7046.61
8	9489.52	2975.28	743.82	8745.70	6903.94
9	9489.52	2642.00	660.50	8829.02	6766.71
10	9489.52	2291.67	572.92	8916.61	6634.79
11	9489.52	1923.41	480.85	9008.67	6508.06
12	9489.52	1536.31	384.08	9105.45	6386.38
13	9489.52	1129.41	282.35	9207.17	6269.64
14	9489.52	701.69	175.42	9314.10	6157.72
15	9489.52	252.09	63.02	9426.50	6050.51
SUM	142342.85	42342.85	10585.71	131757.14	104322.54

Table 2-1-3 $100,000 home loan for a 15-year term and a fixed annual interest rate of 5% with monthly payments

It is interesting to observe in Table 2-1-3 that after it is all said and done, out of pocket present value of money is $104,323 for a $100,000 home loan. Similar results are presented in graphic form in Figure 2-1-1.

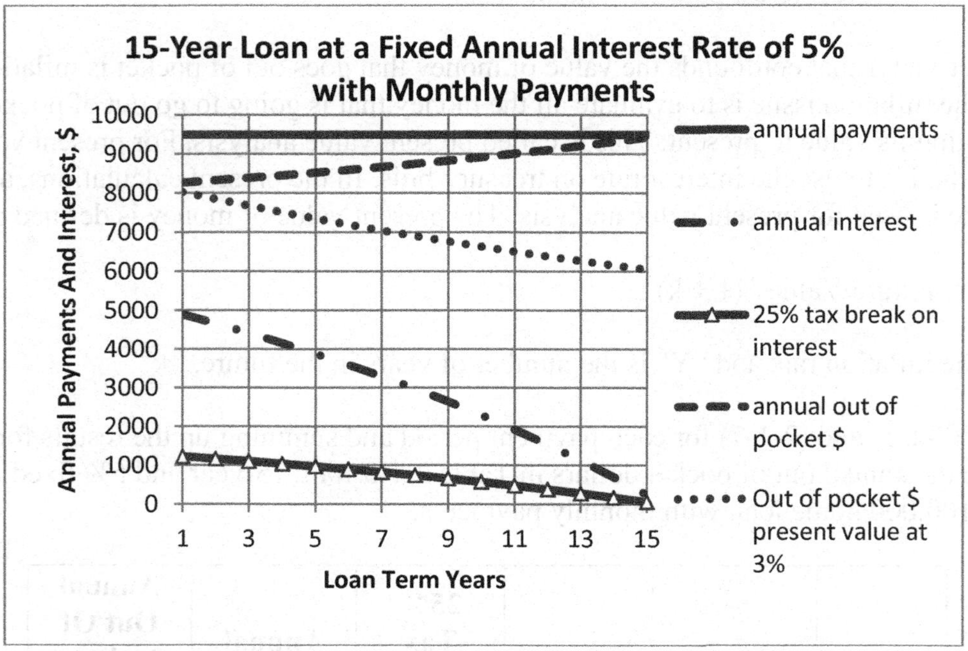

Figure 2-1-1 $100,000 home loan at a 15-year term and a fixed annual interest rate of 5% with monthly payments

By using Eqs. (2-1-5) and (2-1-7) for different loan terms and payment periods and summing up the results for each loan term year give the annual out of pocket dollars in Table 2-1-4 for a $100,000 home loan at a 5% fixed annual interest rate.

$100,000 Home Loan At A Fixed Annual Interest Rate Of 5%				
Loan Term In Years	**Number Of Payment Periods**	**Loan Payments "M"**	**Total Out Of Pocket $**	**Total Out Of Pocket $ Present Value At 3%**
7.5	180 (bi-monthly)	$666.79	115,016.5	101,446.3
15	180 (monthly)	$790.79	131,757.1	104,322.5
15	360 (bi-monthly)	$395.12	131,682.6	104,262.7
15	780 (weekly)	$182.29	131,496.5	104,124.3
30	360 (monthly)	$536.82	169,941.8	109,349.6
30	720 (bi-monthly)	$268.29	169,876.9	109,305.6
30	1560 (weekly)	$123.80	169,580.1	109,142.4

Table 2-1-4 $100,000 Home Loan At A Fixed Annual Interest Rate Of 5%

In Table 2-1-4, if total out of pocket dollars at present value are compared for 7.5, 15, and 30-year term loans with bi-monthly payments, the differences are small. The difference between the total out of pocket dollars at present value for 7.5 and 15-year term loans with bi-monthly payments is $2,816.4. The difference between the total out of pocket dollars at present value for 15 and 30-year term loans with bi-monthly payments is $5,042.9.

Chapter 2-2 Life Cycle Cost of Owning a Car

Owning a car is the second largest investment in one's life after a home. Owning a car affects one's finances, forces one to make many decisions before purchasing a car, and becomes a significant part of one's life.

One of the ways to compare the cost of a car to others is to look at its life cycle costs. These costs covers all the out of pocket expenses that will be encountered during the life of a car. These costs can be categorized as the purchase cost, taxes, registration costs, maintenance costs, insurance costs, and energy costs. There is also a resale or a salvage value income at the end of a car's life. For each year that the car is to be owned, the costs can be summed as follows:

$$Annual\ Cost_j = \sum_{i=1}^{n} cost_{ij} \qquad (2\text{-}2\text{-}1)$$

where Annual $Cost_j$ is the sum of "n" categorized car expenses and incomes for the j'th year. $Cost_{ij}$ are the individual category of expenses or incomes for the j'th year. $Cost_{ij}$ has to include expected inflationary and car age related adjustments. For example, maintenance costs increase as the car ages and registration costs decrease as the car ages. When all the Annual $Cost_j$'s are added up as shown in Eq. (2-2-2), total cost of the car is obtained throughout its life in expected future values of money.

$$Total\ Cost = \sum_{j=1}^{m} \sum_{i=1}^{n} cost_{ij} \qquad (2\text{-}2\text{-}2)$$

Each year's annual cost, except the first year, has to be brought back to its present value for an assumed annual inflation rate. First year's annual cost is not brought to its present value since majority of the costs of a car during the first year are up front. Sum of present value of net annual cost for every year during the expected life of the car divided by the total miles to be driven during its life will provide the actual cost per mile.

Cost per mile at present value is a good metric in comparing different car purchase scenarios and car types.

In the first example given in Table 2-2-1, a car which has $50,000 sale price is considered. The car is paid all up front with no car loan. The life of the car is ten years. The car is to be driven 15,000 miles per year. Average fuel economy for the car is 30 miles per gallon of premium gasoline which is at $3 per gallon at the present time.

YEARS	1	2	3	4	5	6	7	8	9	10	Totals
Purchase cost, $	50,000										50,000
Initial out the door expenses, $	5,000										5,000
Maintenance costs, $	0	500	500	1,000	1,000	1,000	1,500	2,000	2,000	2,000	11,500
Insurance costs, $	2,000	2,000	2,000	2,000	2,000	1,500	1,500	1,500	1,500	1,500	17,500
Annual registration, $	750	675	600	525	450	375	300	225	150	75	4,125
Energy costs, $	1,500	1,575	1,654	1,736	1,823	1,914	2,010	2,111	2,216	2,327	18,867
Salvage cost, $										-2,500	-2,500
Total annual costs, $	59,250	4,750	4,754	5,261	5,273	4,789	5,310	5,836	5,866	3,402	104,492
PRESENT VALUE OF ANNUAL COSTS, $	59,250	4,308	4,106	4,329	4,132	3,574	3,774	3,950	3,781	2,089	93,292
Total miles driven in 10 years = 150,000					Cost per mile in present value = 0.622 $/mile						

Table 2-2-1 Life cycle cost of a car purchased by cash

Out the door price of the car is the sum of first two rows of costs, namely $55,000. Initial out the door expenses include sales tax, transportation charges, etc. Maintenance costs start at zero for the first year, are low during the initial warranty years, and goes up during the final years of the car. Insurance costs start high and goes down as the car ages. Annual registration costs also start high and go down linearly as the car ages. Energy costs start at $3 per gallon and increase at a 5% annual inflation rate. Large jumps in energy due to drastic changes in supply and demand cycles are not considered. Also tax benefits for energy efficient cars are not considered. However, if tax benefits are known, they can be added easily to the calculations. At the end of its ten-year life, the car is assumed to have a $2,500 resale value.

Annual costs are brought to present value by the following Eq. (2-2-3).

$$Present\ Value\ of\ Annual\ Cost_j = \frac{Annual\ Cost_j}{[1+\frac{Annual\ Interest\ Rate}{100}]^j} \tag{2-2-3}$$

where "j" is the age of the car except j=1.

If the cash to purchase the car is not available, the car can be bought by getting a loan. Table 2-2-2 provides the results for the same car bought by a loan at 7% annual percentage rate for a term of 5 years for 60 monthly payments.

YEARS	1	2	3	4	5	6	7	8	9	10	Totals
Purchase cost, $	11,881	11,881	11,881	11,881	11,881						59,404
Initial out the door expenses, $	5,000										5,000
Maintenance costs, $	0	500	500	1,000	1,000	1,000	1,500	2,000	2,000	2,000	11,500
Insurance costs, $	2,000	2,000	2,000	2,000	2,000	1,500	1,500	1,500	1,500	1,500	17,500
Annual registration, $	750	675	600	525	450	375	300	225	150	75	4,125
Energy costs, $	1,500	1,575	1,654	1,736	1,823	1,914	2,010	2,111	2,216	2,327	18,867
Salvage cost, $										- 2,500	-2,500
Total annual costs, $	21,131	16,631	16,634	17,142	17,154	4,789	5,310	5,836	5,866	3,402	113,895
PRESENT VALUE OF ANNUAL COSTS, $	21,131	15,085	14,369	14,103	13,441	3,574	3,774	3,950	3,781	2,089	95,296
Total miles driven in 10 years = 150,000					Cost per mile in present value = 0.635 $/mile						

Table 2-2-2 Life cycle cost of a car purchased by a 5-year loan at 7% annual interest rate

Table 2-2-2 shows that buying the car with a loan is going to cost 0.013 $ per mile more or a total of $2,004 more.

Another option is to lease the same car for three years for $600 per month lease payments. Results for the lease option are shown in Table 2-2-3.

YEARS	1	2	3	Totals
Purchase cost, $	7,200	7,200	7,200	21,600
Initial out the door expenses, $	5,000			5,000
Maintenance costs, $	0	500	500	1,000
Insurance costs, $	2,000	2,000	2,000	6,000
Annual registration, $	750	675	600	2,025
Energy costs, $	1,500	1,575	1,654	4,729
Salvage cost, $				0
Total annual costs, $	16,450	11,950	11,954	40,354
PRESENT VALUE OF ANNUAL COSTS, $	16,450	10,839	10,326	37,615
Total miles driven in 3 years = 45,000	Cost per mile in present value = 0.836 $/mile			

Table 2-2-3 Life cycle cost of a $50,000 car leased for three years for $600 per month

Table 2-2-3 shows that leasing a car is the most expensive option. It has a present value cost per mile of $0.859 which is 0.214 $ per mile more expensive than purchasing the car by cash.

Chapter 2-3 Profits from Book Sales

When there is scattered data available for analysis, it is good to understand the behavior or trend of that data. As an example consider the data for the number of a certain book sold versus its unit price as shown in Table 2-3-1.

Quantity Sold	Book Unit Price, $
1200	17
500	20
200	24
1750	14
2000	12
3000	9
3900	6
10	26
1200	15
350	22
55	25

Table 2-3-1 Quantity sold versus book unit price

Data given in Table 2-3-1 can be plotted to see how it is behaving as shown in Figure 2-3-1.

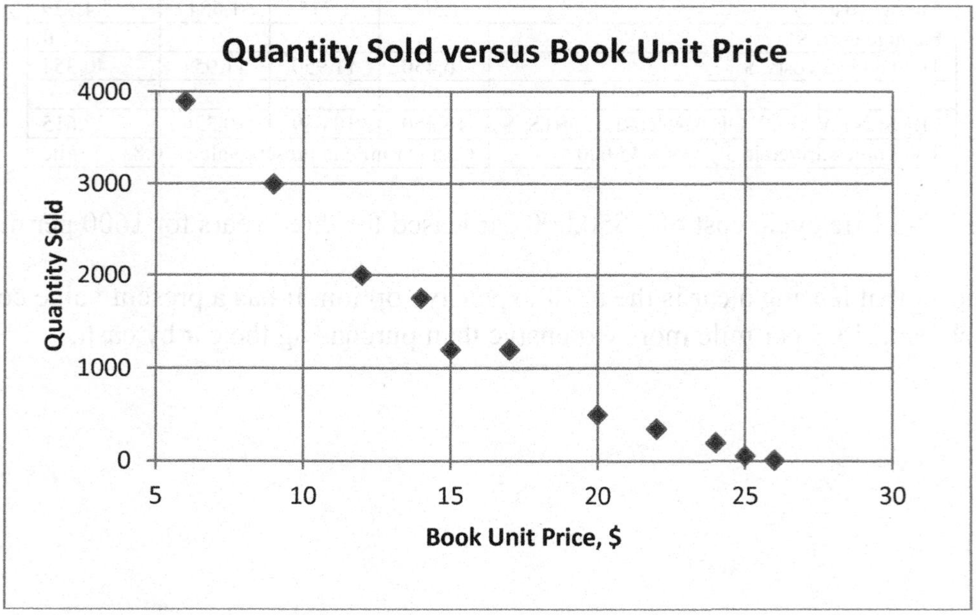

Figure 2-3-1 Data in Table 2-3-1 for book unit price versus quantity sold

Data is showing an increasing trend in quantity sold as the book unit price decreases. This trend can be characterized by a best fit linear curve to this data. This best fit linear curve can be written as

$$Q_{best\ fit} = a \times P + b \qquad (2\text{-}3\text{-}1)$$

where $Q_{best\ fit}$ is book quantity sold according to best fit linear curve, P is the book unit price, a is the slope of the best fit linear curve, and b is the intercept of the best fit linear curve.

Least squares regression method determines the best fit linear curve by minimizing the squares of the residuals between the actual data and the prediction given by the best fit linear curve. Sum of squares of residuals can be written as

$$RES^2 = \sum_{i=1}^{N}(Q_i - Q_{best\ fit})^2 \qquad (2\text{-}3\text{-}2)$$

or

$$RES^2 = \sum_{i=1}^{N}(Q_i - a \times P_i - b)^2 \qquad (2\text{-}3\text{-}3)$$

In order to find the best fit linear curve, constants a and b in Eq. (2-3-1) should be determined by minimizing RES^2 . This minimization is achieved by differentiating Eq. (2-3-3) with respect to a and setting it to zero while holding b constant and also by differentiating Eq. (2-3-3) with respect to b and setting it to zero while holding a constant.

$$\frac{\partial RES^2}{\partial a} = \sum_{i=1}^{N} -2(Q_i - a \times P_i - b)P_i = 0 \qquad (2\text{-}3\text{-}4)$$

$$\frac{\partial RES^2}{\partial b} = \sum_{i=1}^{N} -2(Q_i - a \times P_i - b) = 0 \qquad (2\text{-}3\text{-}5)$$

Eqs. (2-3-4) and (2-3-5) provide two linear simultaneous equations in a and b .

$$a\sum_{i=1}^{N} P_i^2 + b\sum_{i=1}^{N} P_i = \sum_{i=1}^{N} P_i Q_i \qquad (2\text{-}3\text{-}6)$$

$$a\sum_{i=1}^{N} P_i + bN = \sum_{i=1}^{N} Q_i \qquad (2\text{-}3\text{-}7)$$

Eqs. (2-3-6) and (2-3-7) can be solved for a and b to obtain the following results.

$$a = \frac{N\sum_{i=1}^{N} P_i Q_i - \sum_{i=1}^{N} P_i \sum_{i=1}^{N} Q_i}{N\sum_{i=1}^{N} P_i^2 - (\sum_{i=1}^{N} P_i)^2} \qquad (2\text{-}3\text{-}8)$$

$$b = \left(\frac{\sum_{i=1}^{N} Q_i}{N}\right) - a\left(\frac{\sum_{i=1}^{N} P_i}{N}\right) \qquad (2\text{-}3\text{-}9)$$

For the data given in Table 12-1, $a = -184.89$ and $b = 4481.2$. Best fit linear curve Eq. (2-3-1) becomes

$$Q_{best\ fit\ linear\ curve} = -184.89 \times P + 4481.2 \qquad (2\text{-}3\text{-}10)$$

Linear regression analysis shown above can also be applied to power functions in the form of $Q_{best\ fit} = mP^n$. First the power function has to be linearized by taking the logarithm of both sides to any base, say 10. Then the power function takes the following form

$$log_{10}(Q_{best\ fit}) = log_{10}(m) + n \times log_{10}(P) \tag{2-3-11}$$

Using the logarithm to base 10 of raw data, a and b are obtained from Eqs. (2-3-8) and (2-3-9). For the best fit power function, $n = a$ and $m = 10^b$.

Best fit linear curve to data given in Table 2-3-1 is plotted in Figure 2-3-2. Also 95.4% probability of data being within upper and lower boundaries is shown in Figure 2-3-2. Best linear curve plus twice the standard error of estimate provides the upper boundary curve and best linear curve minus twice the standard error of estimate provides the lower boundary curve. The standard error of estimate is determined from the sum of squares of residuals given in Eq. (2-3-2).

$$Standard\ Error\ Of\ Estimate = \sqrt{\frac{RES^2}{N-2}} \tag{2-3-12}$$

If data is assumed to be normally distributed around the best fit curve, then any random data taken will lie within the shown upper and lower boundaries in Figure 2-3-2 with 95.4% probability.

The upper boundary linear equation, namely best fit linear curve plus two times standard error of estimate, becomes

$$Q_{best\ fit\ linear\ curve+2*standard\ error\ of\ estimate} = -184.89 \times P + 5120.4 \tag{2-3-13}$$

The lower boundary linear equation, namely best fit linear curve minus two times standard error of estimate, becomes

$$Q_{best\ fit\ linear\ curve+2*standard\ error\ of\ estimate} = -184.89 \times P + 3842.1 \tag{2-3-14}$$

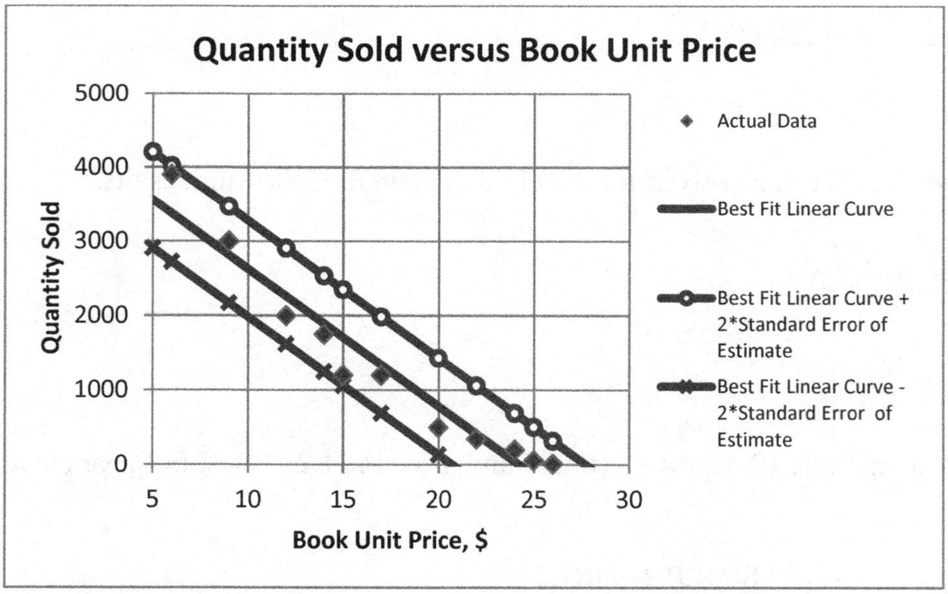

Figure 2-3-2 Quantity sold versus book unit price data and best fit linear curve to data with 95.4% probability upper and lower boundaries

Goodness of fit for the best fit curve to data is defined in literature as follows, see Reference [8]:

$$r = \sqrt{1 - \frac{RES^2}{S^2}} \qquad (2\text{-}3\text{-}15)$$

where RES^2 is sum of squares of residuals as defined in Eq. (2-3-2) and S^2 is sum of squares of each data point's difference to their average or

$$S^2 = \sum_{i=1}^{N}(Q_i - Q_{average})^2 \qquad (2\text{-}3\text{-}16)$$

When RES^2 is zero, all data points lie on the best fit curve and $r = 1$. This is a perfectly fit curve to data. As the value of RES^2 approaches the value of S^2, then $r \to 0$ which means that the best fit curve does not capture the variability in data and the best fit curve is not a good one. In practice, if $r > 0.9$, then the best fit curve represents the data variability very well.

For the example given above, the best fit linear curve, Eq. (2-3-10), fits the data given in Table 2-3-1 with a correlation coefficient value of 0.97 which is considered as a very good fit.
Let us take this best fit linear curve and use it in a book sale profit maximization case. Total cost of the book to the publisher has fixed costs and variable costs per quantity sold.

$$Total\ Cost = Fixed\ Costs + Variable\ Costs \times Q \qquad (2\text{-}3\text{-}17)$$

Profit expected from book sales can be written as

$$Profit = Q \times P - Total\ Cost \qquad (2\text{-}3\text{-}18)$$

Eq. (2-3-10) can be rewritten as book unit price being the dependent variable and book quantity sold as the independent variable.

$$P = -0.00541 \times Q + 24.237 \qquad (2\text{-}3\text{-}19)$$

Combining Eqs. (2-3-17), (2-3-18), and (2-3-19) gives $Profit$ as a function of Q.

$$Profit = -0.00541 \times Q^2 + (24.237 - Variable\ Costs) \times Q - Fixed\ Costs \qquad (2\text{-}3\text{-}20)$$

Profit is maximized when $\frac{d(Profit)}{dQ} = 0$, and if $\frac{d^2(Profit)}{dQ^2} < 0$ assuming that $Variable\ Costs$ and $Fixed\ Costs$ are constants.

$$\frac{d(Profit)}{dQ} = -0.01082 \times Q + 24.237 - Variable\ Costs = 0 \qquad (2\text{-}3\text{-}21)$$

and

$$\frac{d^2(Profit)}{dQ^2} = -0.01082 \qquad (2\text{-}3\text{-}22)$$

If $Variable\ Costs = \$5\ per\ book\ sold$, then using Eq. (2-3-21), the book sale quantity that will maximize profit is 1,778. The optimum book sale price from Eq. (2-3-19) is \$14.62. If

$Fixed\ Costs = \$9{,}000$, then maximum profit that can be achieved from Eq. (2-3-20) is $8,101. Profit versus book quantity sold is plotted for the best linear curve fit, Eq. (2-3-10), if Figure 2-3-3. Eq. (2-3-20) has a parabolic behavior. When profit drops to zero at a quantity of 554, book unit price from Eq. (2-3-19) is $21.24. When profit drops to zero at a quantity of 3,001, book unit price from Eq. (2-3-19) is $8.00.

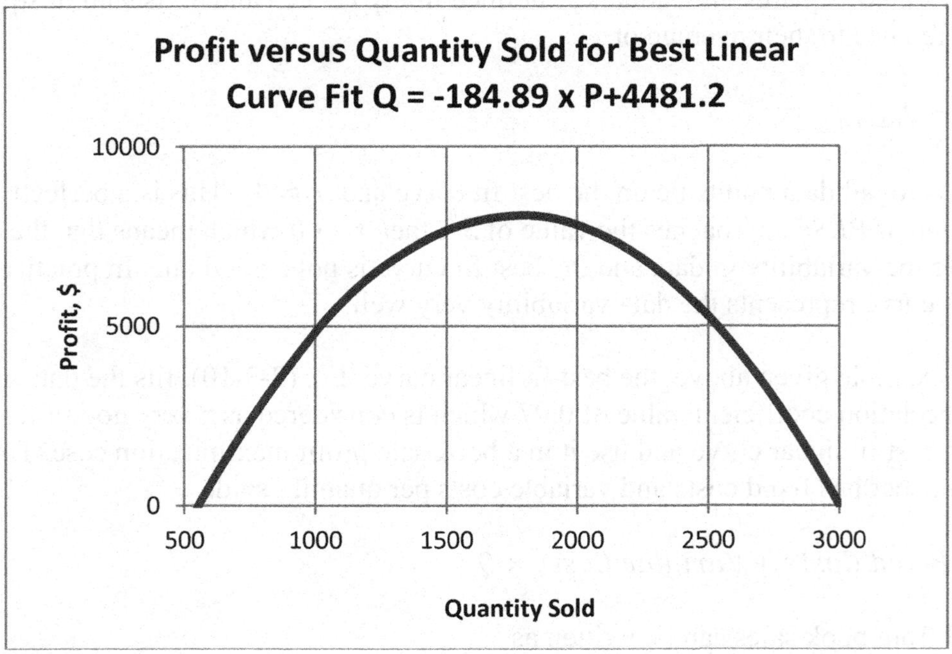

Figure 2-3-3 Profit versus quantity sold for best linear curve fit
$Q = -184.89 \times P + 4481.2$

The question always comes up as to what level of confidence can be assigned to the maximum profit value of $8,101? It was shown above that for a 95.4% probability, upper and lower boundaries of book sale quantities can be determined from Eqs. (2-3-13) and (2-3-14). The upper boundary of the book sale quantity will be 2,418 which will correspond to a profit of $5,885. The lower boundary of the book sale quantity will be 1,138 which will provide the same profit of $5,885. The answer to the above question is "within 95.4% confidence level the book sale profits can vary between $5,885 and $8,101".

Chapter 2-4 Tidal Forces between Two Bodies

Tidal forces result from gravitational forces between two bodies. Newton's law of gravitational attractive forces between two bodies that are treated as points is given by

$$F = G\frac{M_1 M_2}{R^2} \tag{2-4-1}$$

where G is the gravitational constant and has a value $6.67\text{x}10^{-11}$ m^3/s^2-kg in metric units, M_1 is the mass of body 1 in kg, M_2 is the mass of body 2 in kg, and R is the mean distance between bodies 1 and 2 in meters. These gravitational attractive forces have opposite directions for bodies 1 and 2, but have equal magnitudes.

Eq. (2-4-1) applied to bodies that are treated as points, but real planets, moons, satellites have diameters and they are not points. Therefore gravitational attractive forces coming from another planet vary around circumference of the affected planet.

Let us consider the Moon and the Earth as two bodies applying gravitational forces to each other. The Moon will apply stronger gravitational attractive forces to the points on Earth that are nearest to the Moon than to Earth's center. Also, the Moon will apply weaker gravitational attractive forces to the points on Earth that are farthest to the Moon than to Earth's center. The Moon's gravitational attractive forces affecting the Earth are shown in Figure 2-4-1 in a simple one-dimensional fashion.

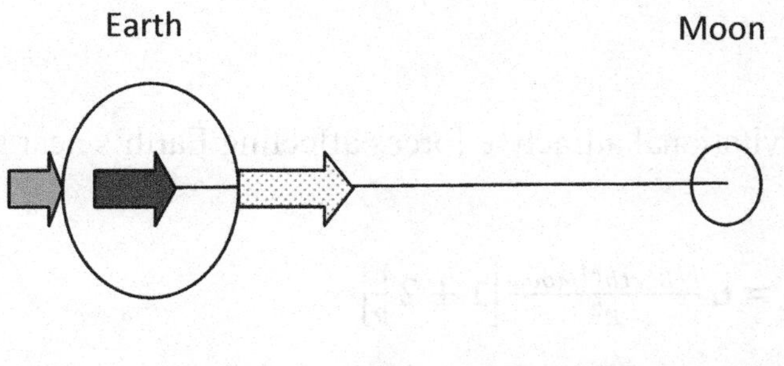

Figure 2-4-1 Moon's gravitational attractive forces affecting Earth's near and far surfaces and its center in a simple one-dimensional fashion

Once the Moon's gravitational attractive force affecting Earth's center is subtracted from near and far surface gravitational attractive forces, tidal forces shown in Figure 2-4-2 are the remaining forces.

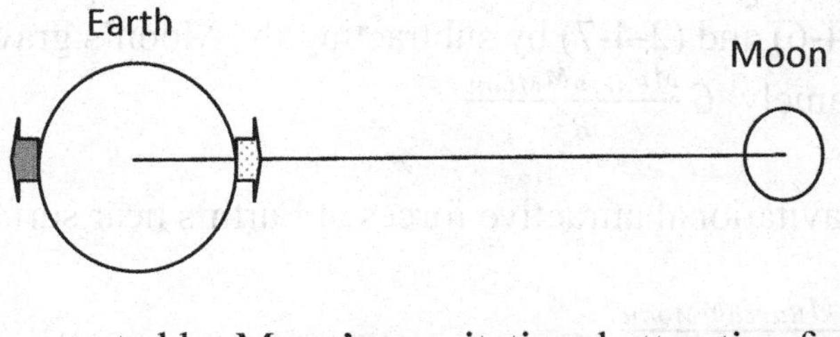

Figure 2-4-2 Tidal forces generated by Moon's gravitational attractive forces affecting Earth's near and far surfaces in a simple one-dimensional fashion

Using Eq. (2-4-1), Moon's gravitational attractive forces affecting Earth's near surface can be written as

$$F_{near\,surface} = G\frac{M_{Earth}M_{Moon}}{(R-r)^2} = G\frac{M_{Earth}M_{Moon}}{R^2}\left[\frac{1}{\left(1-\frac{r}{R}\right)^2}\right] \quad (2\text{-}4\text{-}2)$$

Using Eq. (2-4-1), Moon's gravitational attractive forces affecting Earth's far surface can be written as

$$F_{far\,surface} = G\frac{M_{Earth}M_{Moon}}{(R+r)^2} = G\frac{M_{Earth}M_{Moon}}{R^2}\left[\frac{1}{\left(1+\frac{r}{R}\right)^2}\right] \quad (2\text{-}4\text{-}3)$$

where r is the radius of Earth in meters. Tidal forces can be estimated by expanding the term in brackets in Eqs. (2-4-2) and (2-4-3) as power series, see Reference [3].

Let $= \frac{r}{R}$, then

$$\frac{1}{(1\pm x)^2} = f_n(x) = a_0 + a_1x + a_2x^2 + \cdots + a_nx^n \quad (2\text{-}4\text{-}4)$$

where $f(0) = a_0 = 1$ and $f'(0) = a_1 = \mp 2$.

Since $x = \frac{r}{R} = \frac{6.4\times10^6\ meters}{380\times10^6\ meters} = 0.017$, and it is small, a_2 and higher terms in Eq. (2-4-4) are neglected. Eq. (2-4-5) can be estimated by the following linear relationship.

$$\frac{1}{(1\pm x)^2} = f_n(x) \cong 1 \mp 2x \quad (2\text{-}4\text{-}5)$$

Using Eq. (2-4-5), Moon's gravitational attractive forces affecting Earth's near surface can be rewritten as

$$F_{near\,surface} = G\frac{M_{Earth}M_{Moon}}{(R-r)^2} = G\frac{M_{Earth}M_{Moon}}{R^2}\left[1 + 2\frac{r}{R}\right] \quad (2\text{-}4\text{-}6)$$

Using Eq. (2-4-5), Moon's gravitational attractive forces affecting Earth's far surface can be written as

$$F_{far\,surface} = G\frac{M_{Earth}M_{Moon}}{(R+r)^2} = G\frac{M_{Earth}M_{Moon}}{R^2}\left[1 - 2\frac{r}{R}\right] \quad (2\text{-}4\text{-}7)$$

Tidal forces generated by Moon's gravitational attractive forces affecting Earth's near and far surfaces can be obtained from Eqs. (2-4-6) and (2-4-7) by subtracting the Moon's gravitational attractive force affecting the Earth's center, namely $G\frac{M_{Earth}M_{Moon}}{R^2}$.

Tidal forces due to Moon's gravitational attractive forces at Earth's near surface can be estimated from

$$F_{tidal\,force\,near\,surface} = G\frac{2rM_{Earth}M_{Moon}}{R^3} \quad (2\text{-}4\text{-}8)$$

Tidal forces due to Moon's gravitational attractive forces at Earth's far surface can be estimated from

$$F_{tidal\,force\,far\,surface} = -G\frac{2rM_{Earth}M_{Moon}}{R^3} \quad (2\text{-}4\text{-}9)$$

Tidal forces are equal in magnitude, but opposite in direction as shown in Figure 2-4-2. The magnitude of the tidal force felt by Earth due the Moon can be calculated from either Eq. (2-4-8) or Eq. (2-4-9).

$M_{Earth} = 5.98 \times 10^{24}kg$ and $M_{Moon} = 7.36 \times 10^{22}kg$

$$F_{tidal\,force\,magnitude\,on\,Earth's\,surface\,due\,to\,Moon} = \frac{(6.67\times10^{-11})\times2\times(6.4\times10^{6})\times(5.98\times10^{24})\times(7.36\times10^{22})}{(380\times10^{6})^3} = 6.85\times10^{18}Newtons \quad (2\text{-}4\text{-}10)$$

Tidal force effect on Moon's surface due to Earth can be calculated from either Eqs. (2-4-8) or (2-4-9) by replacing Earth's radius by Moon's radius.

$$F_{tidal\,force\,magnitude\,on\,Moon's\,surface\,due\,to\,Earth} = \frac{(6.67\times10^{-11})\times2\times(1.74\times10^{6})\times(5.98\times10^{24})\times(7.36\times10^{22})}{(380\times10^{6})^3} = 1.86\times10^{18}Newtons \quad (2\text{-}4\text{-}11)$$

It should be noted that these tidal forces affect all of Earth's or Moon's surfaces. Tidal forces effects are seen on water since water shears easily, but all of Earth's and Moon's surfaces are sheared or bulged to a much lesser extent.

Since every planet affects every other planet in our universe due to gravitational attractive forces and since the Sun is the most massive planet close to Earth, let us also consider the effects of our Sun's gravitational attractive forces on Earth's tidal forces.

$$F_{tidal\,force\,magnitude\,on\,Earth's\,surface\,due\,to\,Sun} = G\frac{2rM_{Earth}M_{Sun}}{R^3_{Earth-Sun}} \quad (2\text{-}4\text{-}12)$$

where $M_{Sun} = 1.99 \times 10^{30}\ kg$ and $R_{Earth-Sun} = 150 \times 10^{9}\ meters$.

$$F_{tidal\,force\,magnitude\,on\,Earth's\,surface\,due\,to\,Sun} = 3.01 \times 10^{18}\ Newtons \quad (2\text{-}4\text{-}13)$$

The Sun's effect on tidal forces on Earth is less than half the Moon's effect on tidal forces on Earth. Other planets in the universe also affect the tidal forces on the surfaces of the Earth, but they are negligible when compared to the Moon and the Sun. Large tides that are called Spring Tides are seen on Earth when the Sun and the Moon are lined up with the Earth at new and full phases of the Moon and the tidal forces add up. Small tides that are called Neap Tides are seen on Earth when the Sun and the Moon are lined up at 90 degrees apart to Earth and the tidal forces cancel each other.

Chapter 2-5 Center Of Mass for Vehicles and Their Effect on Rollover Accidents

Center of mass of a system of bodies is defined as the point where there is no moment about it. Namely, masses of all bodies in the system are balanced at the center of mass, like balancing playground seesaw. In a two dimensional rectangular coordinate system, center of mass in the X direction is defined as

$$\bar{X} = \frac{\sum_{i=1}^{N} M_i X_i}{\sum_{i=1}^{N} M_i} \tag{2-5-1}$$

where M_i is the mass of each body and X_i is the corresponding distance from $X = 0$ in the X direction. After $\bar{X}$ is determined, if Eq. (2-5-1) is reapplied to the system of bodies where this time X_i is the corresponding distance from $X = \bar{X}$, then all the moments generated by each body mass should cancel out to get $\bar{X} = 0$. This is a good check for the center of mass in the X direction. Similarly center of mass in the Y direction is defined as

$$\bar{Y} = \frac{\sum_{i=1}^{N} M_i Y_i}{\sum_{i=1}^{N} M_i} \tag{2-5-2}$$

Center of mass is a very critical dimension in the vehicles that we ride in. It affects the stability of the vehicle while cornering. A loaded vehicle with a high center of mass, i.e. an SUV, might be more susceptible to rollover accidents than a low center of mass vehicle, i.e. a sports car, if they both have the same track widths.

Let us consider two vehicles for a rollover analysis during cornering with the parameters given in Table 2-5-1.

Vehicle Type	Vehicle Curb Weight (Vehicle Plus Fluids), pounds	Vehicle Curb Weight Center Of Gravity Height, H_v, feet	Vehicle Track Width, T, feet	Passenger In Riding Position Center Of Gravity Height, H_p, feet	Number Of Passengers Riding In Vehicle	Each Passenger Weight, pounds
SUV	4,000	2.5	5.0	3.0	6	200
4-Door Sedan	3,000	1.7	4.0	2.0	4	200

Table 2-5-1 Vehicle and passenger parameters
Center of gravity heights are from road level

The parameters given in Table 2-5-1 are typical ones for an SUV and for a 4-door sedan vehicle. Center of gravity heights are given from the road level. Track widths are defined as the distance between the two front tire thread centers of parallel wheels. For these vehicles, it is assumed that the front and the rear track widths are the same. Center of gravity of the vehicles loaded with passengers can be calculated using Eq. (2-5-1) where $X = 0$ is the road level.

For the SUV:

$$H_{SUV} = \frac{4000\times2.5+6\times200\times3.0}{4000+6\times200} = 2.62\ feet \quad (2\text{-}5\text{-}3)$$

For the 4-door sedan:

$$H_{4-door\ sedan} = \frac{3000\times1.7+4\times200\times2.0}{3000+4\times200} = 1.76\ feet \quad (2\text{-}5\text{-}4)$$

When a vehicle enters a curve on a flat road, the forces and accelerations acting on the vehicle whose tires have enough traction on the road are shown in Figure 2-5-1. The vehicle in Figure 2-5-1 is not sliding and it is in control, but it is starting to rollover.

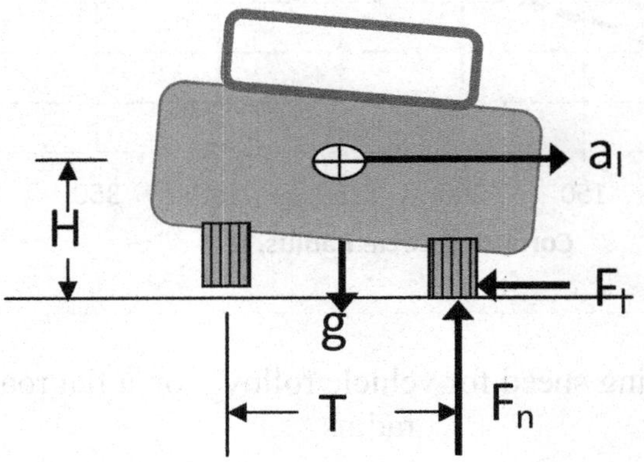

Figure 2-5-1 Forces and accelerations acting on a vehicle that is starting to roll

If moments of the forces acting on the vehicle are taken at the right tire interface with the road, following relationship is obtained.

$$0.5T \times Mg - H \times Ma_l = 0 \quad (2\text{-}5\text{-}5)$$

where g is the gravitational acceleration, 32.2 ft/s^2 , and a_l is the lateral acceleration of the vehicle given by $\frac{V^2}{R}$. V is the cornering speed of the vehicle in ft/s and R is the circle radius in ft that the vehicle is trying to corner. For rollover to occur

$$\frac{a_l}{g} > \frac{0.5T}{H} \quad (2\text{-}5\text{-}6)$$

where $\frac{0.5T}{H}$ is called the vehicle's static stability factor. The larger the static stability factor, it is less likely that it will rollover when we are cornering. For passenger loaded SUV vehicle in Table 2-5-1 and Eq. (2-5-3), the static stability factor is 0.96. For passenger loaded 4-door sedan vehicle in Table 2-5-1 and Eq. (2-5-4), the static stability factor is 1.28. The 4-door sedan is 33% more stable for rollovers during cornering than the SUV. Eq. (2-5-6) can be rewritten as

$$V > \sqrt{Rg\frac{0.5T}{H}} \quad (2\text{-}5\text{-}7)$$

During cornering, speed of the vehicle has to be greater than $\sqrt{Rg\frac{0.5T}{H}}$ for rollover to start occurring. For the two passenger loaded vehicles in Table 2-5-1, maximum cornering speed versus cornering circle radius on a flat road is shown in Figure 2-5-2.

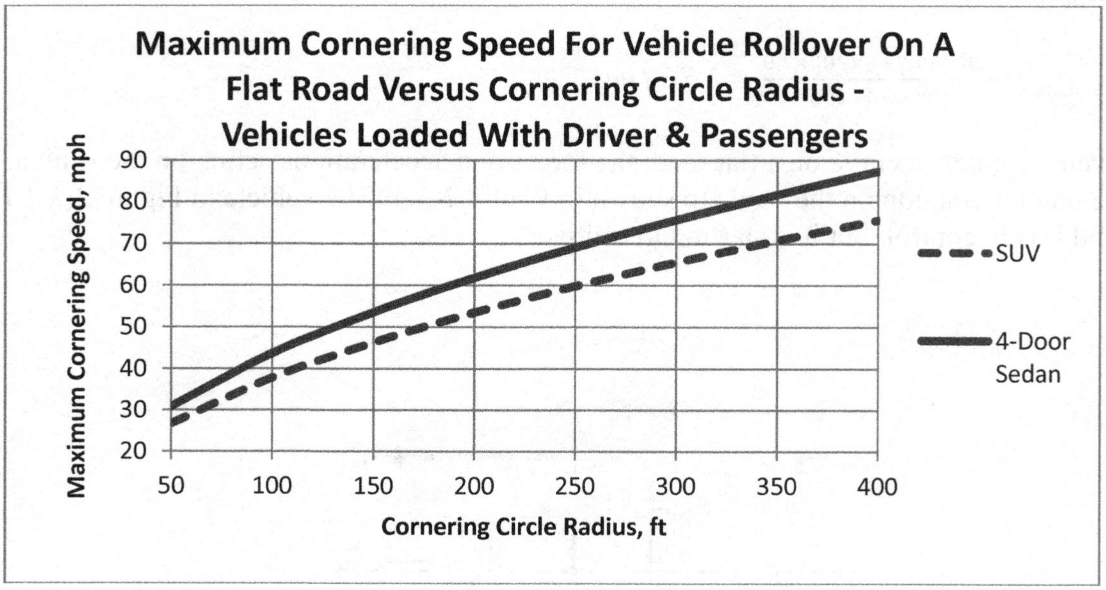

Figure 2-5-2 Maximum cornering speed for vehicle rollover on a flat road versus cornering circle radius

For a given cornering circle radius, maximum cornering speed for an SUV is 16% less than a 4-door sedan.

Cornering speed can be increased or the rollover potential at a given speed can be decreased by introducing an inward facing bank (towards the center of the cornering circle radius) to the cornering segment of the road. Figure 2-5-3 depicts a vehicle cornering on a 10° banked road.

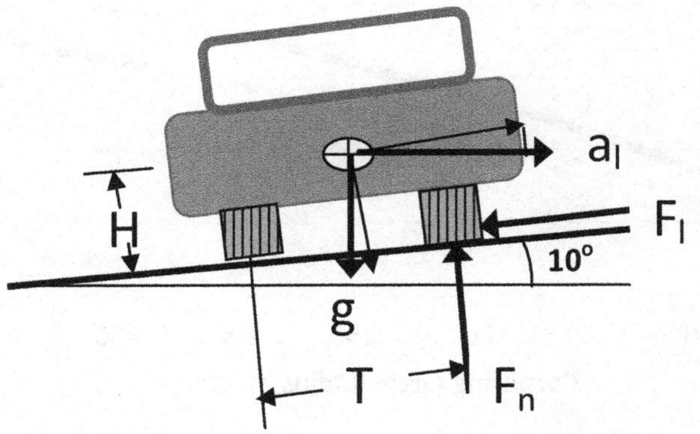

Figure 2-5-3 Forces acting on a vehicle cornering on a 10° banked road

Gravitational and lateral acceleration forces acting on the vehicle in Figure 2-5-3 can be broken into components that are parallel and perpendicular to the banked road. Moments of these forces are taken at the right tire interface with the road to give the following relationship.

$$0.5T \times Mg\cos 10° + H \times Mg\sin 10° - H \times Ma_l\cos 10° + 0.5T \times Ma_l\sin 10° = 0 \qquad (2\text{-}5\text{-}8)$$

or for rollover to start occurring

$$V > \sqrt{Rg\frac{(0.5T\cos 10° + H\sin 10°)}{(H\cos 10° - 0.5T\sin 10°)}} \qquad (2\text{-}5\text{-}9)$$

For the passenger loaded vehicles in Table 2-5-1, maximum cornering speed versus cornering circle radius on a 10° banked road is shown in Figure 2-5-4.

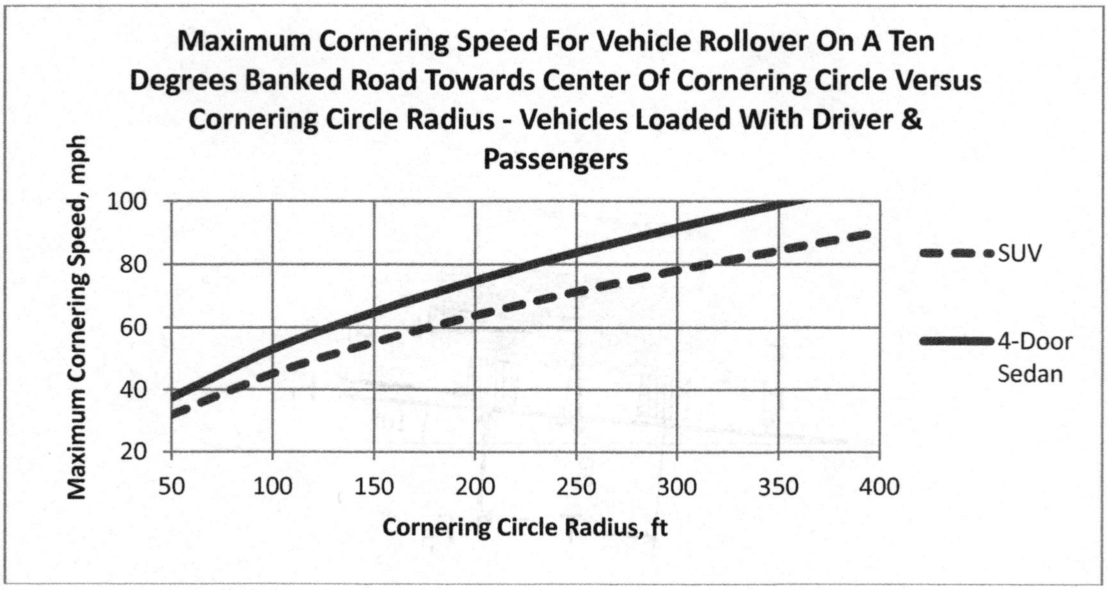

Figure 2-5-4 Maximum cornering speed for vehicle rollover on a ten degrees banked road

A 10° bank on the road while cornering provides a 19% improvement in SUV cornering speed and a 21% improvement in 4-door sedan cornering speed.

Chapter 2-6 Manufacturing Cost versus Hours

In a labor intensive manufacturing environment, manufacturing cost per unit can be correlated to the man-hours its takes to manufacture that unit. In such a case, a logarithmic relationship is determined as best fit curve to available data as follows.

$$MC = 560 \times \log_{10}(1 + 5 \times MH) \qquad (2\text{-}6\text{-}1)$$

where MC is the manufacturing cost in \$/unit and MH is the manufacturing man-hours in hours/unit. Management always wants to keep things simple. So they asked the manufacturing engineer who came up with Eq. (2-6-1) to simplify the $\log_{10}$ relationship in Eq. (2-6-1) and come up with a linear relationship between the manufacturing cost and manufacturing man-hours per unit.

Since $\log_{10}(1 + 5 \times MH)$ is continuous for $MH > 0$ and it has derivatives that are continuous for $MH > 0$, a Taylor series expansion, see Reference [1], of $\log_{10}(1 + 5 \times MH)$ around $MH = 0$ is possible. Taylor series expansion around $MH = 0$ is

$$f(MH) = \frac{f(0)}{0!} + \frac{f'(0)}{1!}(MH - 0) + \frac{f''(0)}{2!}(MH - 0)^2 + \cdots \qquad (2\text{-}6\text{-}2)$$

Only first two terms of the Taylor series expansion is used for a linear approximation.

$$\log_{10}(1 + 5 \times MH) = \frac{\log_{10}(1)}{0!} + \frac{\left[\frac{5}{(1+5\times 0)}\log_{10}(e)\right]}{1!}(MH - 0) \qquad (2\text{-}6\text{-}3)$$

where $\log_{10}(e) = 0.434$. Linear approximation for $\log_{10}(1 + 5 \times MH)$ becomes

$$\log_{10}(1 + 5 \times MH) = 2.17 \times MH \qquad (2\text{-}6\text{-}4)$$

Linear approximation for Eq. (2-6-1) becomes

$$MC = 1216 \times MH \qquad (2\text{-}6\text{-}5)$$

Eqs. (2-6-1) and (2-6-5) are shown in Figure 2-6-1. Linear approximation always over estimates the best fit curve, Eq. (2-6-1). As the manufacturing man-hours increase so does the over estimation by linear approximation. Over estimation by Eq. (2-6-5) is 4.9% when $MH = 0.02$ hours per unit. Over estimation by Eq. (2-6-5) increases to 23.3% when $MH = 0.1$ hours per unit.

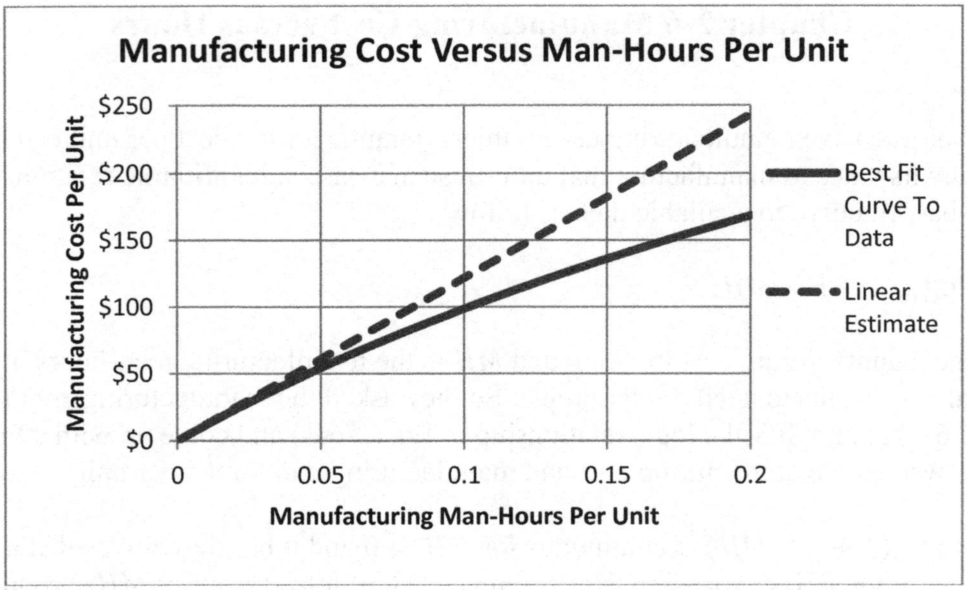

Figure 2-6-1 Manufacturing cost versus manufacturing man-hours per unit

Chapter 2-7 Saving Money

If there is a sequence of numbers in which each number can be obtained by adding a constant to the preceding one, then this sequence of numbers is called to be in arithmetic progression. For example, if you want to save money in an increasing fashion by depositing \$1 the first week, \$3 dollars the second week, \$5 dollars the third week and so on, how much money would you save at the end of 52 weeks?

This weekly savings sequence is an arithmetic progression where the first number in the sequence is \$1 and the constant of progression is \$2. We can write the total savings, S, after 52 weeks as a sum.

$$S = \sum_{N=1}^{52}(2N - 1) = 1 + (1 + 2) + (1 + 2 \times 2) + \cdots + [1 + (52 - 1)2] \qquad \text{(2-7-1)}$$

Eq. (2-7-1) can be generalized by calling the first week's savings W_1 and the constant increase in savings amount from the previous week's amount C. Total savings for m weeks can be written as

$$S = W_1 + (W_1 + C) + (W_1 + 2C) + \cdots + [W_1 + (m - 1)C] \qquad \text{(2-7-2)}$$

Let us call the $m'th$ week's savings amount $W_m = [W_1 + (m - 1)C]$ and rewrite the sum in Eq. (2-7-2) backwards starting with the $m'th$ week and coming back to first week.

$$S = W_m + (W_m - C) + (W_m - 2C) + \cdots + W_1 \qquad \text{(2-7-3)}$$

After adding Eqs. (2-7-2) and (2-7-3) and dividing them by 2, we get

$$S = 0.5 \times m \times (W_1 + W_m) \qquad \text{(2-7-4)}$$

or

$$S = m \times W_1 + 0.5 \times m \times (m - 1)C \qquad \text{(2-7-5)}$$

Eq. (2-7-5) gives the sum of a sequence of m numbers that are in arithmetic progression where W_1 is the first number in the sequence and C is the constant difference between the numbers.

For the case given above, $m = 52$, $W_1 = \$1$, and $C = \$2$. Using Eq. (2-7-5), total money saved after 52 weeks is

$$S = 52 \times \$1 + 0.5 \times 52 \times (52 - 1) \times \$2 = \$2{,}704$$

Section 3 – Everyday Problems Solved Using Differential and Integral Calculus

Main focus of the Section 3 is to be able to solve everyday problems using differential and integral calculus tools.

In nine everyday problems, differential calculus is used to find the maximum or the minimum of a function, namely when the first derivative of governing function with respect to its independent variable is zero. Also second derivative tests are done to governing functions to verify the maximum or the minimum points. These optimization problems cover minimizing cost of piping through a ranch to maximizing area of a rectangular building on an odd shaped lot. Everyday optimization problems that are solved using differential calculus are in chapters 3-5, 3-8, 3-9, 3-10, 3-11, 3-20, 3-22, 3-27, and 3-28.

In fourteen everyday problems, first order linear differential equations are solved using applicable boundary or initial condition. Some of the solutions included both the transient and the steady-state portions. Examples of first order linear differential equations solutions are linear distance, speed, acceleration, and jerk and loss of memory by time. Applications to first order linear differential equations can be found in chapters 3-1, 3-2, 3-3, 3-4, 3-6, 3-7, 3-12, 3-13, 3-14, 3-15, 3-16, 3-17, 3-30, and 3-31.

In four everyday problems, first order non-linear differential equations are solved using applicable boundary and initial conditions. Examples of first order non-linear differential equations solutions are emptying a tank that is full of water and Earth's escape speed. The problems can be found in 3-18, 3-19, 3-21, 3-23.

In one everyday problem, Chapter 3-25, a second order partial differential equation, wave equation, is treated to determine the string tensions in musical instruments.

In four everyday problems, numerical integration methods are used to approximate the area under the applicable integral functions. Numerical integration methods are applied to star temperatures, to Sun protection factor, to surface area of a leaf, and to motion of a falling object. These problems can be found in chapters 3-3, 3-24, 3-29, and 3-32.

Sensitivities of governing functions to independent variables are determined by differential calculus in two everyday problems, namely in chapters 3-26 and 3-34. Sensitivity analysis is performed on the effects of drag forces on a bicycle cyclist and on savings to annual income.

Closed form integration and differentiation solutions can be found in every problem in this section. For difficult integrations and differentiations, standard integration and differentiation tables are used.

Chapter 3-1 Distance, Speed, Acceleration, and Jerk

Let us investigate relationships between distance, speed, acceleration, and jerk for one-dimensional motion.

Speed "v" is defined as the linear distance "$s_2 - s_1$" traveled between two points divided by the duration of time "$t_2 - t_1$" it takes for the trip.

$$v = \frac{(s_2 - s_1)}{(t_2 - t_1)} \qquad (3\text{-}1\text{-}1)$$

For small distance and time differences, speed is the derivative of distance with respect to time.

$$v = \frac{ds}{dt} \qquad (3\text{-}1\text{-}2)$$

If distance-time history of an object is known, then we can calculate its speed-time history by differentiation.

Or if speed-time history of an object is known, we can calculate its distance-time history by integration (finding the area under the speed-time history curve).

$$s = \int v\, dt \qquad (3\text{-}1\text{-}3)$$

If speed-time history of an object is known, then we can calculate its acceleration-time history by differentiation.

$$a = \frac{dv}{dt} = \frac{d(^{ds}/_{dt})}{dt} = \frac{d^2 s}{dt^2} \qquad (3\text{-}1\text{-}4)$$

If acceleration-time history is know, we can calculate its speed-time history by integration (finding the area under the acceleration-time history curve).

$$v = \int a\, dt \qquad (3\text{-}1\text{-}5)$$

If acceleration-time history of an object is known, then we can calculate its jerk-time history by differentiation.

$$j = \frac{da}{dt} = \frac{d^3 s}{dt^3} \qquad (3\text{-}1\text{-}6)$$

If jerk-time history is know, we can calculate its acceleration-time history by integration (finding the area under the jerk-time history curve).

$$a = \int j\, dt \qquad (3\text{-}1\text{-}7)$$

Let us do an example using the above one-dimensional equations for a short distance runner. A 100-meter runner starts a race from the blocks with a jerk as defined in the following jerk-time history in Figure 3-1-1.

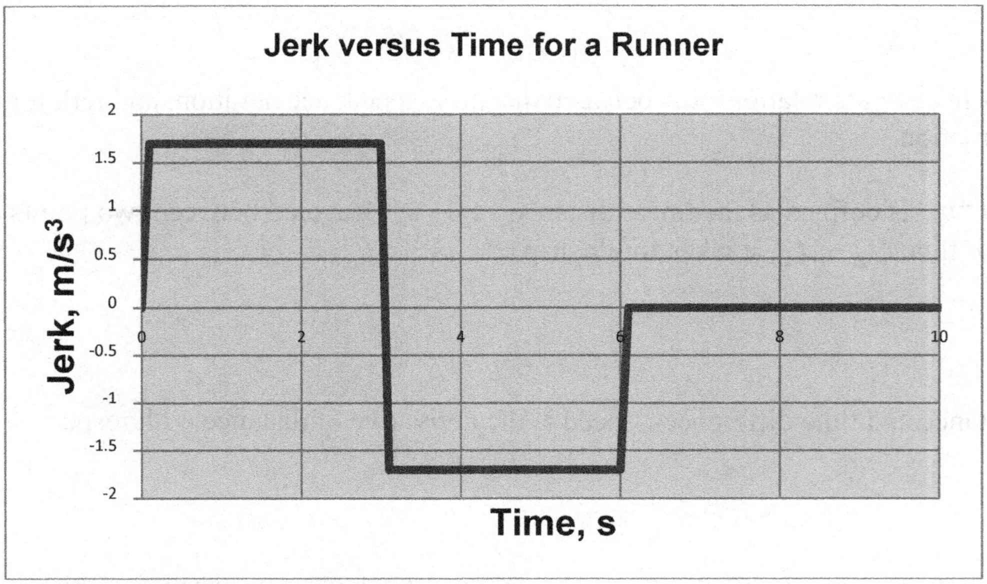

Figure 3-1-1 Jerk versus time for a runner

His maximum jerk is 1.7 m/s^3 or (1.7/9.8)=0.17 g/s.
From Figure 3-1-1, jerk equations can be written as three different continuous equations in m/s^3.

For $t \leq 3$ s $\quad j = 1.7$
For 3 s $< t \leq 6$ s $\quad j = -1.7$ (3-1-8)
For $t > 6$ s $\quad j = 0$

Acceleration equations in m/s^2 can be obtained by integrating the above jerk equations using Eq. (3-1-7).

For $t \leq 3$ s $\quad a = 1.7t$
For 3s $< t \leq 6$ s $\quad a = 1.7(6-t)$ (3-1-9)
For $t > 6$ s $\quad a = 0$

Integration constants for the acceleration equations are obtained by considering initial condition of $a = 0$ m/s^2 at $t = 0$ and acceleration continuity conditions at time $t = 3$ and 6 second interfaces, namely $a = 5.1$ m/s^2 at $t = 3$ s and $a = 0$ m/s^2 at $t = 6$ s.

For $t \leq 3$ s $\quad c_1 = 0$
For 3s $< t \leq 6$ s $\quad c_2 = 1.7*6$ (3-1-10)
For $t > 6$ s $\quad c_3 = 0$

Acceleration of the runner, Eqs. (3-1-9), is shown in Figure 3-1-2.

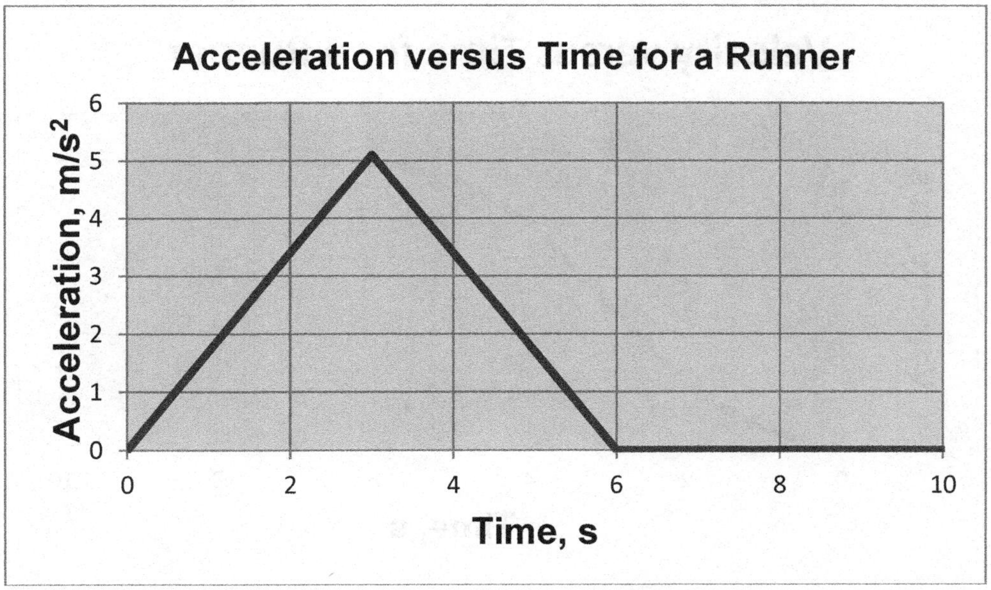

Figure 3-1-2 Acceleration versus time for a runner

Up to 3 s, the runner's acceleration increases linearly to a maximum of 5.1 m/s^2 and then decreases linearly to zero at 6 s.

Velocity equations for the runner in m/s can be obtained by integrating the above acceleration Eqs. 3-1-9.

For $t \leq 3$ s	$v = (1.7/2)\, t^2$	
For 3s $< t \leq 6$ s	$v = 1.7(-0.5\, t^2 + 6\, t - 9)$	(3-1-11)
For $t > 6$ s	$v = 1.7 * 9 = 15.3$	

Integration constants for the velocity equations, Eqs. 3-1-11, are obtained by considering initial condition of $v = 0$ at $t = 0$ and velocity continuity conditions at time $t = 3$ and 6 second interfaces, namely $v =$ 7.65 m/s at $t = 3$ s and $v =$ 15.3 m/s at $t = 6$ s.

For $t \leq 3$ s	$c_4 = 0$	
For 3s $< t \leq 6$ s	$c_5 = -1.7*9 = -15.3$	(3-1-12)
For $t > 6$ s	$c_6 = 1.7*9 = 15.3$	

Velocity of the runner behaves as a parabolic function of time during the first 6 seconds and then it becomes a constant. The runner reaches to a speed of 15.3 m/s or 55 km/hr as shown in Figure 3-1-3.

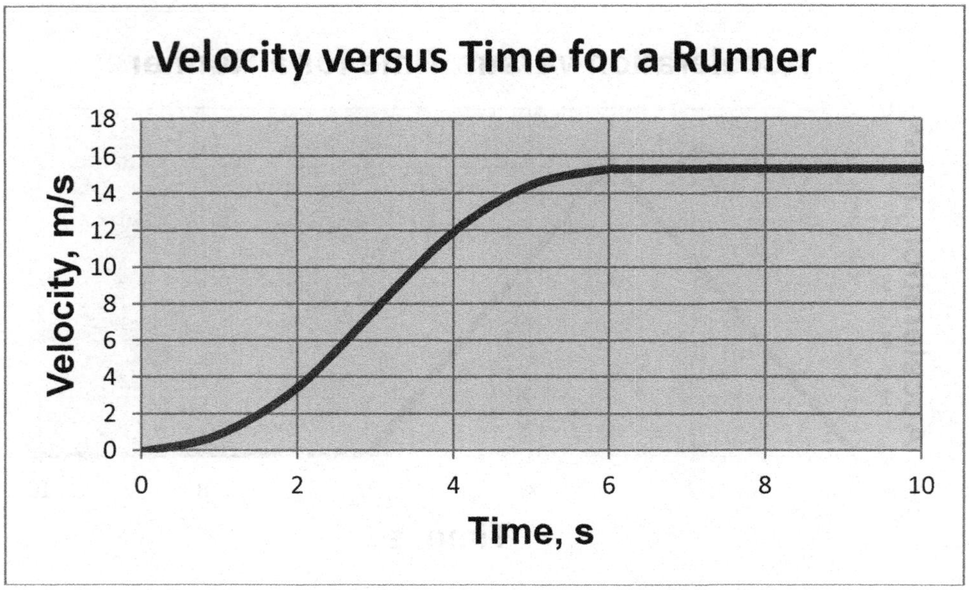

Figure 3-1-3 Velocity versus time for a runner

Distance equations are obtained by integrating the velocity equations, Eqs. 3-1-11, with respect to time and by determining the integration constants.

For $t \leq 3$ s $\quad s = (1.7/6)\, t^3$
For 3s $< t \leq 6$ s $\quad s = 1.7[-(0.5/3)\, t^3 + 3\, t^2 - 9\, t + 9]$ (3-1-13)
For $t > 6$ s $\quad s = 1.7(9\, t - 27)$

where integration constants are obtained by considering initial condition of $s = 0$ at $t = 0$ and distance continuity conditions at time $t = 3$ and 6 second interfaces, namely $s =$ 7.65 m at $t = 3$ s and $s =$ 45.9 m at $t = 6$ s.

For $t \leq 3$ s $\quad c_7 = 0$
For 3s $< t \leq 6$ s $\quad c_8 = 1.7*9 = 15.3$ (3-1-14)
For $t > 6$ s $\quad c_9 = -1.7*27 = -45.9$

As shown in Figure 3-1-4, the runner covers the distance of 100 meters in 9.53 seconds which is a new world record.

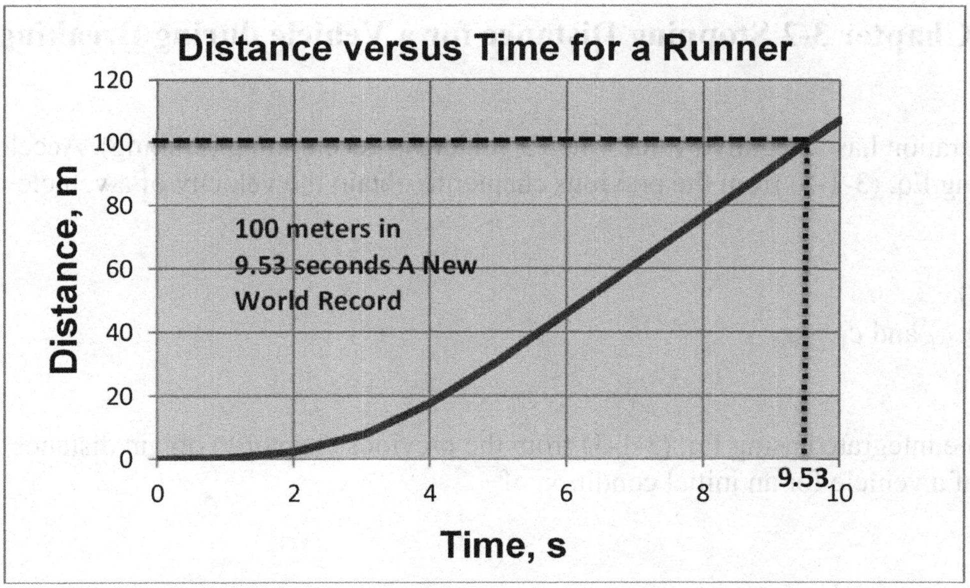

Figure 3-1-4 Distance versus time for a runner

Chapter 3-2 Stopping Distance for a Vehicle during Breaking

Acceleration has a negative value when a vehicle is decelerating (braking). Acceleration can be integrated using Eq. (3-1-5) from the previous chapter to obtain the velocity of a vehicle for an initial condition of

$v = v_0$ at $t = 0$
Then $v = at + c_1$ and $c_1 = v_0$
or $v = at + v_0$ (3-2-1)

Velocity can be integrated using Eq. (3-1-3) from the previous chapter to obtain distance during deceleration of a vehicle for an initial condition of

$s = 0$ at $t = 0$.
Then $s = 0.5at^2 + v_0t + c_2$ and $c_2 = 0$
or $s = 0.5at^2 + v_0t$

"t" in the distance equation can be replaced by the time-velocity relationship Eq. (3-2-1), $t = (v - v_0)/a$, or

$$s = (0.5/a)(v^2 - v_0^2) \quad (3\text{-}2\text{-}2)$$

To come to a full stop, namely $v = 0$, stopping distance becomes

$$s_{stop} = -(0.5/a)v_0^2 \quad (3\text{-}2\text{-}3)$$

During braking to a full stop, the kinetic energy lost, $0.5mv_0^2$, equals to the work done to stop the vehicle – product of vehicle weight, kinematic coefficient of friction between the tires and the road, and the stopping distance , $mg\mu s$.

Where
m = is the mass of the vehicle in lbm or kg,
g = gravitational acceleration on earth,
32.2 ft/s^2 or 9.8 m/s^2,
mg is the weight of the vehicle in lbf or in Newtons.
μ = kinematic coefficient of friction between the vehicle tires and the ground.

Kinematic coefficient of friction is about 0.7 between a dry asphalt road and new tires, goes down to 0.3 when the road gets wet and further goes down to 0.1 when the road is covered with ice. Kinematic coefficient of friction can be thought of as the ratio of force required to overcome the friction between the wheels of the vehicle and the road to the vehicle's weight to keep a vehicle moving while it is in motion.

It always needs a higher force to start a vehicle moving from a static position or to overcome static coefficient of friction.

Once the vehicle starts moving, the force to keep it moving reduces, namely kinematic coefficient of friction applies. For example, if a vehicle weighs 4,000 pounds, then it will take 400 pounds to push it at ground level in order to keep its at constant speed on an icy road, $mg\mu$ = 4000 x 0.1 = 400 pounds .

Kinetic energy lost equals to the work done during braking can be rewritten as: $0.5mv_0^2 = -mg\mu(0.5/a)v_0^2$ or $a/g = -\mu$, namely breaking deceleration, a/g , absolute value has to be less than μ in order not to slip on the road.

Initial velocity v_0 versus stopping distance can be plotted for different absolute a/g values. First Figure 3-2-1 is in the metric unit system and the second one, Figure 3-2-2, is in the English unit system.

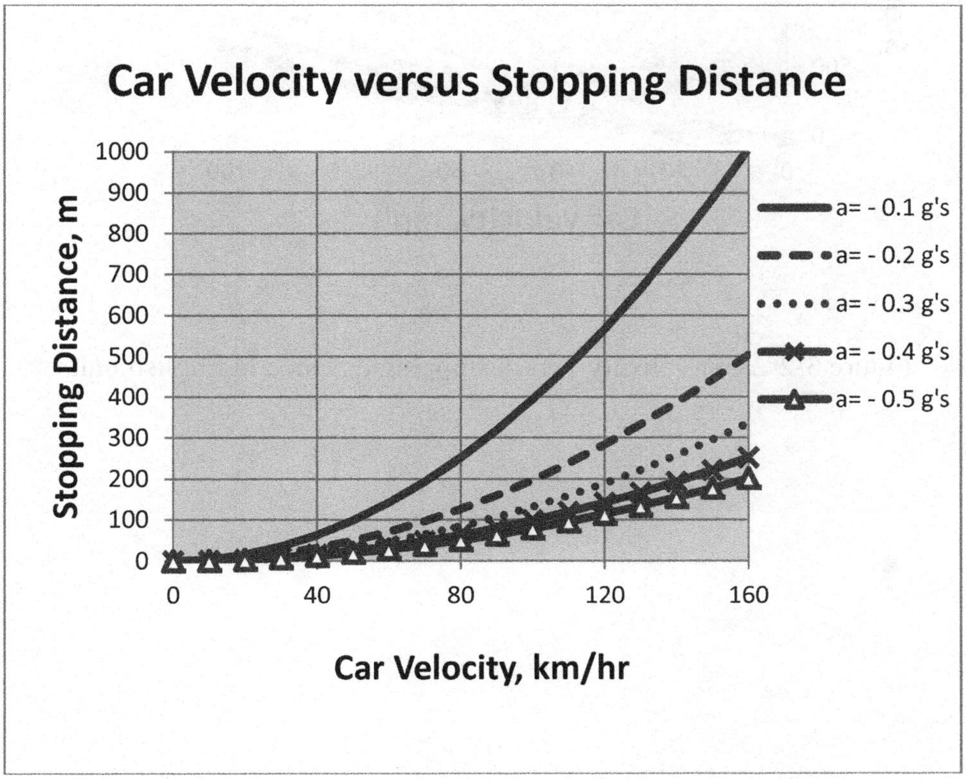

Figure 3-2-1 Car velocity versus stopping distance in metric units

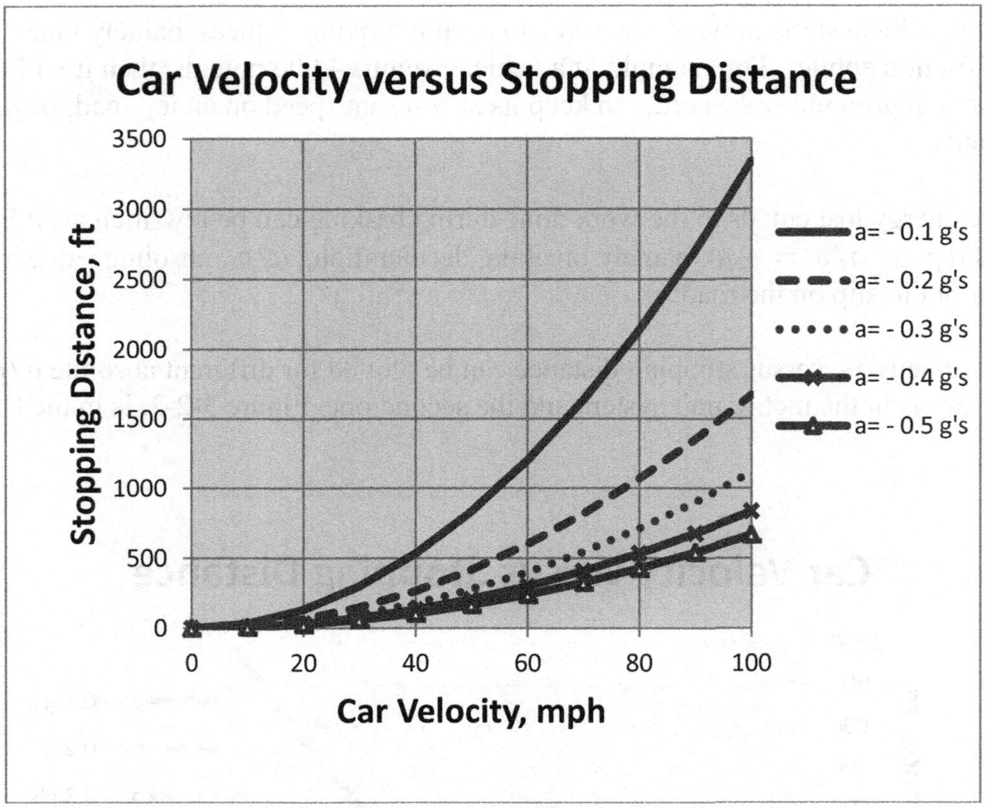

Figure 3-2-2 Car velocity versus stopping distance in English units

Chapter 3-3 Motion of a Falling Object

A falling object in air has an air resistance force that is called drag on it. The drag force on a falling object is defined as being proportional to square of its velocity.

$$drag\ force = cv^2$$

where the proportionality constant , c in kg/m or in lbm/ft, is a function of object's shape, air resisting surface area, and air density. The drag force on a falling object acts against the force of gravity and they can be combined to define the acceleration of an object as follows:

$$ma = mg - cv^2 \quad \text{(3-3-1)}$$

where m is the mass of the object in kg or in lbm and g is earth's gravitational constant, 9.8 m/s^2 or 32.2 ft/s^2. Acceleration of a falling object is

$$a = g - (c/m)v^2 \quad \text{(3-3-2)}$$

Eq. (3-3-2) can be rewritten by using the differential relationship between speed and acceleration, see Eq. (3-1-4) in Chapter 3-1.

$$dv/dt = g - (c/m)v^2 \quad \text{(3-3-3)}$$

Eq. (3-3-3) can be integrated by separating the variables v and t.

$$\int dt = \int \frac{dv}{g-(\frac{c}{m})v^2} \quad \text{(3-3-4)}$$

If we let $u = \sqrt{\frac{c}{mg}}\ v$, Eq. (3-3-4) can be rewritten as

$$\int dt = \sqrt{m/cg} \int \frac{du}{1-u^2} \quad \text{(3-3-5)}$$

From integral tables, see Reference [9], it can be found that $\int \frac{du}{1-u^2} = tanh^{-1}u + C_1$, if $u \leq 1$ and Eq. (3-3-5) is integrated to give

$$\sqrt{cg/m}\, t = tanh^{-1}\left(\sqrt{c/mg}\ v\right) + C_1 \quad \text{or}$$
$$v = \sqrt{mg/c}\, tanh\left(\sqrt{cg/m}\ t\right) + C_1$$

$u \leq 1$ condition requires that $v \leq \sqrt{mg/c}$ which is the maximum falling speed of the object which is called terminal speed, namely $v_{terminal\ speed} = \sqrt{mg/c}$.

The initial speed condition that can be applied to a falling object is $v = 0$ at $t = 0$ and therefore $C_1 = 0$.

$$v = \sqrt{mg/c}\, tanh(\sqrt{cg/m}\ t) \quad (3\text{-}3\text{-}6)$$

Eq. (3-3-6) can be rewritten by using the differential relationship between distance and speed, see Eq. (3-1-3) in Chapter 3-1.

$$s = \int \sqrt{mg/c}\, tanh(\sqrt{cg/m}\ t)\, dt + C_2$$

If we let $x = \sqrt{\frac{cg}{m}}\ t$, above integral can be rewritten as

$$s = (\tfrac{m}{c}) \int tanh(x)dx + C_2$$

From integral tables, see Reference [9], it can be found that $\int tanh(x)dx = ln\ cosh(x) + C_3$. The initial falling distance condition that can be applied to a falling object is $s = 0$ at $t = 0$ and therefore $\left(\frac{m}{c}\right) C_3 + C_2 = 0$.

$$s = ({}^{m}/_{c}) ln\ cosh(\sqrt{{}^{cg}/_{m}}\ \ t) \quad (3\text{-}3\text{-}7)$$

For a 1 kg rock falling down with a drag proportionality constant, c , of 0.005 kg/m can be plotted as an example to see how the falling distance in Figure 3-3-1, speed in Figure 3-3-2, and acceleration in Figure 3-3-3 vary with time. Drag proportionality constant is defined as $c = 0.5 f A \rho$ where f is the drag coefficient that is on the order of one and is obtained experimentally for a falling object. A is the air resisting surface area of the rock in m^2 and it is 0.01 m^2 for this example. ρ is the air density at the location of the falling rock and it is assumed to be 1 kg/m^3 for this example.

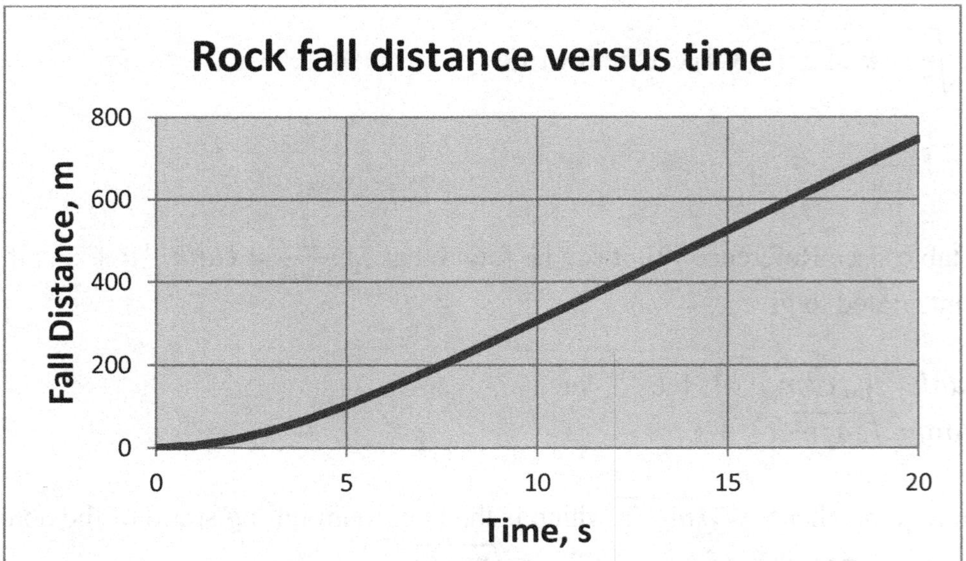

Figure (3-3-1) Rock fall distance versus time

Rock fall distance starts at the initial value of zero and increases 747 meters in 20 seconds.

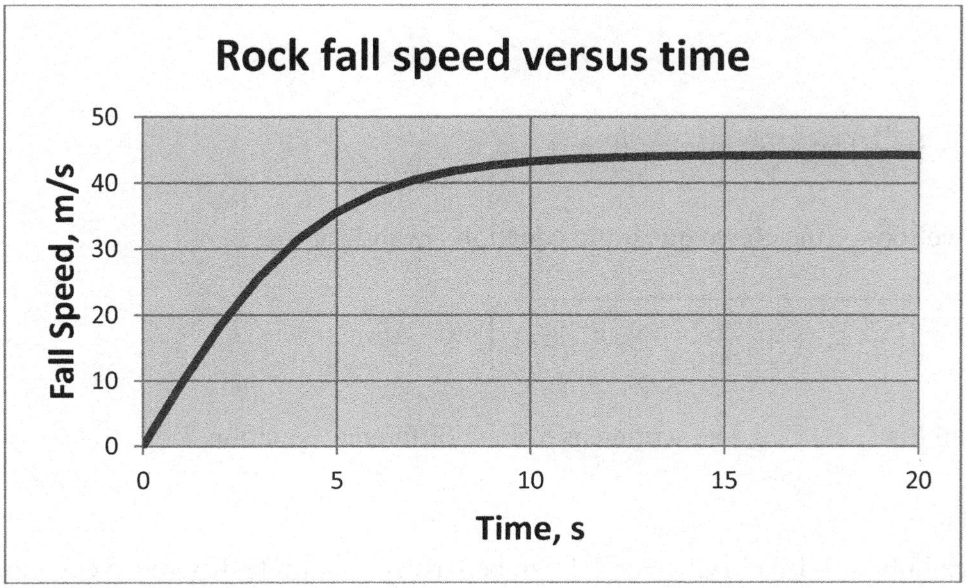

Figure 3-3-2 Rock fall speed versus time

Rock fall speed starts at the initial value of zero and asymptotes to its terminal speed value of 44.3 m/s.

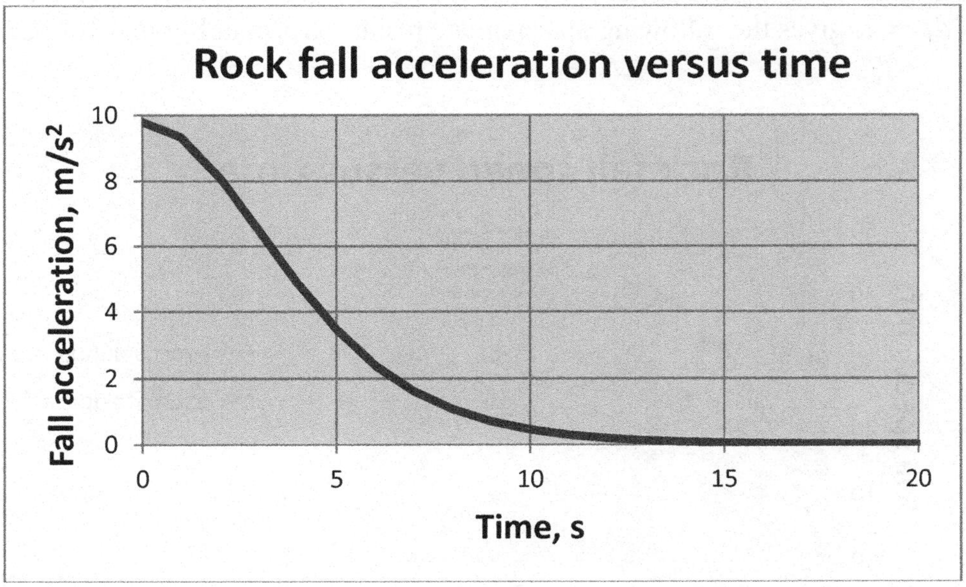

Figure 3-3-3 Object fall acceleration versus time

Rock fall acceleration starts at the initial value of earth's gravitational constant of 9.8 m/s^2 and decreases to zero fast at around 10 seconds when it approaches its terminal velocity.

If the exact solution of an integral is not available, the governing equations can be solved numerically. Numerical solutions can be demonstrated on Eq. (3-3-3), and then Eq. (3-3-1), and Eq. (3-1-3) in Chapter 3-1. Speed at the i'th time can be approximated by the following finite difference equation, see Reference [9].

$$(v_i - v_{i-1})/dt = g - (c/m)v_i^2 \qquad (3\text{-}3\text{-}8)$$

or

$$v_i^2 - \left(\frac{m}{cdt}\right) v_i - \left(\frac{m}{cdt}\right)(v_{i-1} + gdt) = 0$$

Only the positive root of the above quadratic equation is valid for v_i.

$$v_i = \left[-\left(\frac{m}{cdt}\right) + \sqrt{\left(\frac{m}{cdt}\right)^2 + 4\left(\frac{m}{cdt}\right)(v_{i-1} + gdt)}\ \right]/2 \qquad (3\text{-}3\text{-}9)$$

The acceleration Eq. (3-3-2) can be written as a finite difference equation.

$$a_i = g - \left(\frac{c}{m}\right) v_i^2 \qquad (3\text{-}3\text{-}10)$$

and the distance in Eq. (3-1-3) in Chapter 3-1 can be written as a finite difference equation.

$$s_i = s_{i-1} + v_i dt \qquad (3\text{-}3\text{-}11)$$

Numerical approximations for speed, Eq. (3-3-9), for acceleration, Eq. (3-3-10), and for distance, Eq. (3-3-11) are calculated by using the initial conditions of $v = 0$ and $s = 0$ at $t = 0$. A time interval of one second, $dt = 1$, gives the following approximate results shown in Figures 3-3-4, 3-3-5, and 3-3-6.

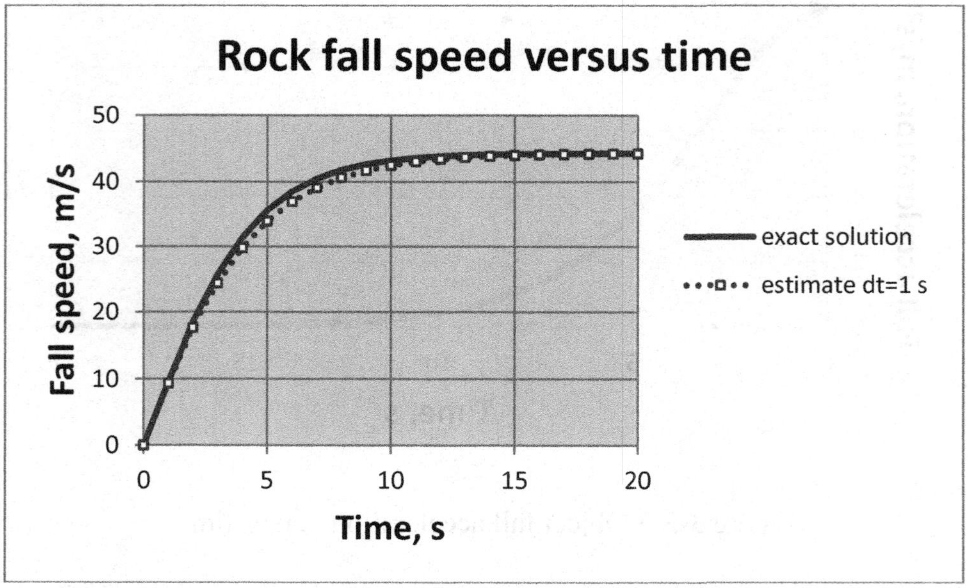

Figure 3-3-4 Rock fall speed versus time

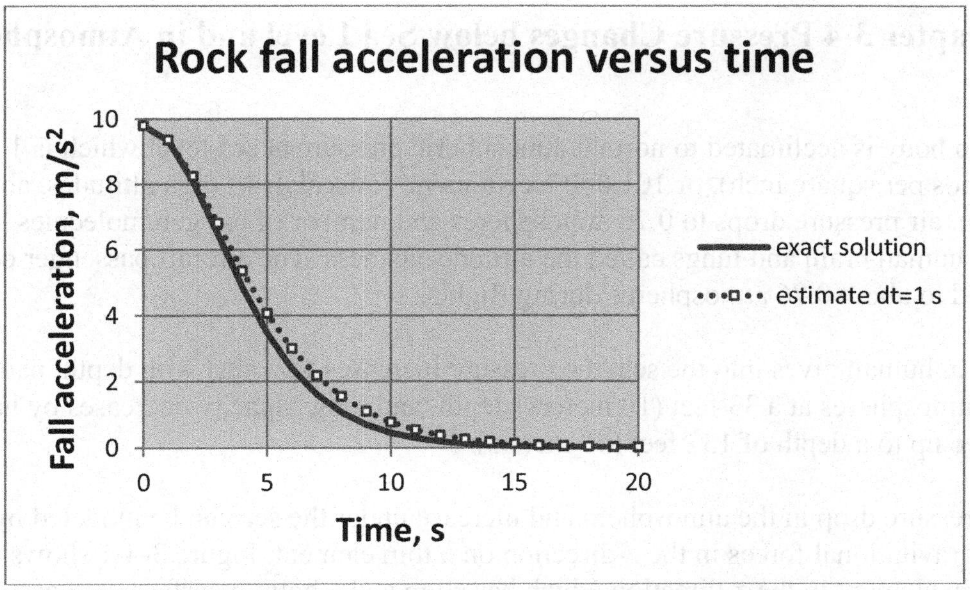

Figure 3-3-5 Rock fall acceleration versus time

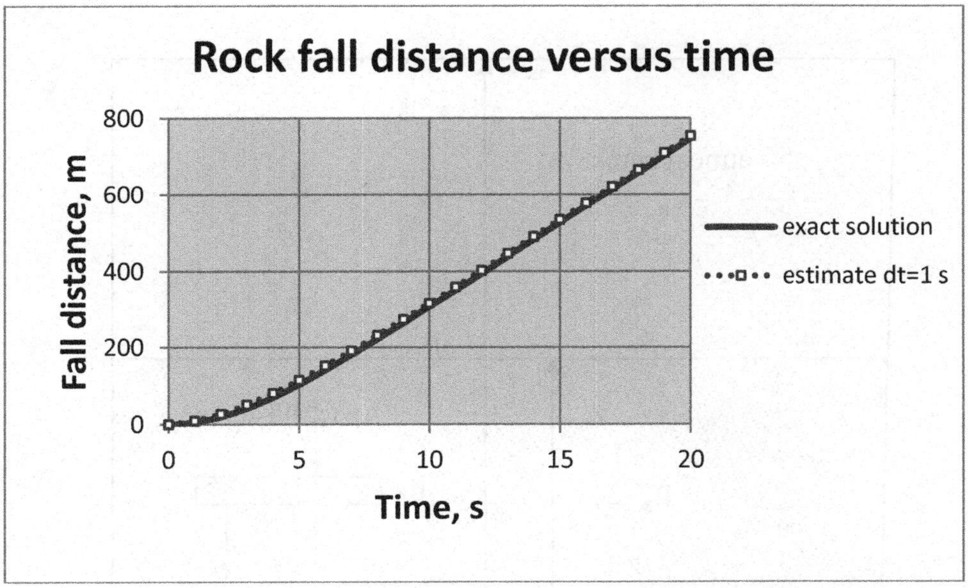

Figure 3-3-6 Rock fall distance versus time

The numerical solutions are close to exact solutions. If the time interval for numerical calculations is smaller than one second, i.e. 0.1 s, then the numerical solutions will be much closer to exact solutions.

Chapter 3-4 Pressure Changes below Sea Level and in Atmosphere

Human body is acclimated to normal atmospheric pressure at sea level which is 1 atmosphere, 14.7 psi (pounds per square inch), or 101,000 Newtons/m^2 (Pascals). At high altitudes, above 8,000 ft (2,400 meters), air pressure drops to 0.76 atmospheres and number of oxygen molecules decrease which affects human brain and lungs called the altitude sickness. The aircraft passenger compartments are pressurized to about 0.76 atmospheres during flights.

When a human dives into the sea, the pressure increases very fast with depth, namely pressure doubles to 2 atmospheres at a 33 feet (10 meters) depth, and lung capacity decreases by half. Advanced divers can dive up to a depth of 132 feet (40 meters).

The pressure drop in the atmosphere and increase under the sea can be modeled by considering pressure and gravitational forces in the z-direction on a thin element. Figure 3-4-1 shows the forces acting on a thin element in the z-direction which has a top and a bottom surface area of "A" and a thickness "dz".

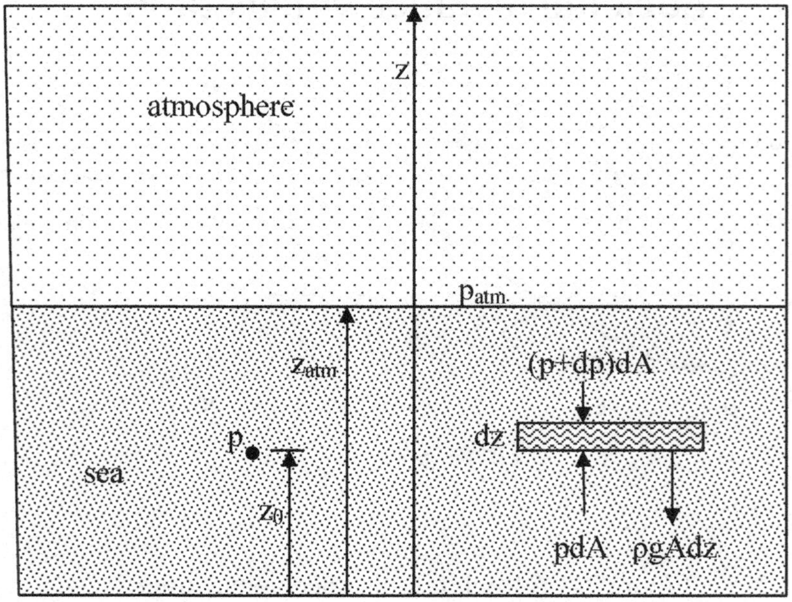

Figure 3-4-1 Pressure and gravitational forces acting in the z-direction on a thin element

In Figure 3-4-1 "dp" represents a change in pressure from the bottom surface to the top surface of the thin element. "ρ" is the density of the thin element and "g" is the gravitational acceleration. The mass of the thin element is its density times its volume, namely "ρgAdz".

In an equilibrium state, sum of pressure and gravitational forces in the z-direction should be zero or

$$pA - (p + dp)A - \rho gAdz = 0 \qquad (3\text{-}4\text{-}1)$$

Eq. (3-4-1) can be rewritten as

$$dp = -\rho gdz \qquad (3\text{-}4\text{-}2)$$

Assuming that the sea water density is not changing with depth, salinity, and temperature, the first order differential equation in Eq. (3-4-2) can be integrated to find pressure, p , at any location, z_o , under the sea surface with respect to the known sea level atmospheric pressure, p_{atm} , at a height z_{atm} , see Figure 3-4-1.

$$\int_{p}^{p_{atm}} dp = -\rho g \int_{z_o}^{z_{atm}} dz \qquad (3\text{-}4\text{-}3)$$

or

$$p_{atm} - p = -\rho g(z_{atm} - z_0) \qquad (3\text{-}4\text{-}4)$$

Let $(z_{atm} - z_o) = d$, the depth below the sea level, then Eq. (3-4-4) becomes

$$p - p_{atm} = \rho gd \qquad (3\text{-}4\text{-}5)$$

Pressure at any depth below sea level can be written as a ratio to p_{atm}.

$$\frac{p}{p_{atm}} = 1 + \left(\frac{\rho g}{p_{atm}}\right) d \qquad (3\text{-}4\text{-}6)$$

In Eq. (3-4-6), $\left(\frac{\rho g}{p_{atm}}\right)$ term is a constant and can be calculated by using a sea water density value $\rho = 1{,}030\frac{kg}{m^3}$, gravitational acceleration $g = 9.8\frac{m}{s^2}$, and $p_{atm} = 101{,}000\frac{N}{m^2}$. Also knowing that one Newton of force, N, is defined as one kg mass accelerating at one $\frac{m}{s^2}$. Then $\left(\frac{\rho g}{p_{atm}}\right) = 0.1\ \frac{1}{m}$.
Eq. (3-4-6) becomes

$$\frac{p}{p_{atm}} = 1 + 0.1d \qquad (3\text{-}4\text{-}7)$$

where the depth d is in meters. Pressure under the sea as a ratio to atmospheric pressure at sea level versus sea depth in meters is shown in Figure 3-4-1.

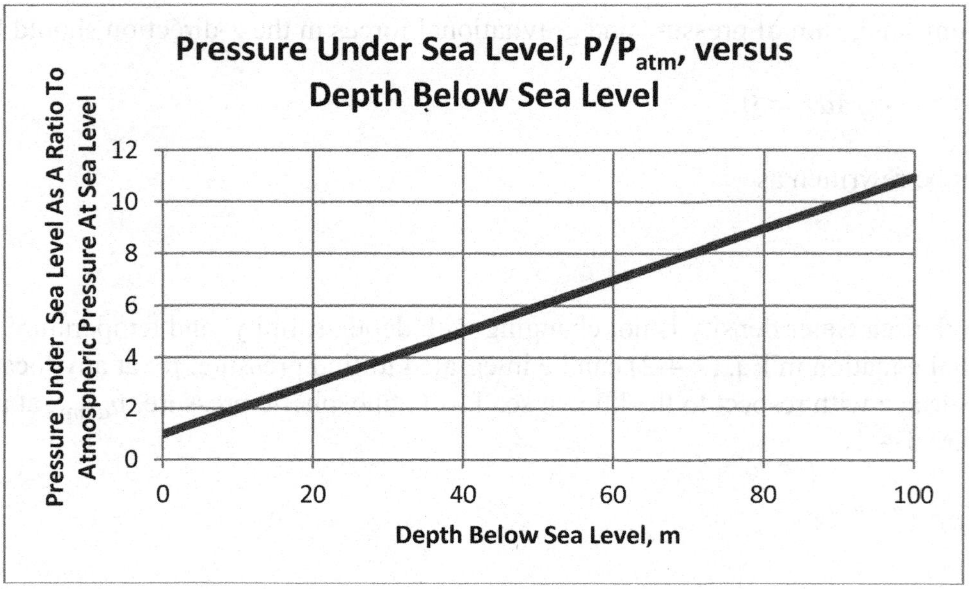

Figure 3-4-1 Pressure under sea level, P/P_{atm}, versus depth below sea level

In English units, Eq. (3-4-7) becomes

$$\frac{p}{p_{atm}} = 1 + 0.03d \tag{3-4-8}$$

where the depth d is in feet. Deepest locations under the sea can be found in the Pacific Ocean. The Marianas Trench is about 10,923 meters (35,838 feet) deep where the pressure reaches up to 1,093 atmospheres or 16,000 psi. Pressure under the sea as a ratio to atmospheric pressure at sea level versus sea depth in feet is shown in Figure 3-4-2.

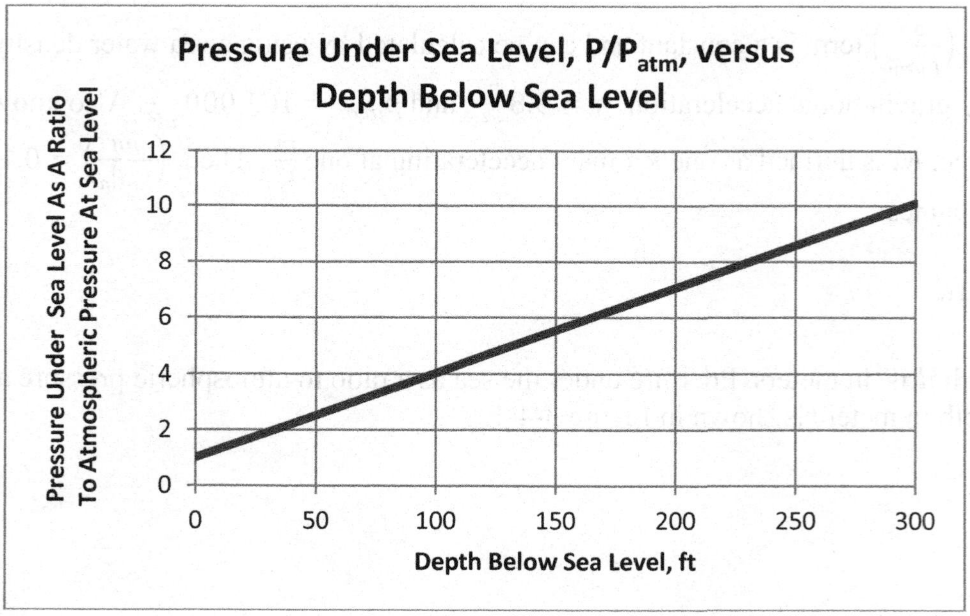

Figure 3-4-2 Pressure under sea level, P/P_{atm}, versus depth below sea level

Eq. (3-4-2) can also be used to estimate the pressure drop in the atmosphere as higher altitudes are reached. In this case the atmospheric air density cannot be assumed to be a constant like it was for sea water above. As the pressure decreases, the air density decreases. Assuming constant air temperature, the following linear relationship between pressure and air density holds pretty well.

$$\frac{p}{p_{atm}} = \frac{\rho}{\rho_{atm}} \qquad (3\text{-}4\text{-}9)$$

Substituting Eq. (3-4-9) into Eq. (3-4-2) gives the following first order differential equation.

$$\frac{dp}{p} = -\left(\frac{\rho_{atm} g}{p_{atm}}\right) dz \qquad (3\text{-}4\text{-}10)$$

By assuming the sea level to be at zero altitude with a known atmospheric pressure of p_{atm}, Eq. (3-4-10) can be integrated between p_{atm} and p from sea level to an altitude h.

$$\int_{p_{atm}}^{p} \frac{dp}{p} = -\left(\frac{\rho_{atm} g}{p_{atm}}\right) \int_0^h dz \qquad (3\text{-}4\text{-}11)$$

Integrating Eq. (3-4-11) gives

$$ln\left(\frac{p}{p_{atm}}\right) = -\left(\frac{\rho_{atm} g}{p_{atm}}\right) h \qquad (3\text{-}4\text{-}12)$$

Eq. (3-4-12) can be rewritten as

$$\frac{p}{p_{atm}} = exp\left[-\left(\frac{\rho_{atm} g}{p_{atm}}\right) h\right] \qquad (3\text{-}4\text{-}13)$$

In Eq. (3-4-13), $\left(\frac{\rho_{atm} g}{p_{atm}}\right)$ term is a constant and can be calculated by using a sea level air density value at 20 C of $\rho_{atm} = 1.21 \frac{kg}{m^3}$, gravitational acceleration $g = 9.8 \frac{m}{s^2}$, and $p_{atm} = 101{,}000 \frac{N}{m^2}$. Then $\left(\frac{\rho_{atm} g}{p_{atm}}\right) = 0.117 \frac{1}{km}$. Eq. (3-4-6) becomes

$$\frac{p}{p_{atm}} = exp[-0.117h] \qquad (3\text{-}4\text{-}14)$$

where the altitude h is in kilometers. Atmospheric pressure, $\frac{p}{p_{atm}}$, versus altitude above sea level in kilometers is shown in Figure 3-4-3.

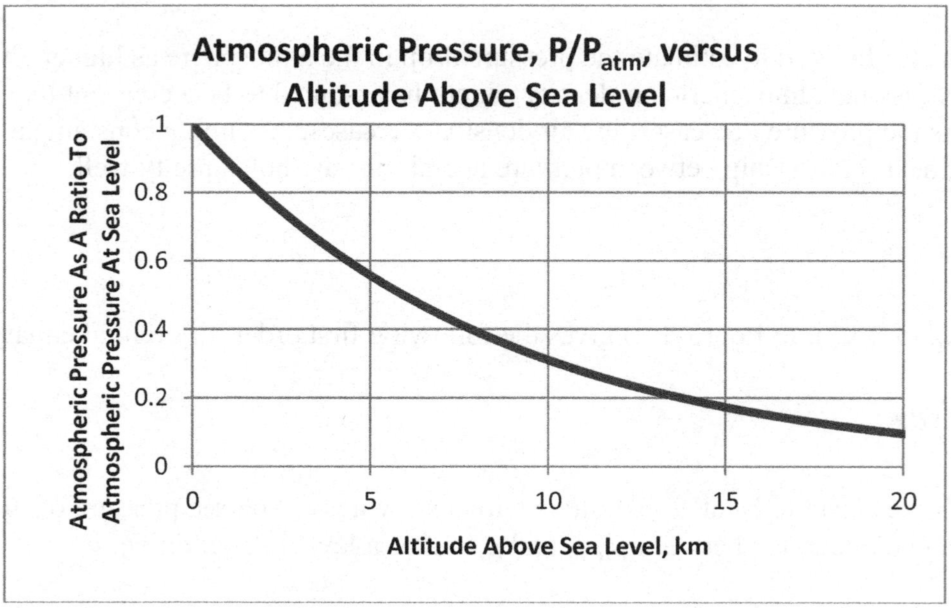

Figure 3-4-3 Atmospheric pressure, P/P_{atm}, versus altitude above sea level in kilometers

In English units, Eq. (3-4-14) becomes

$$\frac{p}{p_{atm}} = exp[-0.188h] \qquad (3\text{-}4\text{-}15)$$

where the altitude h is in miles. Atmospheric pressure and altitude relationships, Eqs. (3-4-14) and (3-4-15), are good approximations even if the density variations due to temperature in the atmosphere are neglected. On top of Mount Everest at 29,029 feet (8,848 meters) altitude, the pressure drops to $0.35p_{atm}$ or $0.52\ psi$. Atmospheric pressure, $\frac{p}{p_{atm}}$, versus altitude above sea level in miles is shown in Figure 3-4-4.

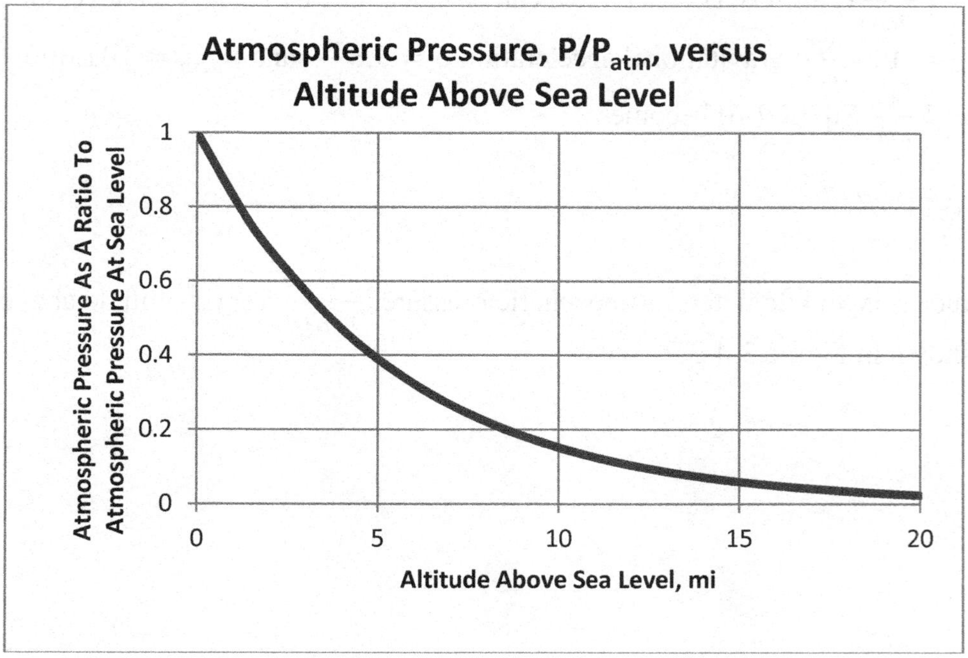

Figure 3-4-4 Atmospheric pressure, P/P_{atm}, versus altitude above sea level in miles

Chapter 3-5 Optimize Piping Costs through a Ranch

Consider a one mile by one mile ranch as shown in Figure 3-5-1. There is a project to run water piping from the north-west corner of the ranch to its south-east corner. On the east boundary of the ranch, the piping can be over the ground, but any other place it has to be under the ground. Underground piping needs trench digging and it costs two to ten times more than the over ground piping from the bids that were received.

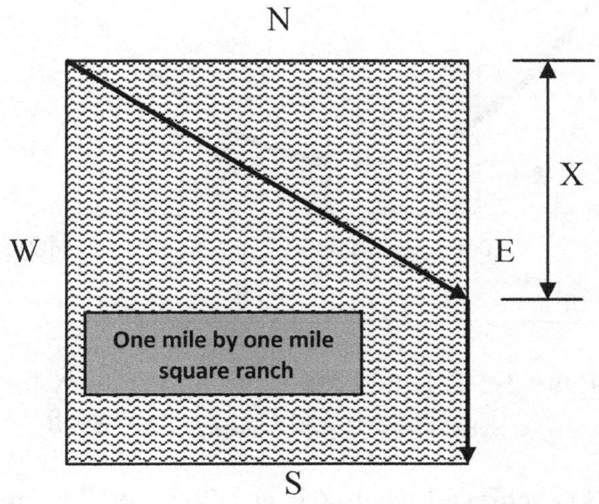

Figure 3-5-1 One mile by one mile ranch water piping project

Total cost of the piping from the north-west corner to the south-east corner has to be formulated. Let us call the over ground piping cost per foot along the eastern border of the ranch $C_{overground}$. Let us call the underground piping cost per foot at any other location on the ranch $C_{underground}$. Total cost of piping from the north-west corner to the south-east corner is

$$Total\ Cost = C_{underground}\sqrt{X^2 + 5280^2} + C_{over\ ground}(5280 - X) \qquad (3\text{-}5\text{-}1)$$

where X is the distance from the north-east corner of the ranch to the location where the underground piping ends and the over ground piping starts at the eastern boundary. One of the bids came in at $C_{overground} = \$2/ft$ and $C_{underground}$ at three times the $C_{overground}$ cost, namely at \$6/ft. Total cost of piping versus the X distance is plotted in Figure 3-5-2.

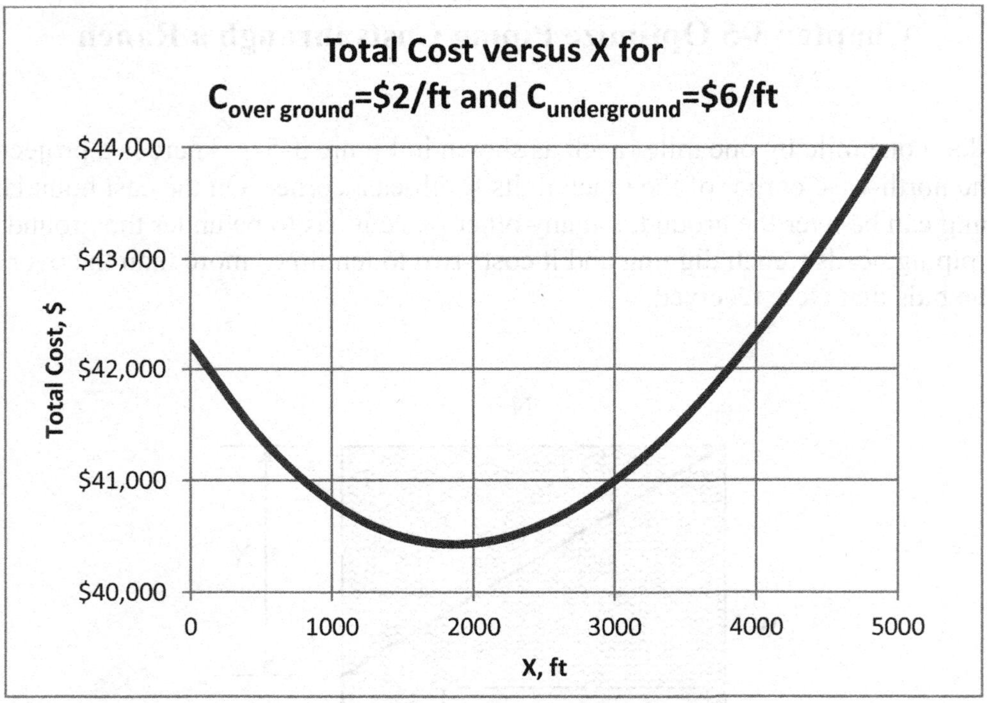

Figure 3-5-2 Total cost of piping versus X for
$C_{over\ ground}$ = \$2/ft and $C_{underground}$ = \$6/ft

Minimum total cost of \$40,428 is achieved when *X* is at 1,867 feet. This minimum value can be obtained by differential calculus, namely by differentiating $Total\ Cost$ in Eq. (3-5-1) with respect to *X* and setting the result equal to zero.

$$\frac{d(Total\ Cost)}{dX} = \frac{C_{underground}\,X}{\sqrt{X^2+5280^2}} - C_{over\ ground} = 0 \qquad (3\text{-}5\text{-}2)$$

or

$$X_{minimum} = +\frac{5280}{\sqrt{\left(\frac{C_{underground}}{C_{over\ ground}}\right)^2 - 1}} \qquad (3\text{-}5\text{-}3)$$

In Eq. (3-5-3) only the positive root is used, since negative $X_{minimum}$ is invalid and $C_{underground}$ has to be greater than $C_{over\ ground}$. The $X_{minimum}$ value in Eq. (3-5-3) provides the minimum $Total\ Cost$. This can be shown by taking the second derivative of $Total\ Cost$ with respect to *X* in Eq. (3-5-2) and verifying that the results are always positive in the region of interest.

$$\frac{d^2(Total\ Cost)}{dX^2} = \frac{5280^2}{(X^2+5280^2)^{3/2}} > 0 \qquad (3\text{-}5\text{-}4)$$

In Eq. (3-5-3), if $\sqrt{2} > \frac{C_{underground}}{C_{over\ ground}} > 1$, then $X_{minimum} > 5280$ feet and $X_{minimum}$ intersects the southern boundary of the ranch. There is no over ground piping as shown in Figure 3-5-3.

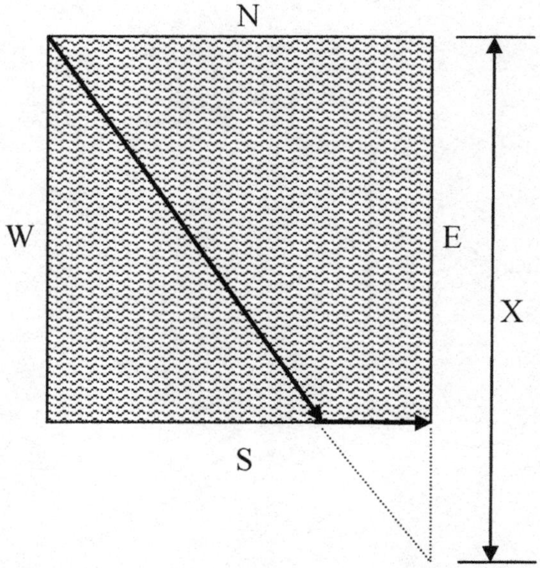

Figure 3-5-3 One mile by one mile ranch water piping project
$\sqrt{2} > \frac{C_{underground}}{C_{over\ ground}} > 1$ and $X_{minimum} > 5280$ feet case

$X_{minimum}$ is plotted for different ratios of $\frac{C_{underground}}{C_{over\ ground}}$ using Eq. (3-5-3) and $Total\ Cost$ for the piping project is determined using Eq. (3-5-1) and $X_{minimum}$ for a case of $C_{over\ ground} = \$2/ft$. The results are shown in Figure 3-5-4.

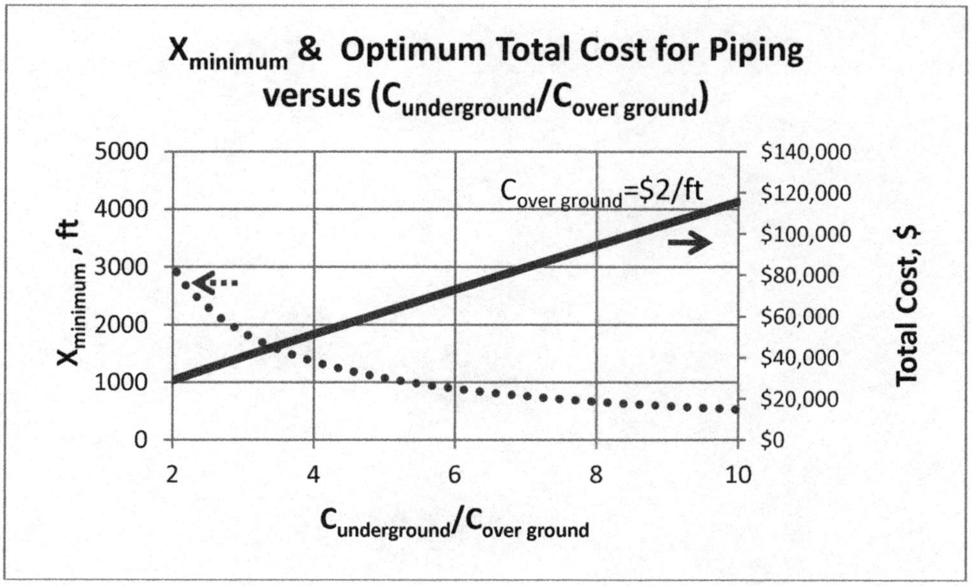

Figure 3-5-4 $X_{minimum}$ versus $\frac{C_{underground}}{C_{over\ ground}}$ and $Total\ Cost$ for piping using $X_{minimum}$ for a case of $C_{over\ ground} = \$2/ft$

As the underground piping costs increase with respect to the over ground piping costs, the crossing point at the eastern boundary of the ranch gets closer to the northern boundary or $X_{minimum}$ gets smaller.

Chapter 3-6 Cooling of Tea With and Without Adding Milk

Cooling of a cup of tea can be modeled by using the first law of thermodynamics which states that the heat loss from the tea to the environment is proportional the temperature change of the tea with respect to time. This statement can be written in an equation form as follows.

$$\frac{dT}{d\theta} = -K(T - T_{environment}) \tag{3-6-1}$$

where T is the temperature of the tea in degrees Fahrenheit, θ is time in minutes when the tea is cooling down, and K is a proportionality constant in 1/minutes. The minus sign in front of the proportionality constant comes from the fact that heat is lost from the tea to the environment. $\frac{1}{K}$ is in minutes and it is called the thermal time constant for this problem. If $T_{environment}$ is assumed to be a constant, then Eq. (3-6-1) can be rearranged to be

$$\frac{d(T-T_{environment})}{(T-T_{environment})} = -Kd\theta \tag{3-6-2}$$

Eq. (3-6-2) is a first order differential equation which can be integrated from $\theta = 0$ when the tea initial temperature is $T_{initial}$ to an arbitrary tea temperature T at time θ.

$$\int_{T_{initial}-T_{environment}}^{T-T_{environment}} \frac{d(T-T_{environment})}{(T-T_{environment})} = \int_0^{\theta} -Kd\theta \tag{3-6-3}$$

or

$$ln(T - T_{environment}) - ln(T_{initial} - T_{environment}) = -K\theta \tag{3-6-4}$$

Eq. (3-6-4) can be rewritten as

$$\frac{(T-T_{environment})}{(T_{initial}-T_{environment})} = e^{(-K\theta)} \tag{3-6-5}$$

In Eq. (3-6-5), $T_{environment}$ and $T_{initial}$ are known. The only unknown is the proportionality constant K which can easily be determined for the cup of tea in question by a simple experiment. In heat transfer the proportionality constant K is defined as

$$K = \frac{hA}{\rho V c_p}$$

where h is the heat transfer coefficient between tea and the environment, A , is the surface area of tea from which heat is being transferred to the environment, ρ , is the density of tea, V is the volume of tea, and c_p , is the specific heat of tea at constant pressure. According to Eq. (3-6-5), temperature of the tea decreases exponentially towards $T_{environment}$ and temperature potential difference that affects the tea cooling process is $T_{initial} - T_{environment}$. $\frac{1}{K}$ is the thermal time constant for the cooling process, namely tea cools down by $0.63(T_{initial} - T_{environment})$ when $\theta = \frac{1}{K}$.

The proportionality constant K is different for every cooling system. Therefore it is also different for a cup of tea and it can be determined by a simple experiment instead of dealing with the heat transfer parameters. For example, tea whose initial temperature is known, say at 80 C, is poured into a cup. A small thermometer that is at the temperature of tea is inserted into the tea and the tea temperature is read after three minutes have elapsed. It is determined that the tea temperature cools down to about 70 C after three minutes. During this experiment, the environment temperature is held as a constant at 25 C. The experiment is repeated several times and an average of 70 C is obtained after three minutes cooling period. Using Eq. (3-6-5), the proportionality constant K can be determined.

$$\frac{(70-25)}{(80-25)} = e^{(-3K)} \qquad (3\text{-}6\text{-}6)$$

or

$$K = \left(\frac{-1}{3}\right) ln\left(\frac{45}{55}\right) = 0.0669 \ \frac{1}{min} \qquad (3\text{-}6\text{-}7)$$

The thermal time constant for this cup of tea is about 15 minutes, namely tea cools down to $80 - 0.63(80 - 25) = 45\ C$ degrees centigrade in 15 minutes. This time constant can be used for different initial and environmental conditions for the same type of tea and its cup. If the initial tea temperature is 90 C and the environment is 10 C, tea will cool down to 40 C, $90 - 0.63(90 - 10) = 40\ C$, in 15 minutes.

Using Eq.s (3-6-5) and (3-6-7), $T_{environment} = 25\ C$ and $T_{initial} = 90\ C$, the cooling of tea in a cup is shown in Figure 3-6-1.

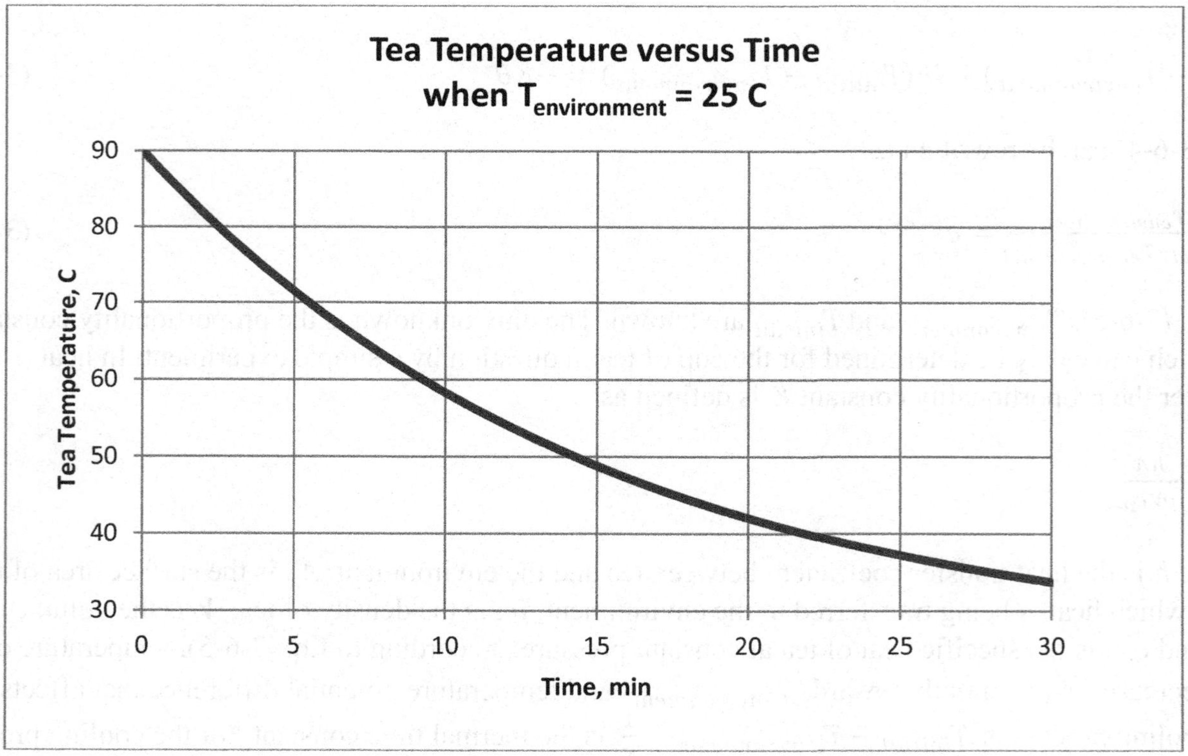

Figure 3-6-1 Tea temperature versus time when $T_{environment}$=25 C

Next the question comes up when milk is added to tea. Should the milk be added to tea initially or sometime after, say 5 minutes, for a warmer mixture to drink?

Intuitively from Eq. (3-6-5), temperature potential difference, $T_{initial} - T_{environment}$, that affects tea cooling process is larger, if milk is not added to tea initially, then tea-milk mixture will cool faster. If a warmer tea-milk mixture is desired, the milk should be added to tea initially so the temperature potential difference, $T_{initial} - T_{environment}$, is smaller.

Using Eq. (3-6-5) and also by determining the temperature drop due to the addition of milk can show that if a warmer tea-milk mixture is desired for drinking, the milk should be added to tea initially. Temperature drop due to addition of cold milk to tea is estimated from the following equation that assumes densities for hot tea and cold milk to be equal and also assumes specific heats at constant pressure for hot tea and cold milk to be equal.

$$(V_{tea} + V_{milk})T_{tea-milk} = V_{tea}T_{tea} + V_{milk}T_{milk} \tag{3-6-8}$$

If T_{milk} is at 10 C and T_{tea} is at 90 C initially, namely when $\theta = 0$, and if the amount of milk that is added is half of tea volume, $V_{milk} = 0.5V_{tea}$, then from Eq. (3-6-8), $T_{tea-milk}$ becomes 63.3 C or the tea-milk mixture cools down by 26.7 C.

Same exercise can be applied when milk is added to tea after 5 minutes. This time T_{milk} is again at 10 C and T_{tea} is at 71.5 C and again if the amount of milk that is added is half of tea volume, $V_{milk} = 0.5V_{tea}$, then from Eq. (3-6-8), $T_{tea-milk}$ becomes 51 C or the tea-milk mixture cools down by 20.5 C.

Adding milk to tea options are shown in Figure 3-6-2.

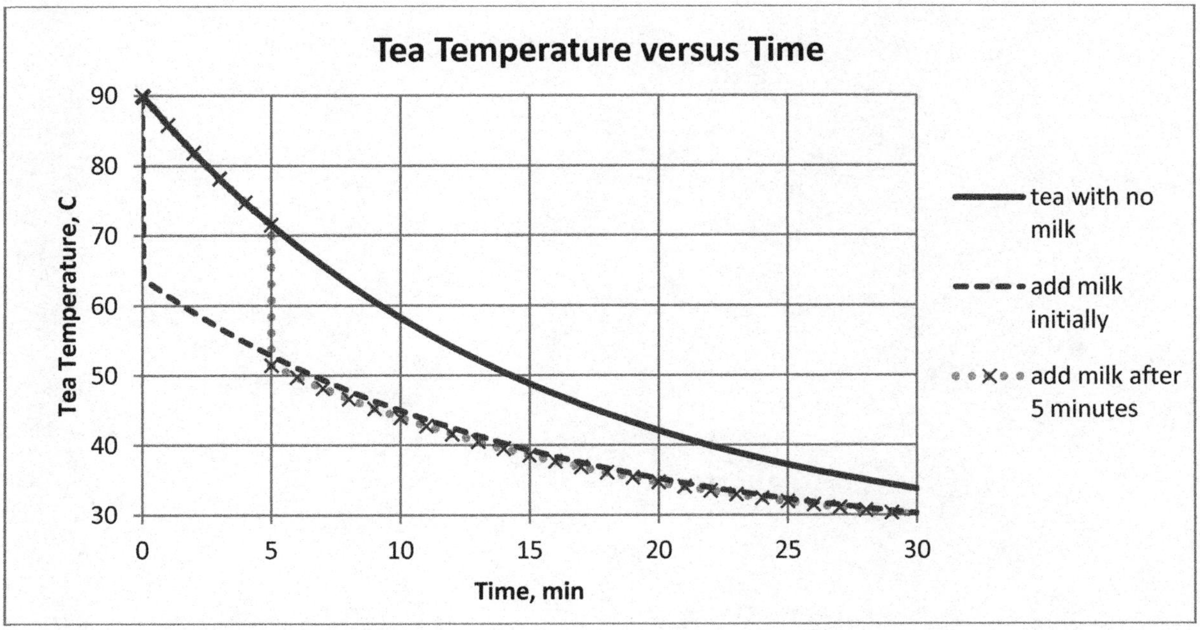

Figure 3-6-2 Tea temperature versus time with no milk case, adding milk initially case, and adding milk after 5 minutes case where milk temperature is at 10 C and added milk volume is half the volume of tea

Figure 3-6-2 shows that adding milk to tea initially will provide a warmer tea-milk mixture than adding milk to tea after 5 minutes. If milk is added to tea earlier than 5 minutes, difference between two cooling curves will get smaller. If milk is added to tea later than 5 minutes, difference between two cooling curves will get larger. Adding milk to tea initially will always provide a warmer tea-milk mixture.

Chapter 3-7 A Basketball Player's Hang Time

A basketball player can jump towards the hoop to dunk the ball and stay airborne for a while. His or her airborne time can be calculated by considering the gravitational acceleration.

$$a = \frac{dv}{dt} = -g \qquad (3\text{-}7\text{-}1)$$

g is earth's gravitational acceleration, 32.2 $\frac{ft}{s^2}$ and it is negative since it is against the upward motion of the basketball player. Eq. (3-7-1) can be integrated with respect to time to find the player's upward speed.

$$v = C_1 - gt \qquad (3\text{-}7\text{-}2)$$

Integration constant C_1 can be determined from the initial condition that the upward speed of the player at $t = 0$ is v_0 . Then Eq. (3-7-2) becomes

$$v = v_0 - gt \qquad (3\text{-}7\text{-}3)$$

Upward displacement of the player is determined by integrating Eq. (3-7-3) with respect to time.

$$z = C_2 + v_0 t - 0.5gt^2 \qquad (3\text{-}7\text{-}4)$$

Integration constant C_2 can be determined from the initial condition that the upward displacement of the player at $t = 0$ is 0 . Then Eq. (3-7-4) becomes

$$z = v_0 t - 0.5gt^2 \qquad (3\text{-}7\text{-}5)$$

The player reaches his or her maximum jump height when $v = 0$ in Eq. (3-7-3). Then the time to reach maximum jump height is

$$t_{maximum\ jump\ height} = \frac{v_0}{g} \qquad (3\text{-}7\text{-}6)$$

Maximum jump height of the player can be obtained by inserting the maximum jump height time from Eq. (3-7-6) into Eq. (3-7-5).

$$z_{maximum\ jump\ height} = 0.5\frac{v_0^2}{g} \qquad (3\text{-}7\text{-}7)$$

The basketball player's hang time can be obtained from Eq. (3-7-5) when the jump height is zero.

$$0 = v_0 t - 0.5gt^2 = t(v_0 - 0.5gt) \qquad (3\text{-}7\text{-}8)$$

Eq. (3-7-8) is satisfied when $t = 0$ or when $t = 2\frac{v_0}{g}$. Jump height is zero initially when the player starts to jump and it is again zero when he or she comes down to the floor. So the hang time is the latter solution to Eq. (3-7-8).

$$t_{hang\ time} = 2\frac{v_0}{g} \qquad (3\text{-}7\text{-}9)$$

Initial upward speed of the player can be determined from Eq. (3-7-7), if his or her maximum jump height is known. Figure 3-7-1 shows initial upward speed of the player versus his or her maximum jump height.

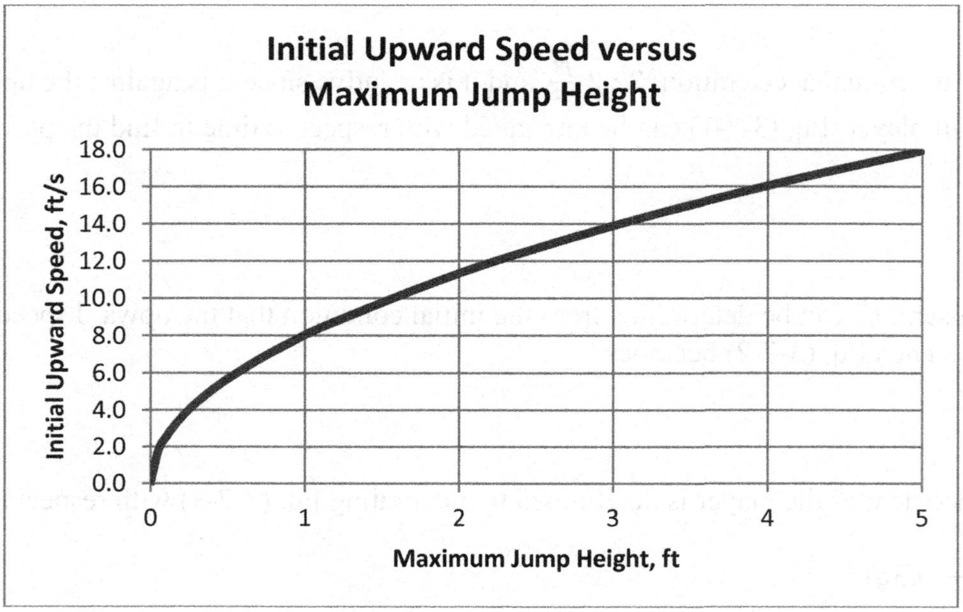

Figure 3-7-1 Initial upward speed versus maximum jump height

Jump height of the player versus time is determined from Eq. (3-7-5) for different maximum jump heights. Jump height versus time is a parabolic behavior and plots are shown in Figure 3-7-2.

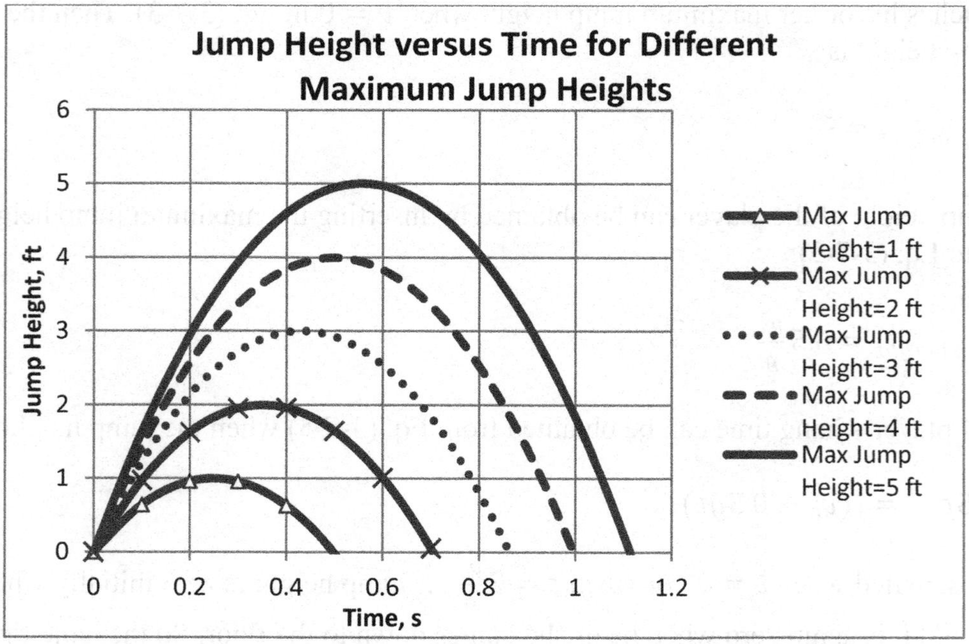

Figure 3=7-2 Jump height versus time for different maximum jump heights

Player's hang time versus maximum jump height is determined using Eqs. (3-7-7) and (3-7-9). Player's hang time varies as the square root of his or her maximum jump height.

$$t_{hang\ time} = \sqrt{\frac{8z_{maximum\ jump\ height}}{g}} \tag{3-7-10}$$

Player's hang time versus his or her maximum jump height is shown in Figure 3-3-3.

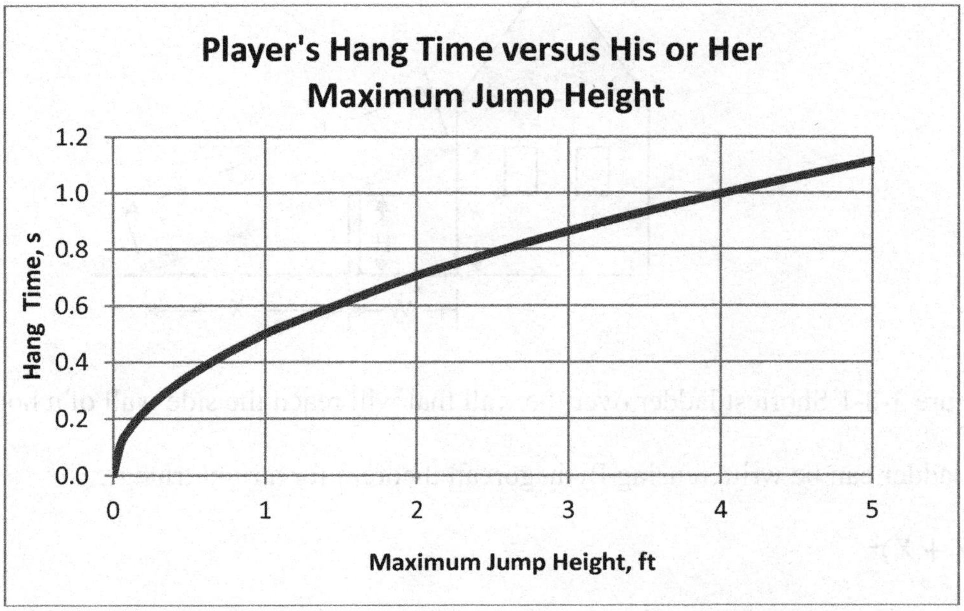

Figure 3-7-3 Player hang time versus maximum jump height

Player's hang time versus his or her initial upward speed is a linear relationship as given in Eq. (3-7-9). Player's hang time versus his or her initial upward speed is shown in Figure 3-7-4.

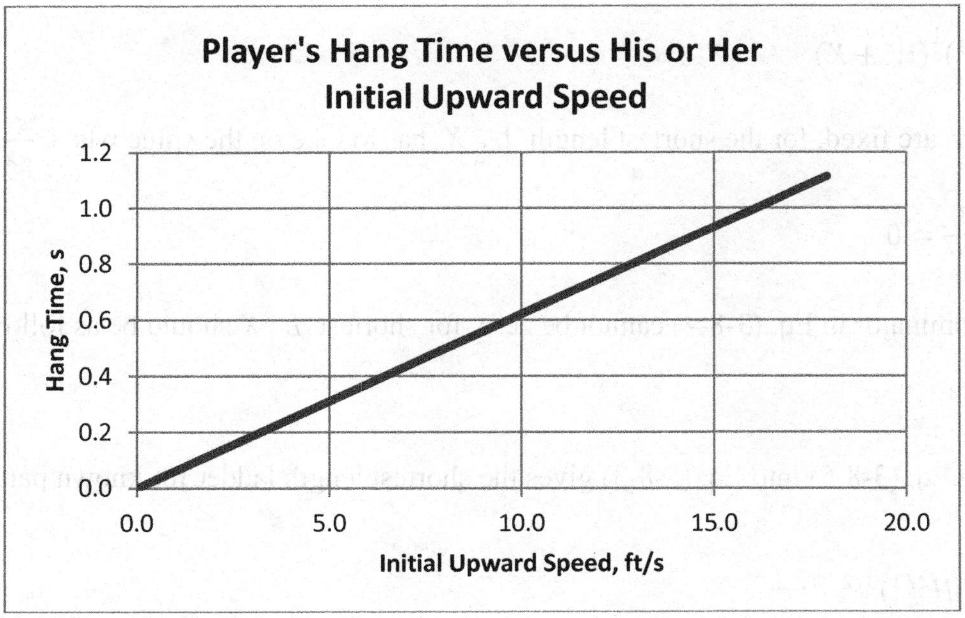

Figure 3-7-4 Player's hang time versus his or her initial upward speed

Chapter 3-8 Shortest Ladder over a Wall

Some optimization problems require the investigation of first and second derivatives of the governing function. To illustrate this optimization process, let us find the shortest ladder that will go over a wall and reach a house's side wall as it is illustrated in Figure 3-8-1.

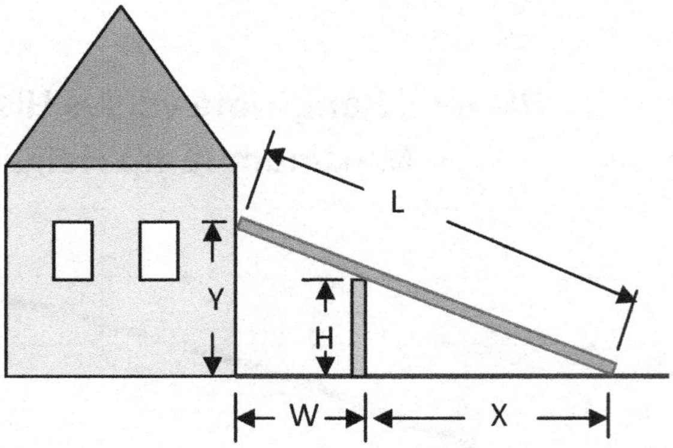

Figure 3-8-1 Shortest ladder over the wall that will reach the side wall of a house

Length of the ladder can be written using Pythagorean theorem for a right triangle.

$$L^2 = Y^2 + (W + X)^2 \tag{3-8-1}$$

We can get rid of Y by using similar triangles

$$\frac{Y}{W+X} = \frac{H}{X} \tag{3-8-2}$$

Substituting Y from Eq. (3-8-2) into Eq. (3-8-1) gives the following equation for L.

$$L = (H^2 + X^2)^{\frac{1}{2}}(W + X) \tag{3-8-3}$$

Since H and W are fixed, for the shortest length L, X has to take on the value when $\frac{dL}{dX}$ is zero.

$$\frac{dL}{dX} = \frac{X^3 - WH^2}{(H^2+X^2)^{1/2}} = 0 \tag{3-8-4}$$

Since the denominator in Eq. (3-8-4) cannot be zero, for shortest L, X should be as follows:

$$X = W^{1/3}H^{2/3} \tag{3-8-5}$$

or substituting Eq. (3-8-5) into Eq. (3-8-3) gives the shortest length ladder for known parameters W and H.

$$L = (W^{2/3} + H^{2/3})^{3/2} \tag{3-8-6}$$

Shortest ladder length, L , in Eq. (3-8-6) is plotted against the garden width, W, for different wall heights, H, in Figure 3-8-1.

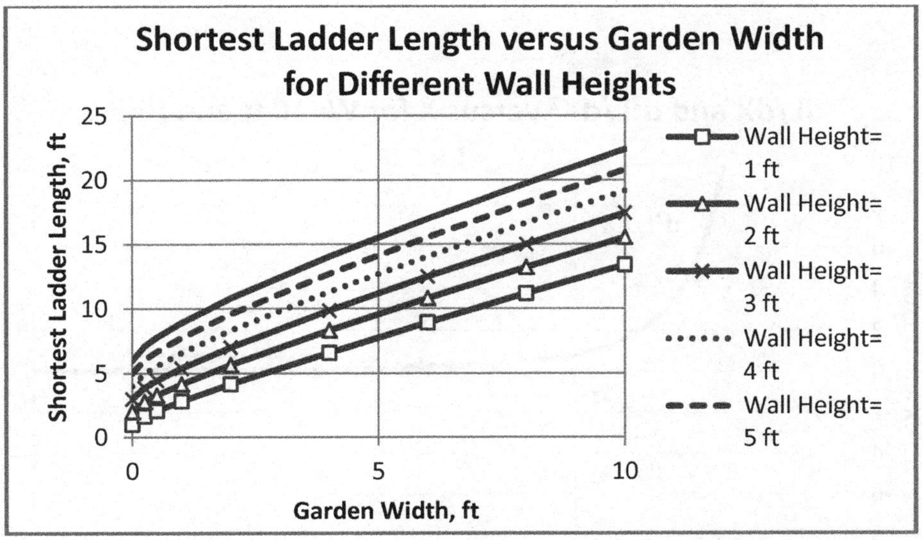

Figure 3-8-1 Shortest ladder length versus garden width for different wall heights

Ladder length L can be plotted as a function of X using Eq. (3-8-3) for a given garden width, W, and wall height, H. Figure 3-8-2 shows the ladder length versus X for a garden width of 10 feet and a wall height of 5 feet. Ladder length decreases as X increases and goes through a minimum value for L at $X = 6.3$ ft $L_{minimum} = 20.81$ ft. After this X value the ladder length starts to increase.

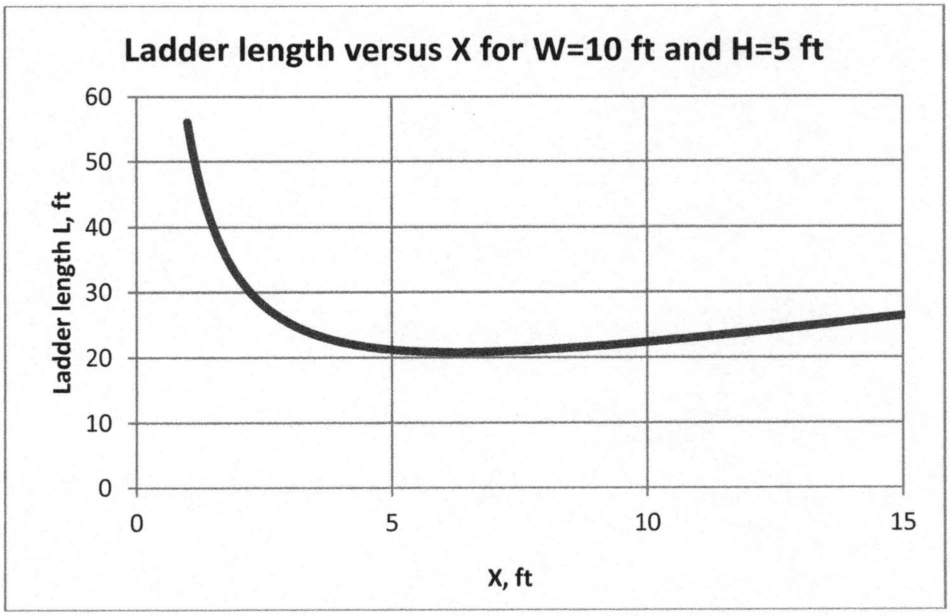

Figure 3-8-2 Ladder length versus X for W=10 ft and H=5 ft

Let us look at the behavior of first and second derivatives of L with respect to X which are shown in Figure 3-8-3. Both first and second derivatives are continuous functions. First derivative, $\frac{dL}{dX}$, is negative at small values of X, goes through zero at $x = 6.3$, and becomes positive. Second derivative, $\frac{d^2L}{dX^2}$, is positive around $x = 6.3$. Therefore the ladder length is a minimum at this value of X.

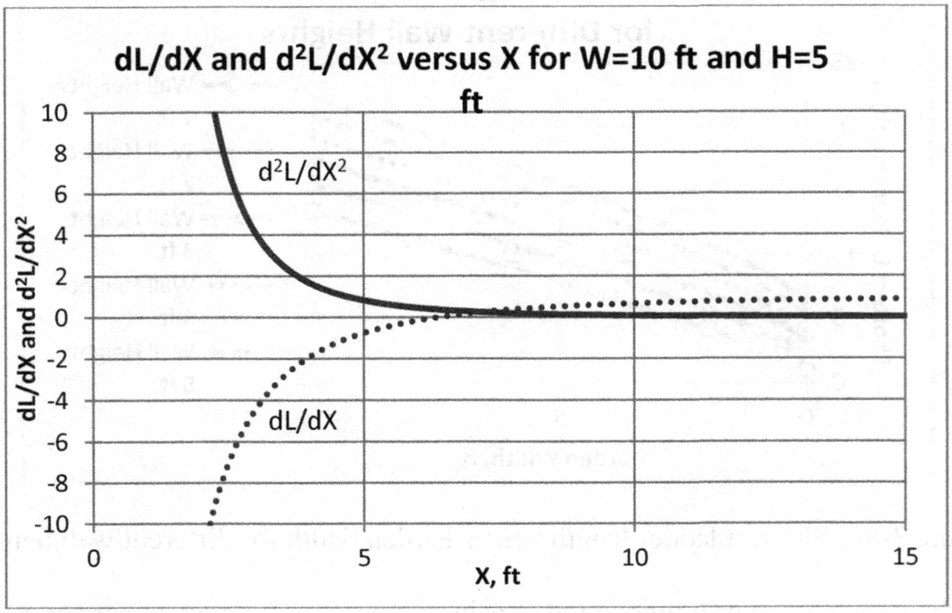

Figure 3-8-3 First and second derivatives of L with respect to X

Chapter 3-9 Maximize Area of a Rectangular Building on an Odd Shaped Lot

Let us consider an irregular lot as shown in Figure 3-9-1 where the lot dimensions are shown in feet. The object is to find the largest rectangular single floor building that will fit on this lot.

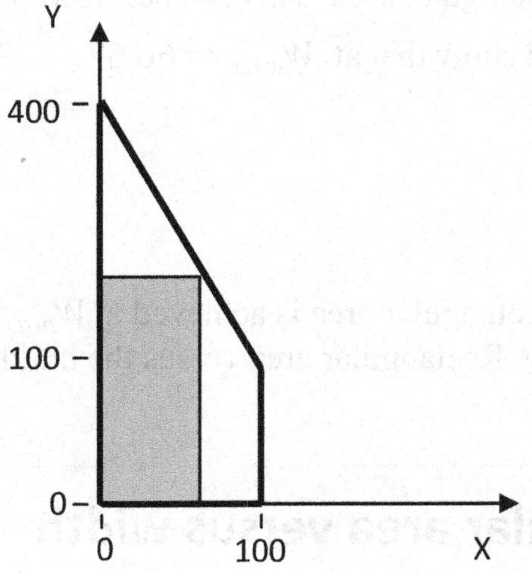

Figure 3-9-1 Rectangular building on an irregular lot

Area of the building has to be maximized in this lot. Let us define this area as

$$Area = W \times L \qquad (3\text{-}9\text{-}1)$$

where W is the width of the building in feet in the X direction and L is the length of the building in feet in the Y direction. For a width value of W, the length L has to lie on the lot line that lays between the points (0,400) and (100,100). The line that lays between the points (0,400) and (100,100) can be written in a general form as follows:

$$Y = A \times X + B \qquad (3\text{-}9\text{-}2)$$

where A is the slope of the line and B is the intercept of the line at $X = 0$ or at the Y axis.

Eq. (3-9-2) has to satisfy the point (0, 400) or when $X = 0$, $Y = 400$ and therefore the constant $B = 400$. Eq. (3-9-2) has to also satisfy the point (100,100) and this requirement gives $A = -3$. Eq. (3-9-2) becomes

$$Y = -3 \times X + 400 \qquad (3\text{-}9\text{-}3)$$

Since $Y = L$ and $X = W$ for this problem, Eq. (3-9-1) can be rewritten as

$$Area = W \times (-3 \times W + 400) = -3 \times W^2 + 400 \times W \qquad (3\text{-}9\text{-}4)$$

In order to maximize the rectangular area, the derivative of $Area$ with respect to W is taken and first derivative is set to zero.

$$\frac{d(Area)}{dW} = -6 \times W + 400 = 0 \qquad (3\text{-}9\text{-}5)$$

or the width of the building that will maximize its rectangular area is

$$W_{max} = 66.67\ ft \tag{3-9-6}$$

The width in Eq. (3-9-6) maximizes its rectangular area. This can be shown by taking the second derivative of $Area$ with respect to W and show that at $W_{max} = 66.67\ ft$, $\frac{d^2(Area)}{dW^2} < 0$. From Eq. (3-9-5),

$$\frac{d^2(Area)}{dW^2} = -6 \tag{3-9-7}$$

Another way to show that the maximum rectangular area is achieved at $W_{max} = 66.67\ ft$ is by graphing $Area$ versus W using Eq. (3-9-4). Rectangular area versus the building width is shown in Figure 3-9-2.

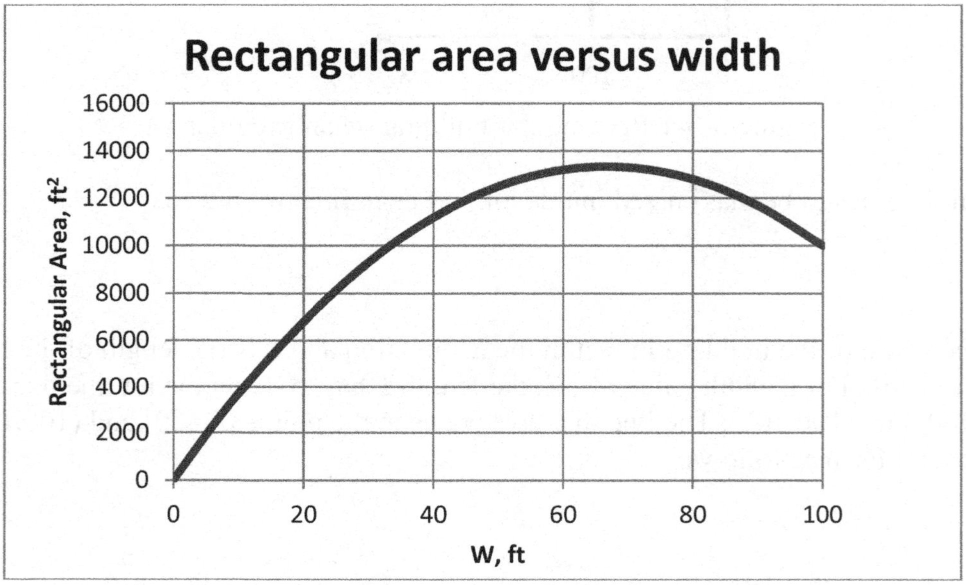

Figure 3-9-2 Rectangular building area versus its width

Figure 3-9-2 shows that the rectangular building area increases as its width increases. The area reaches its maximum at 66.67 feet and then starts to decrease. At its maximum point, the length of the building can found from Eq. (3-9-3) and it is 200 feet. Therefore the maximum rectangular single floor building that can be achieved on this lot is

$$Area_{max} = 66.67 \times 200 = 13{,}333.33\ ft^2 \tag{3-9-8}$$

Chapter 3-10 Minimum Material Cost for a Can

Assume that a can that is a right circular cylinder as shown in Figure 3-10-1. It has a radius R and a height H . The volume of the can is

$$Volume_{can} = \pi R^2 H \tag{3-10-1}$$

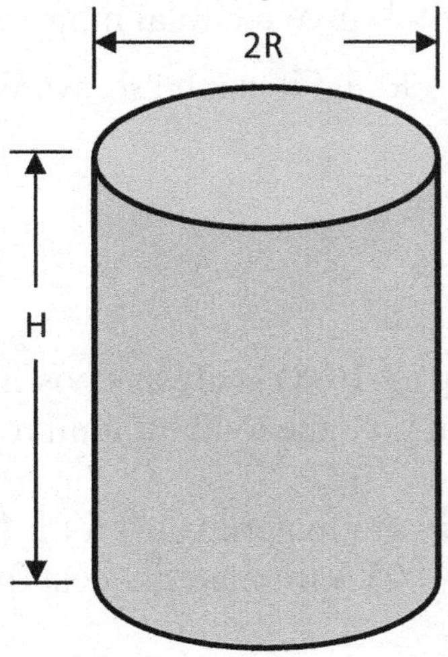

Figure 3-10-1 A right circular cylinder

The material cost for top and bottom surfaces is $A\ \$/in^2$ and for side surfaces it is $B\ \$/in^2$. The total material cost for the can be written as

$$Total\ Material\ Cost = A \times 2\pi R^2 + B \times 2\pi RH \tag{3-10-2}$$

Eq. (3-10-2) has two independent variables for can dimensions, namely R and H. One of the independent variables, H, can be eliminated by substituting Eq. (3-10-1) into Eq. (3-10-2).

$$Total\ Material\ Cost = A \times 2\pi R^2 + B \times \frac{2V}{R} \tag{3-10-3}$$

For a given volume, total minimum material cost of a can occurs at

$$\frac{d(Total\ Material\ Cost)}{dR} = 0 \tag{3-10-4}$$

or

$$A \times 4\pi R - B \times \frac{2V}{R^2} = 0 \tag{3-10-5}$$

A can with a known volume has a minimum total material cost at a radius shown in Eq. (3-10-6).

$$R_{\substack{total\ minimum \\ material\ cost}} = \left[\frac{B \times V}{2\pi \times A}\right]^{1/3} \qquad (3\text{-}10\text{-}6)$$

If Eq. (3-10-6) is substituted into Eq. (3-10-1) and solved for the can height, then a can with a known volume has a total minimum material cost at a height shown in Eq. (3-10-7).

$$H_{\substack{total\ minimum \\ material\ cost}} = \left(\frac{2A}{B}\right)^{2/3} \times \left(\frac{V}{\pi}\right)^{1/3} \qquad (3\text{-}10\text{-}7)$$

In order to test if Eqs. (3-10-6) and (3-10-7) give the total minimum material cost, second derivative of Eq. (3-10-3) can be taken with respect to R and it can be shown that $\frac{d^2(Total\ Material\ Cost)}{dR^2} > 0$ in all cases.

$$\frac{d^2(Total\ Material\ Cost)}{dR^2} = 4\pi \times A + \frac{4B \times V}{R^3} \qquad (3\text{-}10\text{-}8)$$

Since $A, B, V, and\ R$ are all positive, Eq. (3-10-8) is always positive. Therefore Eqs. (3-10-6) and (3-10-7) provide the can dimensions that give the total minimum material cost.

As an example, let us find the total minimum material cost for a 1 ft^3 can which has 0.1 $/in^2 material cost for its top and bottom surfaces and 0.05 $/in^2 material cost for its side surfaces.

Radius of the can for total minimum material cost from Eq. (3-10-6) is

$$R_{\substack{total\ minimum \\ material\ cost}} = \left[\frac{0.05 \times 1 \times 12^3}{2\pi \times 0.1}\right]^{1/3} = 5.16\ in \qquad (3\text{-}10\text{-}9)$$

12^3 is introduced into Eq. (3-10-9) to convert ft^3 of volume into in^3 and to be able to obtain the can radius in inches. Height of the can for total minimum material cost from Eq. (3-10-1) is

$$H_{\substack{total\ minimum \\ material\ cost}} = \frac{V}{\pi R^2} = \frac{1 \times 12^3}{\pi \times 5.16^2} = 20.65\ in \qquad (3\text{-}10\text{-}10)$$

Height of the can for total minimum material cost can also be obtained from Eq. (3-10-7).

Total minimum material cost for the can is obtained from Eq. (3-10-3).

$$Total\ Material\ Cost = 0.1 \times 2\pi \times 5.16^2 + 0.05 \times \frac{2 \times 1 \times 12^3}{5.16} = \$50.22 \qquad (3\text{-}10\text{-}11)$$

$16.74 or 33% of the total minimum material cost is for the top and bottom surfaces of the can. $33.48 or 67% of the total minimum material cost is for the side surfaces of the can.

Chapter 3-11 Maximize Viewing Angle That Covers Height of a TV Screen

There are no set rules as to from where to watch a TV screen. A recommended distance from TV screen is 30 degrees to cover the screen width as shown in Figure 3-11-1. This is a rule of thumb criteria used in home theater designs. Distance from TV to achieve the 30 degrees angle is

$$\frac{(\frac{W}{2})}{L} = \tan 15° = 0.268 \qquad (3\text{-}11\text{-}1)$$

or

$$L = 1.87 \times W \qquad (3\text{-}11\text{-}2)$$

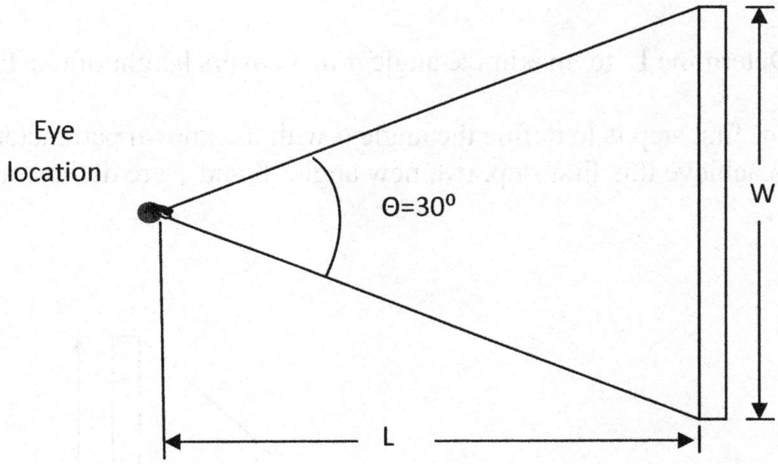

Figure 3-11-1 30° is recommended to cover TV screen width adequately

There must be an optimum distance from TV that maximizes the angle that covers height of the TV screen too. This maximum angle that covers height of the TV screen is visualized in Figure 3-11-2.

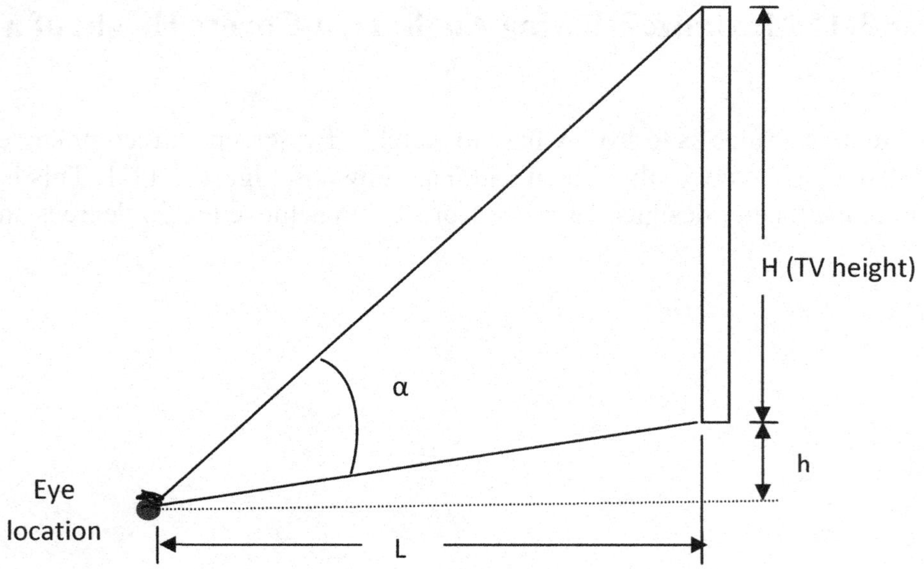

Figure 3-11-2 Determine L to maximize angle α that covers height of the TV screen H

To maximize the angle α, first step is to define the angle α with the known parameters H and h and the unknown variable L. To achieve this first step, two new angles β and φ are defined as shown in Figure 3-11-3.

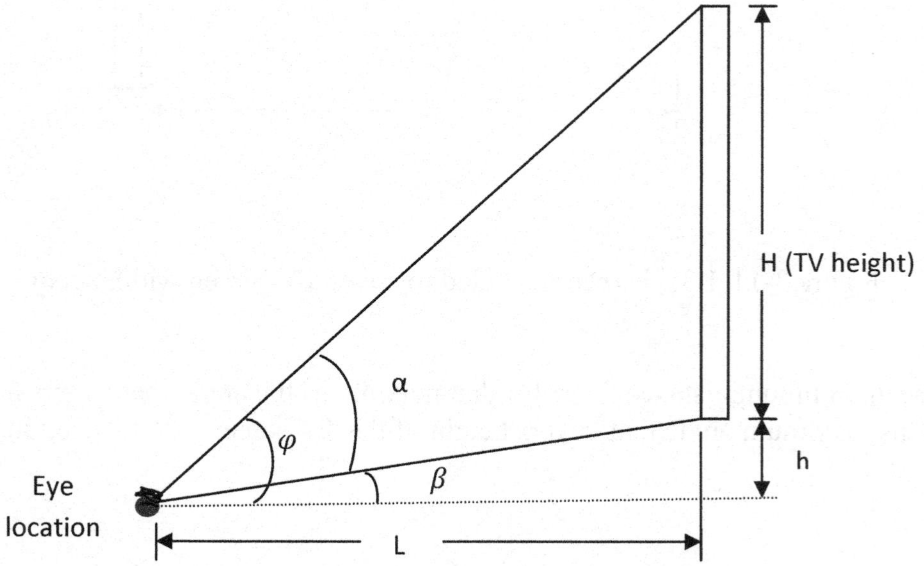

Figure 3-11-3 Definition of angles β and φ

From Figure 3-11-3,

$$\alpha = \varphi - \beta \tag{3-11-3}$$

$$\tan\beta = \frac{h}{L} \quad \text{or} \quad \beta = \tan^{-1}\frac{h}{L} \tag{3-11-4}$$

$$\tan\varphi = \frac{H+h}{L} \quad \text{or} \quad \varphi = \tan^{-1}\frac{H+h}{L} \tag{3-11-5}$$

Eq. (3-11-3) can be rewritten as follows:

$$\alpha = \tan^{-1}\frac{H+h}{L} - \tan^{-1}\frac{h}{L} \tag{3-11-6}$$

Eq. (3-11-6) gives the angle α in radians. In order to convert it to degrees, the results has to be multiplied by $\frac{180}{\pi}$. To maximize α with respect to L, $\frac{d\alpha}{dL}$ has to be determined from Eq. (3-11-6) and set equal to zero to solve for L.

From differentiation tables, Reference [9], one can find for $y = \tan^{-1} x$ that the differential function $\frac{dy}{dx}$ is $\frac{1}{1+x^2}$. Then

$$\frac{d\alpha}{dL} = \frac{1}{1+\frac{(H+h)^2}{L^2}}\left[\frac{-(H+h)}{L^2}\right] - \frac{1}{1+\frac{h^2}{L^2}}\left(\frac{-h}{L^2}\right) = 0 \tag{3-11-7}$$

Solving for L in Eq. (3-11-7) and using only the positive solution, the length L that maximizes α is

$$L = \sqrt{h(H+h)} \tag{3-11-8}$$

Eq. (3-11-8) is plotted for different TV heights, H. The optimum distance to TV, L, is plotted as a function of height, h, from eye level to bottom of TV.

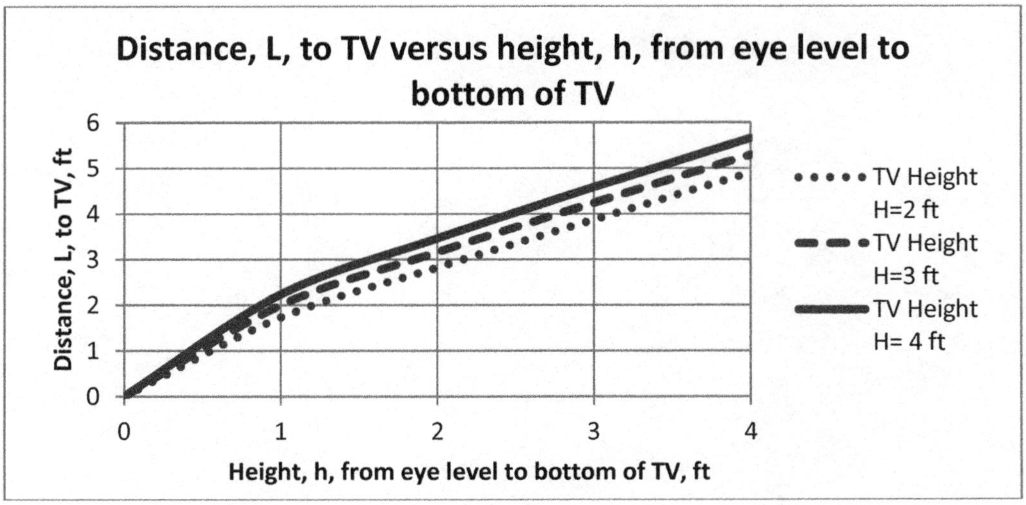

Figure 3-11-1 Distance, L, to TV versus height, h, from eye level to bottom of TV that maximizes the viewing angle α

For $H = 4\,ft$ and for $h = 1\,ft$, the TV viewing angle α is plotted as a function of distance L to TV using Eq. (3-11-6) in Figure 3-11-2. TV viewing angle starts at zero and increases to a maximum at $L = 2.24\,feet$ and then starts to decrease. The slope of the curve α versus L starts positive and decreases to zero at $L = 2.24\,feet$ and becomes negative. The slope $\frac{d\alpha}{dL}$ is always decreasing and therefore $\frac{d^2\alpha}{dL^2}$ is always negative which verifies that the result that is obtained at $L = 2.24\,feet$ is the maximum TV viewing angle.

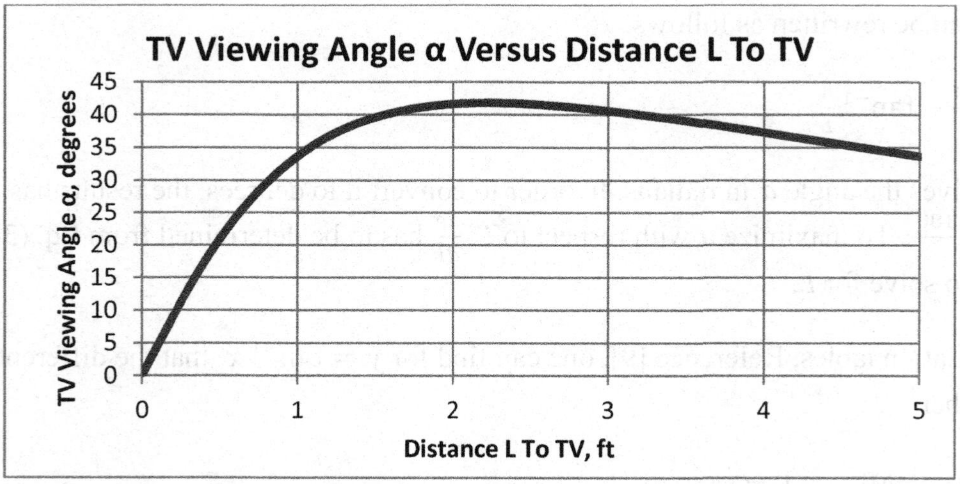

Figure 3-11-2 TV viewing angle α versus distance L to TV for H=4 feet and h=1 foot

Depending upon the home theater design, either Eq. (3-11-2) or Eq. (3-11-8) can be used to determine the optimum distance L to TV.

Chapter 3-12 Population Growth or Decline

Natural population increases and decreases in a community can be formulated by a first order differential equation barring effects from sudden occurrences such as natural disasters, epidemics, migrations, wars, etc.

Under natural circumstances, population changes at a rate during a small increment of time that is proportional to the population present at the beginning of a small increment of time. Increment of population change, ΔP, is equal to the product of change rate, r, population, P, present at the beginning of a small increment of time, and small increment of time, Δt.

$$\Delta P = rP\Delta t \qquad (3\text{-}12\text{-}1)$$

As $\Delta t \to 0$, Eq. (3-12-1) can be expressed by the following first order and linear differential equation.

$$\frac{dP}{dt} = rP \qquad (3\text{-}12\text{-}2)$$

Change rate, r, has the dimensions change per unit time, for example 1% increase per year. For 1% population increase per year, r has a value of +0.01. For 0.5% population decrease per year, r has a value of -0.005.

In order to be able to solve Eq. (3-12-2), an initial condition at $t = 0$ is required so that the integration constant can be determined. If the population at $t = 0$ is P_0, then the initial condition is

$$P = P_0 \ at\ t = 0 \qquad (3\text{-}12\text{-}3)$$

Before integration, variables in Eq. (3-12-2) are separated and rewritten as follows:

$$\frac{dP}{P} = rdt \qquad (3\text{-}12\text{-}4)$$

Integrating both sides of Eq. (3-12-4) gives

$$\int \frac{dP}{P} = \int rdt \qquad (3\text{-}12\text{-}5)$$

$$\ln(P) = rt + C \qquad (3\text{-}12\text{-}6)$$

where C is the integration constant. Then both sides of Eq. (3-12-6) is raised to the power e, the exponential function.

$$P = e^{\ln(P)} = e^{rt+C} = e^{C}e^{rt} = C_1 e^{rt} \qquad (3\text{-}12\text{-}7)$$

Using the initial condition in Eq. (3-12-3), C_1 can be found to be P_0. The natural population increase or decrease equation becomes

$$P = P_0 e^{rt} \qquad (3\text{-}12\text{-}8)$$

As an example, consider the population of a country that doubles every 20 years. From this information change rate r can be determined using Eq. (3-12-8).

$$2P_o = P_o e^{20r} \qquad (3\text{-}12\text{-}9)$$

or

$$r = \frac{\ln(2)}{20} = 0.0347 \qquad (3\text{-}12\text{-}10)$$

The population of this particular country increases 3.47% per year. Population growth with respect to P_o is

$$\frac{P}{P_o} = e^{0.0347t} \qquad (3\text{-}12\text{-}11)$$

Eq. (3-12-11) is plotted against time in years in Figure 3-12-1. Population of this country doubles in 20 years, quadruples in forty years, becomes 8 fold in 60 years, becomes 16 fold in 80 years, and becomes 32 fold in 100 years.

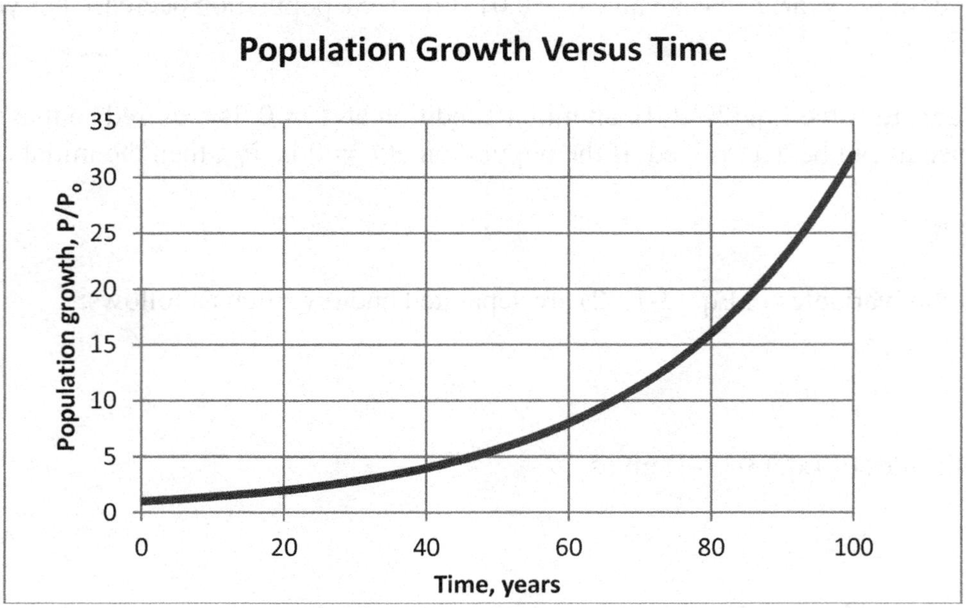

Figure 3-12-1 Population growth versus time for an increase rate of 3.46% per year

In another example, a town of 40,000 has a population decline of 1% per annum. What will the town's population after 50 years? Eq. (3-12-8) can be rewritten for this problem.

$$P = 40{,}000e^{-0.01t} \qquad (3\text{-}12\text{-}12)$$

When $t = 50$ years, $P = 24{,}261$. Population decline of this town, Eq. (3-12-12), is plotted against time in Figure 3-12-2.

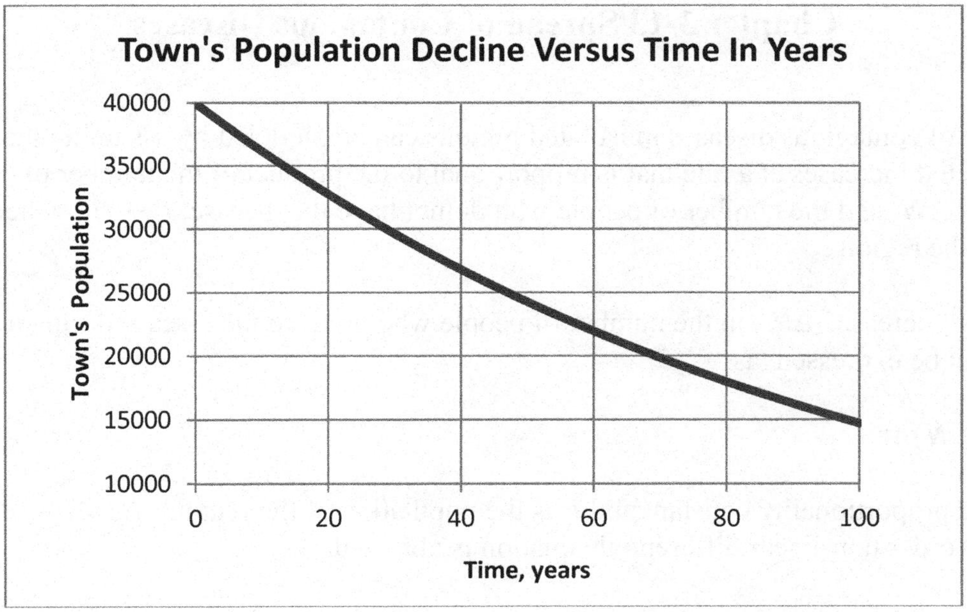

Figure 3-12-2 Town's population decline versus time in years

If the slopes of curves in Figures 3-12-1 and 3-12-2 are analyzed, one can see that slopes also behave as exponential functions. The derivative of Eq. (3-12-11) is

$$\frac{dP}{dt} = 0.0347P_o e^{0.0347t} \qquad (3\text{-}12\text{-}13)$$

The slope represented by Eq. (3-12-13) never goes through zero in positive time, always increases with time, and goes to infinity or the curve in Figure 3-12-1 becomes parallel to population axis. On the other hand, the derivative of Eq. (3-12-12) is

$$\frac{dP}{dt} = -400e^{-0.01t} \qquad (3\text{-}12\text{-}14)$$

The slope represented by Eq. (3-12-14) is always negative and approaches zero as time goes to infinity.

Chapter 3-13 Spread of Contagious Diseases

Spread of contagious diseases in isolated regions can be modeled by assuming that the number of people infected increases at a rate that is proportional to the product of the number of people who do have the disease, N, and the number of people who do not have the disease, $P - N$, where P is the population of the region .

An incremental increase, ΔN , in the number of people who do have the disease during a small time duration Δt can be expressed as

$$\Delta N = kN(P - N)\Delta t \tag{3-13-1}$$

where k is the proportionality constant and P is the population of the region. As $\Delta t \rightarrow 0$, the following first order non-linear differential equation is obtained.

$$\frac{dN}{dt} = kN(P - N) \tag{3-13-2}$$

After separating the variables, Eq. (3-13-2) can be written in integral form.

$$\frac{1}{P}\int \left(\frac{1}{N} + \frac{1}{P-N}\right) dN = \int k dt \tag{3-13-3}$$

Integrating Eq. (3-13-3) and introducing the integration constant C give the following relationship for spreading of a contagious disease.

$$N = \frac{P}{1+e^{-P(kt+C)}} \tag{3-13-4}$$

The integration constant C in Eq. (3-13-4) is obtained from the initial condition of the contagious disease at $t = 0$.

The proportionality constant k is obtained from the disease spread data after $t = 0$.

As an example, consider an isolated region which has a population of 40,000 people. Initially there are 5 people who are infected by the contagious disease. Disease control center finds out that one more person gets infected during the first week. How fast this contagious disease will spread in this community?

First the integration constant C can be determined from Eq. (3-13-4) at $t = 0$.

$$C = \frac{ln\left(\frac{40000-5}{5}\right)}{-40000} = -2.25 \times 10^{-4} \tag{3-13-5}$$

Then the proportionality constant k can be determined from the condition that one more person gets infected during the first week.

$$k = \frac{ln\left(\frac{40000-6}{6}\right)}{-40000} - C = 4.56 \times 10^{-6} \tag{3-13-6}$$

Once C and k are known, spread of contagious disease for this community can be plotted as a function of time. Plots for one, two and three infected persons during the first week are shown in Figure 3-13-1 where time is in weeks.

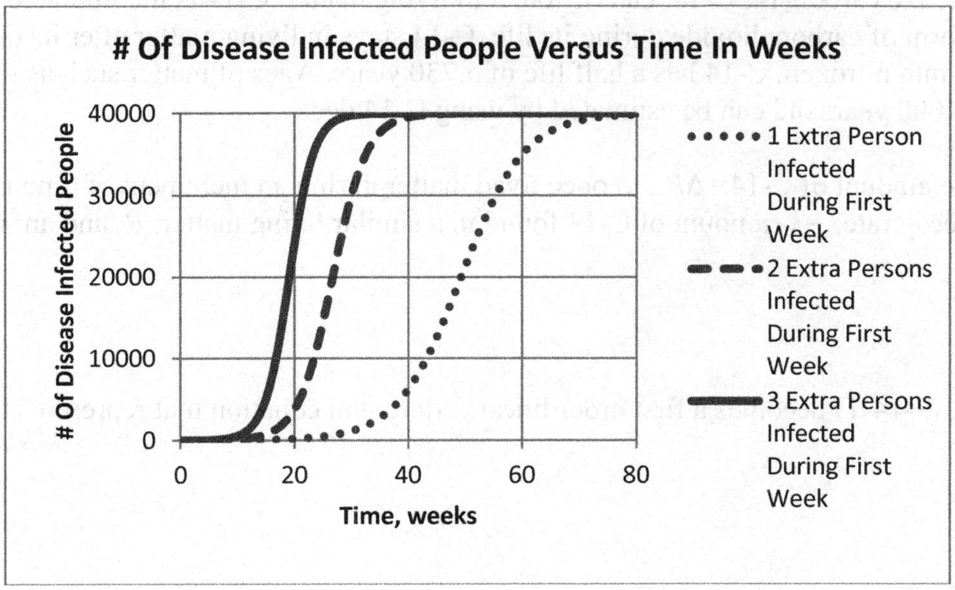

Figure 3-13-1 Number of disease infected people versus time in weeks for three different spread rates

With low spread rate, i.e. one person got infected during first week, the whole community gets infected in two years. However, if two persons get infected during the first week, spread of contagious disease accelerates, k gets larger, 8.41×10^{-6} , and the whole community gets infected in a little over a year. If three persons get infected during the first week, spread of contagious disease accelerates even more as expected, $k = 11.75 \times 10^{-6}$, and the whole community gets infected in eight months.

Chapter 3-14 Radioactive Carbon Dating

Radioactive carbon-14, C-14, can be found in living matter. C-14 is incorporated into living matter in the form of carbon dioxide during its life. C-14 stays in living matter after its death, but C-14 decays slowly into nitrogen. C-14 has a half life of 5,730 years. Ages of matter such as bones and wood up to 60,000 years old can be estimated by using C-14 decay.

Decrease in the amount of C-14, ΔR , in once lived matter during an increment of time is equal to the product of a decay rate, $-k$, amount of C-14 found in a similar living matter, R, and an increment of time, Δt.

$$\Delta R = -kR\Delta t \qquad (3\text{-}14\text{-}1)$$

As $t \rightarrow 0$, Eq. (3-14-1) becomes a first order linear differential equation that represents the C-14 decay.

$$\frac{dR}{dt} = -kR \qquad (3\text{-}14\text{-}2)$$

Eq. (3-14-2) can be integrated after separating the variables R and t. Integration constant can be obtained from the initial condition

$$R = R_0 \; at \; t = 0 \qquad (3\text{-}14\text{-}3)$$

Integrating Eq. (3-14-2) and applying the initial condition Eq. (3-14-3) gives

$$R = R_0 e^{-kt} \qquad (3\text{-}14\text{-}4)$$

The scientists have determined that the half like of C-14 is 5,730 years. From this information the decay rate, $-k$, can be found.

$$\frac{1}{2} R_0 = R_0 e^{-5730k} \qquad (3\text{-}14\text{-}5)$$

or after canceling R_0 and taking natural logarithm of both sides in Eq. (3-14-5) give

$$k = \frac{\ln(0.5)}{-5730} = 0.000121 \qquad (3\text{-}14\text{-}6)$$

For radioactive carbon dating, Eq. (3-14-4) becomes

$$\frac{R}{R_0} = e^{-0.000121t} \qquad (3\text{-}14\text{-}7)$$

where time, t , is in years. Radioactive carbon decay versus years, Eq. (3-14-7), is plotted in Figure 3-14-1.

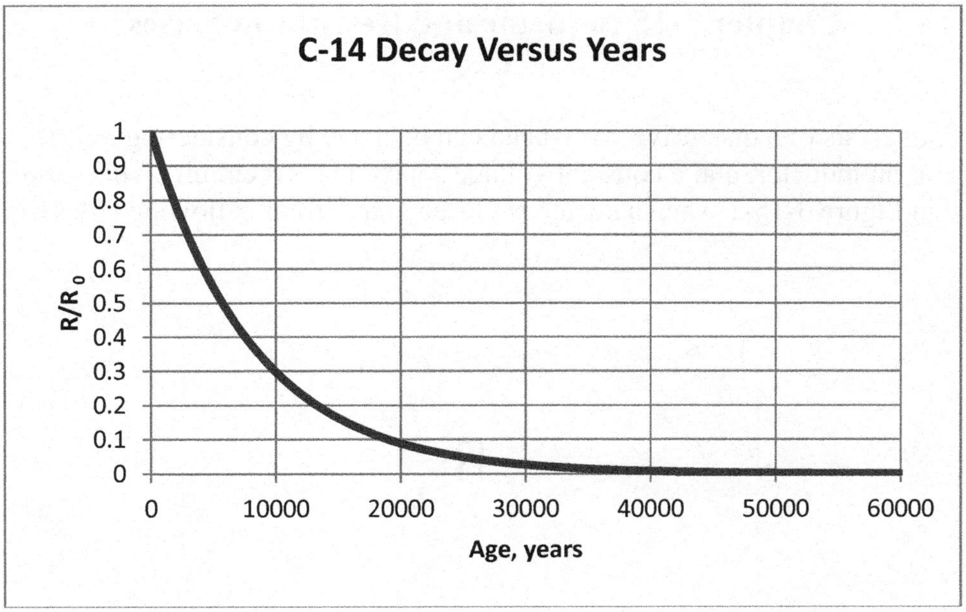

Figure 3-14-1 Radioactive carbon decay versus years

Figure 3-14-1 shows that half of C-14 decays into nitrogen in 5,730 years and 90% of C-14 decays into nitrogen in 19,035 years. 1% of C-14 is left in the fossils after 38,069 years and 0.1% of C-14 is left after 57,104 years. This is the reason the scientists limit the radioactive carbon dating method of once lived matter to 60,000 years.

Chapter 3-15 Inductor and Resistor in Series

A simple analysis of a disk drive write head can be made by considering a series circuit utilizing a resistor, an inductor, and a constant voltage source (an RL circuit). This simple RL series circuit is shown in Figure 3-15-1 when a switch is closed and current is flowing clockwise.

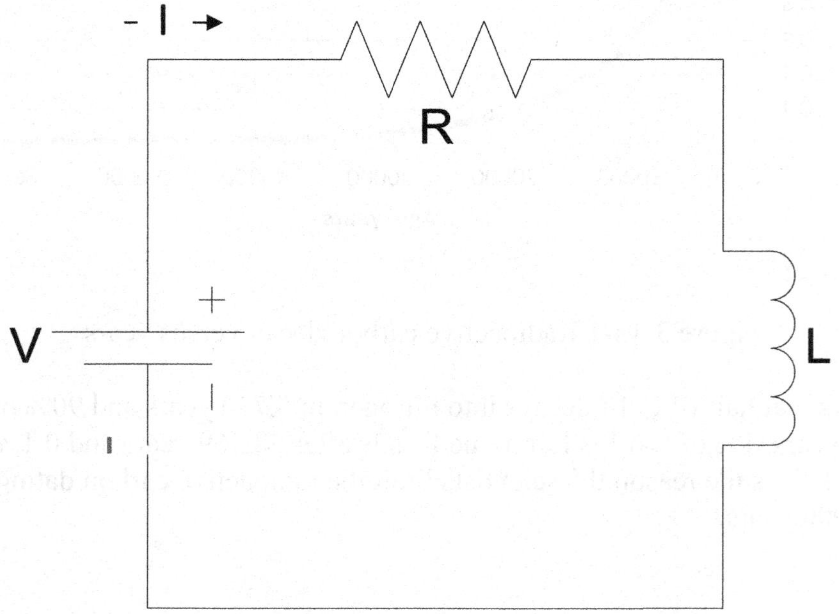

Figure 3-15-1 Resistance and inductance with a constant voltage, *RL* series circuit

When the RL series circuit in Figure 3-15-1 is closed and current I starts to flow and to increase, there is a voltage drop across the resistor R which amounts to $-IV$. As the current goes through the inductor L , magnetic field generated in the inductor opposes the changes in the current. The voltage drop through the inductor is the product of coil inductance and time rate of change in current through the inductor, namely $-L\frac{dI}{dt}$. However, the sum of all the voltages around this circuit must be zero. These conditions generate the following first order differential equation for the *RL* series circuit in Figure 3-15-1.

$$V - IR - L\frac{dI}{dt} = 0 \tag{3-15-1}$$

Eq. (3-15-1) has to satisfy an initial condition of

$$I = 0 \ at \ t = 0 \tag{3-15-2}$$

Separating the variables in Eq. (3-15-1) for ease of integration gives

$$\int \frac{dI}{\left(\frac{V}{R} - I\right)} = \frac{R}{L}\int dt \tag{3-15-3}$$

After integration, Eq. (3-15-3) becomes

$$-\ln\left(\frac{V}{R}-I\right)=\frac{Rt}{L}+C_1 \tag{3-15-4}$$

where C_1 is the integration constant and taking exponential powers of both sides in Eq. (3-15-4) gives

$$I=\frac{V}{R}-e^{\left(-\frac{Rt}{L}-C_1\right)} \tag{3-15-5}$$

Eq. (3-15-5) can be rewritten as

$$I=\frac{V}{R}-C_2 e^{\left(-\frac{Rt}{L}\right)} \tag{3-15-6}$$

Where constant $C_2=e^{-C_1}$. From the initial condition, Eq. (3-15-2), $C_2=\frac{V}{R}$. The final form of current as a function of time through the RL series circuit becomes

$$I=\frac{V}{R}\left[1-e^{\left(-\frac{Rt}{L}\right)}\right] \tag{3-15-7}$$

In Eq. (3-15-7), at $t=0$ Eq. (3-15-2) is satisfied. As time t gets large, the current reaches a steady-state value of $\frac{V}{R}$, time rate of change of current decreases, and the magnetic field generated by the inductor affects the RL series circuit less and less. An important parameter that is used in magnetic disk drive write head designs is $\frac{L}{R}$ which is called the inductive time constant. The designer can put more bits (magnetic flux reversals) onto a disk as the inductive time constant gets smaller. In Eq. (3-15-7), one can observe that the current reaches 63% of its steady-state value of $\frac{V}{R}$ when $t=\frac{L}{R}$ or

$$I=\frac{V}{R}(1-e^{-1})=0.63\frac{V}{R} \tag{3-15-8}$$

As an example, consider a 3.5" diameter disk drive. A write head (an inductor) that is flying very close to the disk surface (several micro inches of flying height) is applied an alternating square write current through its write channel circuit. The write current increases with time and generates a magnetic field in the inductive write head. As the write current rises, the magnetic field in the inductive write head increases in strength and the write head generates a stray magnetic field through its gap that reaches the magnetic disk. This stray magnetic field switches the magnetization direction in the magnetic layer on the disk depending upon the zeros and ones that one wants to record on the disk. There are many circumferential recording tracks on a disk surface, namely between 100,000 to 200,000 tracks per inch. Number of bits that can be squeezed on each track depends on the linear bit density of the head-disk design combination. Inductance of the write head plays a crucial role in determining the final linear bit density.

Consider a 7,600 RPM disk drive that has a linear bit density of 750,000 bits per inch (BPI) at its outer radius track which is at a 1.7 inches radius. The linear velocity S at this outer track is

$$S=2\pi\times 1.7\times\frac{7{,}600}{60}=1{,}353\ in/s \tag{3-15-9}$$

Linear bit density per second becomes

$$Linear\ bit\ density\ per\ second = BPI \times S = 1.03 \times 10^9 \qquad (3\text{-}15\text{-}10)$$

In order to put less magnetic flux reversals onto a track, the disk drive designers use codes that convert the incoming user bits to magnetic flux reversals, see Reference [11]. In this case assume 1,7 coding that reduces magnetic flux reversal by 25% from incoming user bits. Minimum time required between magnetic flux changes is the inverse of 75% of linear bit density per second, namely

$$t_{minimum} = \frac{1}{1.03 \times 10^9 \times \frac{3}{4}} = 1.31\ ns \qquad (3\text{-}15\text{-}11)$$

Also assume that the disk drive has a 5-volt supply and the write head needs a current rise of 40 mA zero-to-peak in order to be able to generate high enough stray magnetic field to reverse the magnetization on the disk track.

If there are two different types of write head designs, as shown in Table 3-15-1 , available for this particular disk drive, which one should be chosen?

Write Head	Inductance, μH	Resistance, Ohm	Inductive Time Constant, L/R, ns
A	0.1	100	1.0
B	0.08	100	0.8

Table 3-15-1 Electrical properties of two write heads

Neglecting other factors in the disk drive circuit design such as head capacitance, damping resistance, etc., a simple *RL* series circuit introduced in Figure 3-15-1 can be used to determine the write current rise time. Using Eq. (3-15-7), write current rise for write heads A and B are shown in Figure 3-15-2.

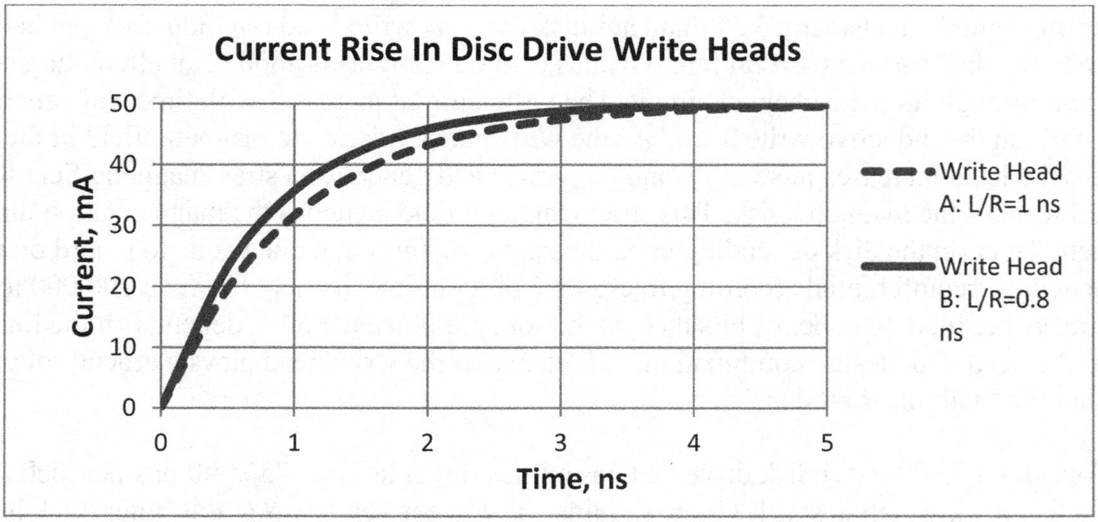

Figure 3-15-2 Current rise in two different disk drive write heads

For write head A, the write current rises to required 40 mA in 1.61 ns which does not meet the minimum time requirement of 1.31 ns between magnetic flux changes. However, the write head B's current rises to required 40 mA in 1.29 ns and meets the minimum time requirement of 1.31 ns between magnetic flux changes. A 20% reduction in inductance from write head A to write head B can increase the linear bit density by as much as 25%.

Chapter 3-16 Loss of Memory by Time

Everyone advises that it is best to learn a foreign language when you are young. As you get older, your memory retention declines faster. Behavior of memory retention behaves like an exponential decay curve which was first pointed out by Hermann Ebbinghaus towards the end of the 19th century.

Increment of memory retention change, ΔM, is equal to the product of memory retention rate, r, memory, M, present at the beginning of a small increment of time, and small increment of time, Δt.

$$\Delta M = -rM\Delta t \tag{3-16-1}$$

The negative sign in Eq. (3-16-1) comes from the fact that we always forget things we learn by time or our memory retention declines. It is more meaningful to relate memory retention strength to a time constant that a certain amount of information is retained. Therefore, memory retention rate r is represented by a time constant τ where $r = \frac{1}{\tau}$. Small τ means forgetting fast and large τ means what is learned is retained in memory a long time. Separating the variables and as $\Delta t \rightarrow 0$, Eq. (3-16-1) becomes

$$\frac{dM}{M} = -\frac{dt}{\tau} \tag{3-16-2}$$

Integrating Eq. (3-16-2) and taking the exponential power of the result provides

$$M = C \times e^{-\frac{t}{\tau}} \tag{3-16-3}$$

where C is the integration constant which can be determined from the initial condition

$$\frac{M}{M_o} = 100 \ at \ t = 0 \tag{3-16-4}$$

or 100% of input to our brain, M_o, is retained at time equals to zero. Eq. (3-16-3) becomes

$$\frac{M}{M_0} = 100 \times e^{-\frac{t}{\tau}} \tag{3-16-5}$$

In Eq. (3-16-5), when the time constant $\tau = 1\ day$, 37% of inputs to memory is retained after one day. If you like the subject you are learning and the subject is easy for you to learn, and you have high memory retention strength, then the time constant $\tau = 1\ month$, 37% of inputs to memory is retained after one month or 67% of inputs to memory is forgotten. Figure 3-16-1 shows percent of memory retained versus time for different memory retention strengths (time constants). As the memory retention strength gets weaker, memory retention declines faster.

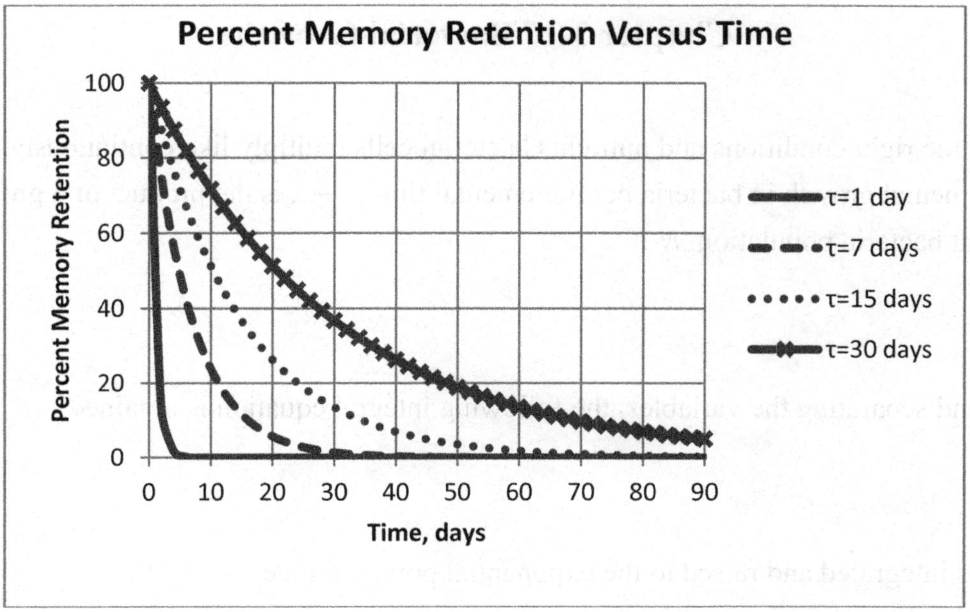

Figure 3-16-1 Percent memory retention versus time for different time constants

A way of improving the memory retention strength is reviewing the material over and over again. After each review of the material, memory retention strength for that material gets higher or the time constant in Eq. (3-16-5) gets longer. Such a case is exhibited in Figure 3-16-2. Learned material at time zero is reviewed after tenth, twentieth, and thirtieth days. After each review, the memory retention strength gets higher and higher for that material. After initial learning, the memory retention strength is 30 days or 63% of the material is forgotten in 30 days. After the third review, the memory retention strength increases to 120 days or 63% of the material is forgotten in 120 days after the third review.

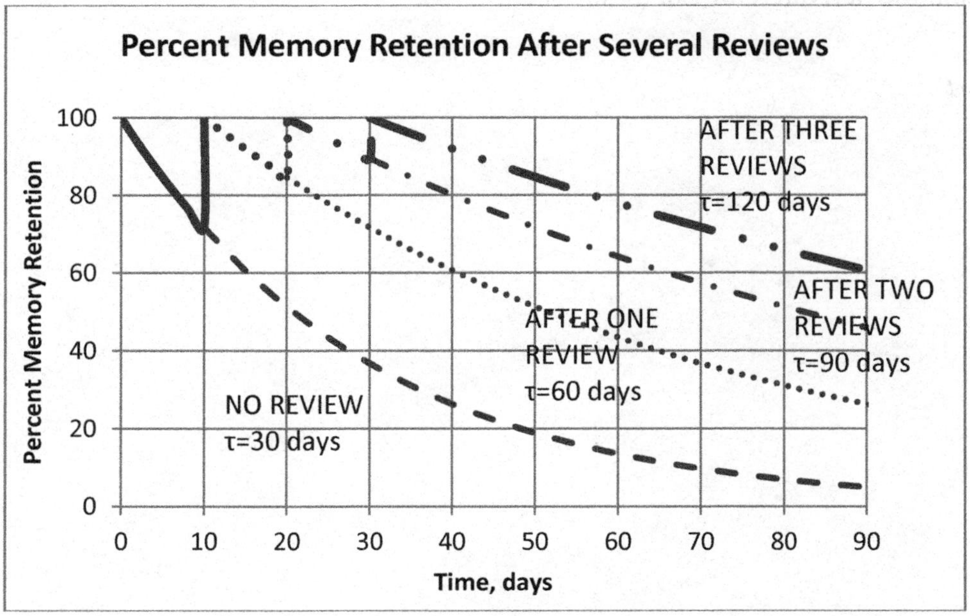

Figure 3-16-2 Percent memory retention after several reviews

Chapter 3-17 Bacterial Growth

Under the right conditions and nutrients bacterial cells multiply like continuously compounded interest. Incremental growth in bacteria per incremental time, $\frac{\Delta N}{\Delta t}$, is the product of a growth rate, r , and the present bacteria population, N .

$$\frac{\Delta N}{\Delta t} = rN \qquad (3\text{-}17\text{-}1)$$

As $\Delta t \to 0$ and separating the variables, the following integral equation is obtained.

$$\int \frac{dN}{N} = \int r dt \qquad (3\text{-}17\text{-}2)$$

Eq. (3-17-2) is integrated and raised to the exponential power to give

$$N = C \times e^{rt} \qquad (3\text{-}17\text{-}3)$$

where C is the integration constant and it is determined from the initial condition that $N = N_o$ at $t = 0$. N_o is the initial bacteria population. Eq. (3-17-3) takes its final form as follows:

$$\frac{N}{N_o} = e^{rt} \qquad (3\text{-}17\text{-}4)$$

The bacterial growth rate r is specified as a percentage growth per unit time, i.e. 1% per hour, 10% per week, etc. Eq. (3-17-4) is plotted for different growth rates in Figure 3-17-1. As the growth rate increases, bacteria population multiplies faster and the growth slope becomes larger with time. This can be seen from the derivative of Eq. (3-17-4).

$$\frac{d\left(\frac{N}{N_o}\right)}{dt} = r \times e^{rt} \qquad (3\text{-}17\text{-}5)$$

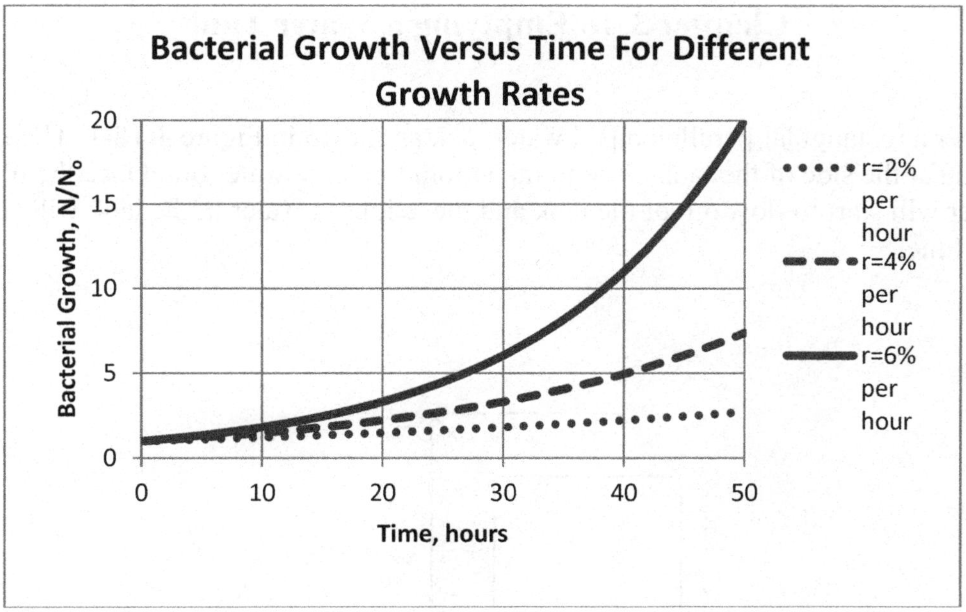

Figure 3-17-1 Bacterial growth rate versus time for different growth rates

In bacterial growth, bacteria population doubling time is an important parameter. Bacteria population doubling time is obtained from Eq. (3-17-4).

$$\frac{2N_o}{N_o} = e^{rt} \qquad (3\text{-}17\text{-}6)$$

Taking natural logarithm of both side in Eq. (3-17-6) and rearranging gives

$$t_{doubling\ time} = \frac{\ln(2)}{r} = \frac{0.693}{r} \qquad (3\text{-}17\text{-}7)$$

Using Eq. (3-17-7), bacteria population doubles in 34.7 hours for $r = 2\%\ per\ hour$ growth rate and doubles in 11.6 hours for $r = 6\%\ per\ hour$ growth rate.

Chapter 3-18 Emptying a Water Tank

Consider a rectangular parallelepiped water tank as shown in Figure 3-18-1. There is a small hole with a plug at the side of the tank close to its bottom to let the water out. Once the plug is removed, water will start to flow out of the tank and the height of water in the tank will start to decrease with time.

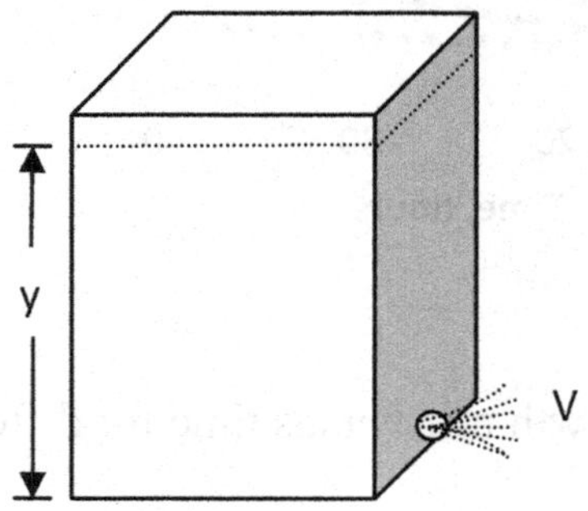

Figure 3-18-1 A rectangular parallelepiped water tank

Assume that the tank has a base area of S and the initial water height is H. The water outlet hole is a circular hole with a diameter d. Assuming a quasi-steady flow with no friction, the Bernoulli's equation, see Reference [5], can be applied between the top surface of the tank and the water jet that is coming out of the hole. The mechanical energies at these two surfaces are equated as follows:

$$y + \frac{p_{atm}}{\rho} = \frac{V^2}{2g} + \frac{p_{atm}}{\rho} \qquad (3\text{-}18\text{-}1)$$

where the pressures are atmospheric at the top surface of the tank and at the water jet surface, ρ is the density of water, y is the water height in the tank at time t, V is the average velocity of the water jet at the outlet hole, and g is the gravitational acceleration 32.2 $\frac{ft}{s^2}$ in English units. Average velocity of the water jet at the outlet hole is

$$V = \sqrt{2gy} \qquad (3\text{-}18\text{-}2)$$

Eq. (3-18-2) is known as the Torricelli's law which relates the speed of water flowing out of a hole to the height of the water above the hole. Eq. (3-18-2) is an upper bound approximation for the speed of water flowing out of a hole, since frictional effects and exit hole configuration effects are neglected. Area of the outlet hole is $A = \pi \frac{d^2}{4}$. Decrease of water volume in the tank with respect to time should be equal to the water volume that is being released from the hole per unit time. The following first order non-linear differential equation is obtained.

$$\frac{d(Sy)}{dt} = -AV = -A\sqrt{2gy} \qquad (3\text{-}18\text{-}3)$$

Separating the variables in Eq. (3-18-3) provides the following integral relationship.

$$\int \frac{dy}{y^{1/2}} = -W \int dt \qquad (3\text{-}18\text{-}4)$$

where $W = \frac{A\sqrt{2g}}{S}$ is a parametric constant with dimensions $\frac{ft^{1/2}}{s}$. Integrating Eq. (3-18-4) gives

$$2y^{1/2} = -Wt + C \qquad (3\text{-}18\text{-}5)$$

The integration constant C is obtained from the initial condition $y = H \; at \; t = 0$ or $C = 2H^{1/2}$. Eq. (3-18-5) can be rewritten as

$$y^{1/2} = H^{1/2} - 0.5Wt \qquad (3\text{-}18\text{-}6)$$

Squaring both sides of Eq. (3-18-6) gives the height of water in the tank as a function of time which is a parabolic relationship.

$$y = 0.25W^2t^2 - WH^{1/2}t + H \qquad (3\text{-}18\text{-}7)$$

Velocity of the water jet as a function of time can be obtained from Eq. (3-18-6).

$$V = \sqrt{2gy} = \sqrt{2g}\left(H^{1/2} - 0.5Wt\right) \qquad (3\text{-}18\text{-}8)$$

When the jet velocity goes to zero, time to empty the tank is obtained. Therefore Eq. (3-18-7) is valid between $t = 0$ and t_{empty}.

$$t_{empty} = \frac{2H^{1/2}}{W} \qquad (3\text{-}18\text{-}9)$$

From Eq. (3-18-7), the decrease of water volume in the tank with respect to time is obtained.

$$WATER\ VOLUME = Sy = S\left(0.25W^2t^2 - WH^{\frac{1}{2}}t + H\right) \qquad (3\text{-}18\text{-}10)$$

Water height in the tank versus time is plotted for three different outlet hole diameters in Figure 3-18-2.

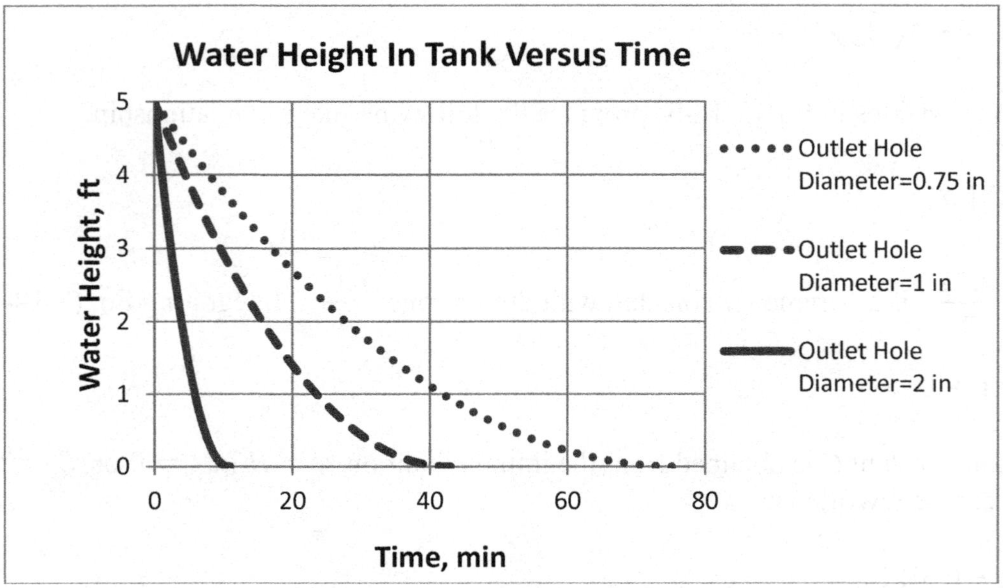

Figure 3-18-2 Water height in tank versus time for different outlet hole diameters

The water height versus time curves in Figure 3-18-2 show parabolic behavior. The curves start to go up in water height after t_{empty} values. The tank empties in 10.6 minutes for the 2 inch diameter outlet hole, in 42.6 minutes for the 1 inch diameter outlet hole, and in 75.7 minutes for the 0.75 inch diameter outlet hole.

Chapter 3-19 Emptying a Water Tank with a Constant Inflow

Consider a rectangular parallelepiped water tank as shown in Figure 3-19-1 with a constant inflow of water into the tank after time equals to zero. This problem is a little more complicated version of the problem in Chapter 3-18. There is also a small hole with a plug at the side of the tank close to its bottom to let the water out. Once the plug is removed at time equals to zero, water will start to flow out of the tank and the height of water in the tank will start to decrease with time to an equilibrium level.

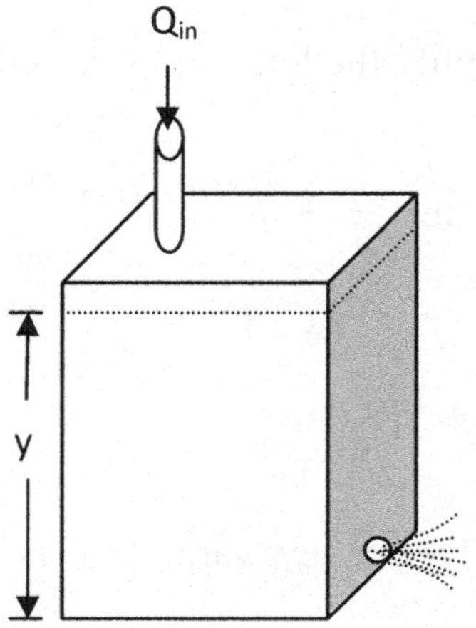

Figure 3-19-1 A rectangular parallelepiped water tank with constant inflow and outflow of water after time equals to zero

Assume that the tank has a base area of S and the initial water height is H. The water outlet hole is a circular hole with a diameter d. After time equals to zero, there is a constant inflow of water into the tank called Q_{in}. Assuming a quasi-steady flow with no friction, the Bernoulli's equation, Reference [5], can be applied between the top surface of the tank and the water jet that is coming out of the hole. The mechanical energies at these to surfaces are equated as follows:

$$y + \frac{p_{atm}}{\rho} = \frac{V^2}{2g} + \frac{p_{atm}}{\rho} \qquad (3\text{-}19\text{-}1)$$

where the pressures are atmospheric at the top surface of the tank and at the water jet surface, ρ is the density of water, y is the water height in the tank at time t, V is the average velocity of the water jet at the outlet hole, and g is the gravitational acceleration 32.2 $\frac{ft}{s^2}$ in English units. Average velocity of the water jet at the outlet hole is

$$V = \sqrt{2gy} \qquad (3\text{-}19\text{-}2)$$

Eq. (3-19-2) is known as the Torricelli's law which relates the speed of water flowing out of a hole to the height of the water above the hole. Eq. (3-19-2) is an upper bound approximation for the speed of water flowing out of a hole, since frictional effects and exit hole configuration effects are neglected.

Area of the outlet hole is $A = \pi \frac{d^2}{4}$. Decrease of water volume in the tank with respect to time should be equal to the water volume that is going in minus the water volume that is being released from the outlet hole per unit time. The following first order non-linear differential equation is obtained.

$$\frac{d(Sy)}{dt} = Q_{in} - AV = Q_{in} - A\sqrt{2gy} \qquad (3\text{-}19\text{-}3)$$

After separating the variables, Eq. (3-19-3) can be written in the following integral form.

$$\int \frac{dy}{\frac{Q_{in}}{S} - \frac{A\sqrt{2g}}{S} y^{1/2}} = \int dt \qquad (3\text{-}19\text{-}4)$$

Let us define two parametric constant to simplify the left hand side integral.

$W = \frac{A\sqrt{2g}}{S}$ *with dimension* $\frac{ft^{1/2}}{s}$ and $B = \frac{Q_{in}}{SW}$ *with dimension* $ft^{1/2}$

Then Eq. (3-19-4) can be rewritten as

$$\frac{1}{W} \int \frac{dy}{B - y^{1/2}} = \int dt \qquad (3\text{-}19\text{-}5)$$

Left hand side integral can be integrated easily, if a new variable x is introduced in the form of $y = x^2$. Then $dy = 2xdx$. Eq. (3-19-5) becomes

$$\frac{2}{W} \int \frac{xdx}{B - x} = \frac{2}{W} \int \left(\frac{B}{B - x} - 1\right) dx = \int dt \qquad (3\text{-}19\text{-}6)$$

Integrating Eq. (3-19-6) and going back to the y variable give

$$\frac{2}{W}\left[-B \times \ln\left(B - y^{1/2}\right) - y^{1/2}\right] = t + C \qquad (3\text{-}19\text{-}7)$$

The integration constant C in Eq. (3-19-7) is obtained from the initial condition $y = H \; at \; t = 0$.

$$C = \frac{2}{W}\left[-B \times \ln\left(B - H^{1/2}\right) - H^{1/2}\right] \qquad (3\text{-}19\text{-}8)$$

Eq. (3-19-7) takes the following final form.

$$t = \frac{2}{W}\left[\left(\sqrt{H} - \sqrt{y}\right) + B \times \ln\frac{(B - \sqrt{H})}{(B - \sqrt{y})}\right] \qquad (3\text{-}19\text{-}9)$$

In Eq. (3-19-9), for a given water height y , time t can be obtained in a straightforward manner. However, for a given time t , water height y can be obtained by trial and error. The trial and error formulation can be written as follows.

$$t - \frac{2}{W}\left[\left(\sqrt{H} - \sqrt{y}\right) + B \times \ln\frac{(B - \sqrt{H})}{(B - \sqrt{y})}\right] = 0 = ERROR \qquad (3\text{-}19\text{-}10)$$

Conditions for the value y and H in the argument of the natural logarithm are

$$y \geq B^2 = \frac{Q_{in}^2}{2gA^2} \text{ and } H \geq B^2 = \frac{Q_{in}^2}{2gA^2} \qquad (3\text{-}19\text{-}11)$$

As time increases, water height decreases and approaches an equilibrium value which is $B^2 = \frac{Q_{in}^2}{2gA^2}$. Let us analyze Eq. (3-19-10) for a 1-inch diameter outlet hole, 25 ft^2 tank base area, 5 feet initial water height, and for three different inflow rates. The results are presented in Figure 3-19-1. As the inflow rate decreases, the equilibrium water height also decreases. Under the given conditions, 35.9 gallons per minute inflow provides an equilibrium water height of 3.34 feet. As the inflow rate decreases to 22.4 gallons per minute, equilibrium water height goes down to 1.3 feet. As the inflow rate decreases further to 9.0 gallons per minute, equilibrium water height goes down to 0.21 feet. As the inflow rate decreases, time to reach equilibrium water height also decreases. At the limit, when there is no inflow rate, the tank empties in 42.6 minutes.

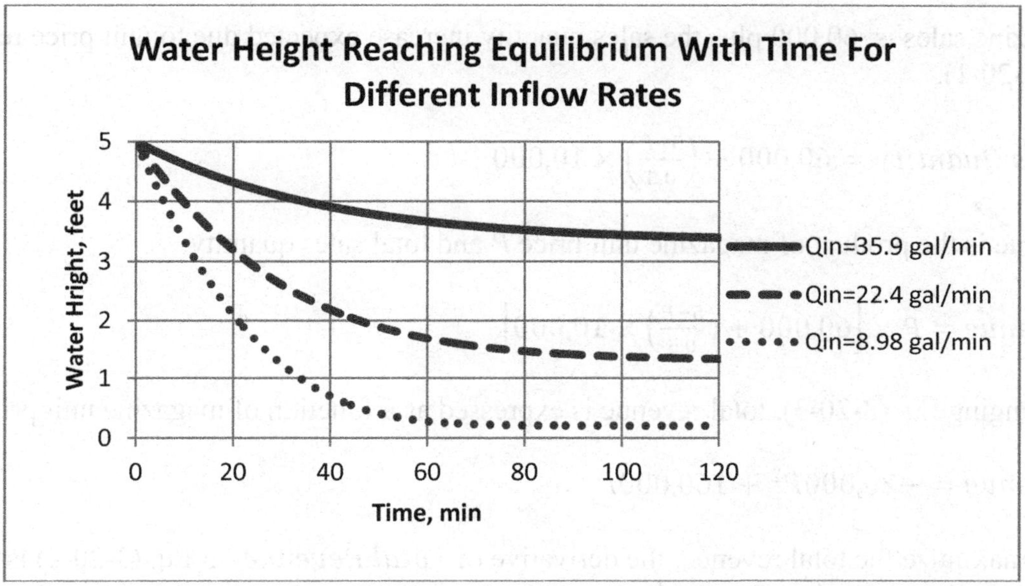

Figure 3-19-1 Water height reaching equilibrium with time for different inflow rates

Another way to solve the differential Eq. (3-19-3) is by numerical methods. Finite difference form, see Reference [10], of Eq. (3-19-3) can be written as

$$y_i = y_{i-1} + \Delta t \times \left(\frac{Q_{in}}{S} - W y_{i-1}^{1/2}\right) \qquad (3\text{-}19\text{-}12)$$

where y_i is the water height at time i and y_{i-1} is the water height at time $i-1$. Δt is the time interval between time i and time $i-1$. Numerical calculations start at $t = 0$, where $y = H = 5\ ft$. If small time intervals are used, i.e. $\Delta t = 1\ s$, then accurate results can be obtained that follows the physical process.

Chapter 3-20 Optimum Magazine Sale Price

Sales volume of a weekly magazine is price sensitive. The magazine sells 60,000 copies a week at \$5.00 unit price. A marketing study determines that \$0.50 decrease in the unit price will boost the magazine sales by 10,000 units. The marketing department wants to maximize revenues from this magazine's sales. What unit price should be assigned to this weekly magazine to maximize revenues?

First step is to find the relationship between the magazine unit price and the revenue. Let us call the magazine unit price P . Every \$0.50 decrease from \$5.00 unit price will increase magazine sales by 10,000 units which can be expressed as follows.

$$\left(\frac{\$5.00-P}{\$0.50}\right) \times 10{,}000 \qquad (3\text{-}20\text{-}1)$$

Total magazine sales is 60,000 plus the sales quantity increase expected due to unit price reduction from Eq. (3-20-1).

$$Total\ Sales\ Quantity = 60{,}000 + \left(\frac{5-P}{0.5}\right) \times 10{,}000 \qquad (3\text{-}20\text{-}2)$$

Total revenue is the product of magazine unit price P and total sales quantity.

$$Total\ Revenue = P \times \left[60{,}000 + \left(\frac{5-P}{0.5}\right) \times 10{,}000\right] \qquad (3\text{-}20\text{-}3)$$

After rearranging Eq. (3-20-3), total revenue is expressed as a function of magazine unit price.

$$Total\ Revenue = -20{,}000P^2 + 160{,}000P \qquad (3\text{-}20\text{-}4)$$

In order to maximize the total revenue, the derivative of $Total\ Revenue$ in Eq. (3-20-4) is taken with respect to magazine unit price P and this first derivative is set to zero.

$$\frac{d(Total\ Revenue)}{dP} = -40{,}000P + 160{,}000 = 0 \qquad (3\text{-}20\text{-}5)$$

The unit price of the magazine that will maximize the total revenue is determined from Eq. (3-20-5).

$$P_{optimum} = \$4.00 \qquad (3\text{-}20\text{-}6)$$

The \$4.00 magazine unit price definitely maximizes the total revenue. This can be shown by taking the second derivative of $Total\ Revenue$ with respect to P in Eq. (3-20-4) and by verifying that at \$4.00 magazine unit price $\frac{d^2(Total\ Revenue)}{dP^2} < 0$.

$$\frac{d^2(Total\ Revenue)}{dP^2} = -40{,}000 \qquad (3\text{-}20\text{-}7)$$

At \$4.00 optimum magazine unit price, total weekly sales quantity from Eq. (3-20-2) is 80,000 and maximum total revenue from Eq. (3-20-4) is \$320,000.

Another way to determine maximum total revenue is by plotting $Total\ Revenue$ versus P as shown in Figure 3-20-1.

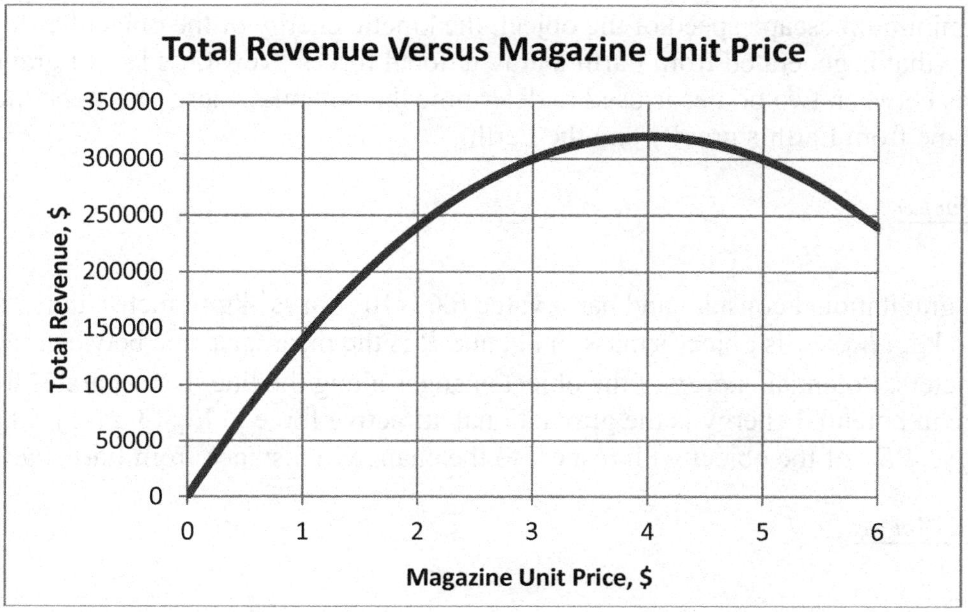

Figure 3-20-1 Total revenue versus magazine unit price

Figure 3-20-1 shows that total revenue increases as the magazine unit price increases. Then total revenue goes through a maximum at the magazine unit price of $4.00 and starts to decrease as the magazine unit price increases further.

Chapter 3-21 Earth's Escape Speed

What is the minimum speed required for an object to escape from Earth's gravity? In order to determine the minimum escape speed of the object, the kinetic energy of the object has to balance its potential energy that is generated from Earth's gravitational forces. Newton's law of gravitational attractive forces between two bodies is used to determine the potential energy between the object that is trying to escape from Earth's gravity and the Earth.

$$F = G\frac{M_{Earth}M_{Object}}{R^2} \qquad (3\text{-}21\text{-}1)$$

where G is the gravitational constant and has a value 6.67×10^{-11} m^3/s^2-kg in metric units, M_{Earth} is Earth's mass in kg, M_{Object} is object's mass in kg, and R is the mean distance between the Earth and the object in meters. Potential energy of the object changes along the line that separates the two bodies and this change in potential energy is the gravitational attractive force in Eq. (3-21-1). Change in potential energy, PE, of the object with respect to the change in distance from Earth can be written as

$$\frac{d(PE)}{dR} = G\frac{M_{Earth}M_{Object}}{R^2} \qquad (3\text{-}21\text{-}2)$$

Integrating Eq. (3-21-2) gives

$$PE = G\int\frac{M_{Earth}M_{Object}}{R^2}dR \qquad (3\text{-}21\text{-}3)$$

or

$$PE = -G\frac{M_{Earth}M_{Object}}{R} + C \qquad (3\text{-}21\text{-}4)$$

The integration constant C in Eq. (3-21-4) can be obtained from the condition that Earth's gravity does not affect the potential energy of the object as $R \to \infty$ and $PE \to 0$. Therefore, $C = 0$.

The kinetic energy of the object, KE, has to balance its potential energy, PE, that is generated from Earth's gravitational attractive forces in order to determine the object's minimum escape speed.

$$KE + PE = 0 \qquad (3\text{-}21\text{-}5)$$

The kinetic energy of the object can be written as

$$KE = \frac{1}{2}M_{Object}V^2 \qquad (3\text{-}21\text{-}6)$$

Combining Eqs. (3-21-4), (3-21-5), and (3-21-6) gives

$$\frac{1}{2}M_{Object}V^2 - G\frac{M_{Earth}M_{Object}}{R} = 0 \qquad (3\text{-}21\text{-}7)$$

The object's minimum escape speed from Earth's gravitational attractive forces is

$$V = \sqrt{\frac{2GM_{Earth}}{R_{Earth}}} \qquad (3\text{-}21\text{-}8)$$

If Earth's mass is 5.98×10^{24} kg and its average radius is 6.37×10^{6} m, then the object's minimum escape speed from Earth's gravitational attractive forces is

$$V = 11{,}191\ \frac{m}{s} = 11.2\ \frac{km}{s} = 25{,}027\ mph \qquad (3\text{-}21\text{-}9)$$

Similarly, the object's minimum escape speed from Moon's gravitational attractive forces can be calculated. If Moon's mass is 7.36×10^{22} kg and its average radius is 1.74×10^{6} m, then the object's minimum escape speed from Moon's gravitational attractive forces is

$$V = 2{,}375\ \frac{m}{s} = 2.4\ \frac{km}{s} = 5{,}312\ mph \qquad (3\text{-}21\text{-}10)$$

The object needs about one-fifth the Earth's minimum escape speed to be able to escape from the Moon.

Similarly, the object's minimum escape speed from Sun's gravitational attractive forces can be calculated. If Sun's mass is 1.99×10^{30} kg and its average radius is 6.96×10^{8} m, then the object's minimum escape speed from Sun's gravitational attractive forces is

$$V = 617{,}589\ \frac{m}{s} = 617.6\ \frac{km}{s} = 1{,}381{,}154\ mph \qquad (3\text{-}21\text{-}11)$$

The object needs about fifty-five times the Earth's minimum escape speed to be able to escape from the Sun.

Chapter 3-22 Maximum Volume of a Right Circular Cylinder in a Sphere

In this maximization problem, largest volume of a right circular cylinder is to be fitted into a sphere which has a radius R . The layout of the sphere and the imbedded right circular cylinder is shown in Figure 3-22-1.

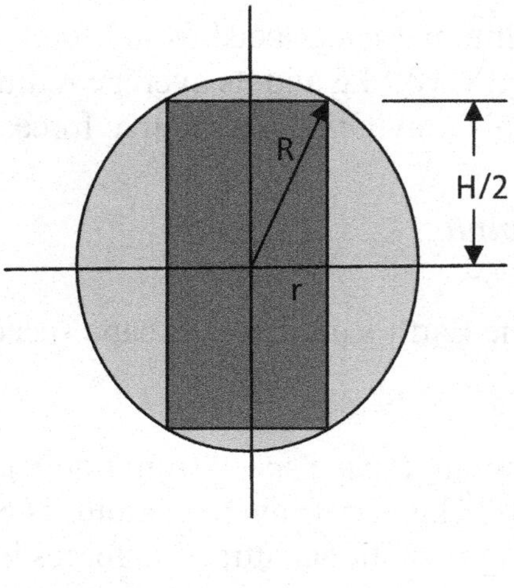

Figure 3-22-1 Largest volume of a right circular cylinder in a sphere

Volume of the right circular cylinder is defined as

$$V_{right\ circular\ cylinder} = \pi r^2 H \tag{3-22-1}$$

In Eq. (3-22-1) the volume of the right circular cylinder depends on two variables, namely its radius, r, and its height, H. The object is to reduce these variables to one by finding a relationship between the radius, r, and the height, H . The height of the right circular cylinder, H, can be written as a function of sphere's radius, R, and right circular cylinder's radius, r as follows.

$$\left(\frac{H}{2}\right)^2 + r^2 = R^2 \tag{3-22-2}$$

Then the volume of the right circular cylinder becomes

$$V_{right\ circular\ cylinder} = 2\pi r^2\sqrt{R^2 - r^2} \tag{3-22-3}$$

Since the volume of a right circular cylinder is being maximized for a fixed sphere's radius, R, taking the derivative of the right circular cylinder's volume with respect to its radius in Eq. (3-22-3) and setting its to zero, should give the value of the radius, r, that maximizes its volume.

$$\frac{dV_{right\ circular\ cylinder}}{dr} = 2\pi\left(2r\sqrt{R^2 - r^2} - \frac{r^3}{\sqrt{R^2-r^2}}\right) = 0 \tag{3-22-4}$$

Multiplying both sides in Eq. (3-22-4) by $\sqrt{R^2 - r^2}$ gives

$$2r(R^2 - r^2) - r^3 = 0 \qquad (3\text{-}22\text{-}5)$$

The radius, r , that maximizes the volume of a right circular cylinder inside a sphere with radius, R, becomes

$$r = \sqrt{\frac{2}{3}}R = 0.816R \qquad (3\text{-}22\text{-}6)$$

The result in Eq. (3-22-6) can be verified by rewriting Eq. (3-22-3) as follows and plotting the non-dimensional volume of a right circular cylinder as a function of $\frac{r}{R}$.

$$\frac{V_{right\ circular\ cylinder}}{R^3} = 2\pi\left(\frac{r}{R}\right)^2\sqrt{1-\left(\frac{r}{R}\right)^2} \qquad (3\text{-}22\text{-}7)$$

Figure 3-22-2 shows the volume of a right circular cylinder as its normalized radius varies between 0 and 1. The maximum of the volume occurs when $\frac{r}{R} = 0.816$.

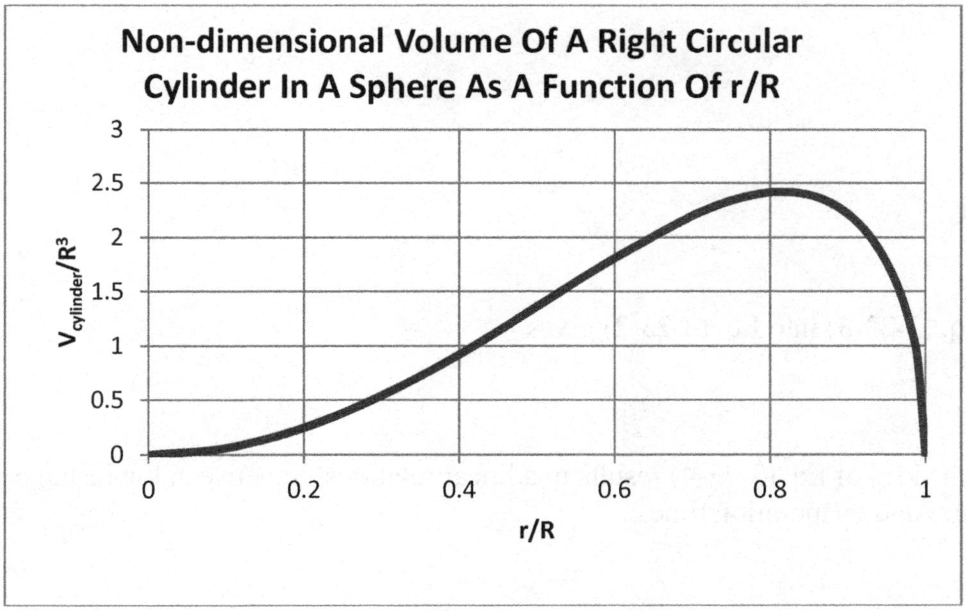

Figure 3-22-2 Non-dimensional volume of a right circular cylinder in a sphere as a function of r/R

Chapter 3-23 Evaporation of a Drop

Evaporation of a water drop depends on atmospheric temperature, pressure, and relative humidity. One can assume that evaporation causes a spherical water drop to decrease its volume at a rate proportional to its surface area. In order to find the time that it will take a water drop with an initial radius of R_i to evaporate entirely, a first order differential equation can be written to describe the physical phenomenon.

$$\frac{dV}{dt} = -C \times 4\pi R^2 \qquad (3\text{-}23\text{-}1)$$

where V is the volume of the water drop that changes with respect to time , C is the proportionality constant of this change to surface area, the minus sign comes from the fact that the volume is always decreasing, and $4\pi R^2$ is the surface area of the spherical water drop at an instant in time with R being its radius.

The volume of the water drop at an instant in time is $V = \frac{4}{3}\pi R^3$ .Eq. (3-23-1) is rewritten after separating the variables R and t .

$$\frac{dR^3}{R^2} = -3Cdt \qquad (3\text{-}23\text{-}2)$$

But

$$\frac{dR^3}{dR} = 3R^2 \qquad (3\text{-}23\text{-}3)$$

Substituting Eq. (3-23-3) into Eq. (3-23-2) gives

$$dR = -Cdt \qquad (3\text{-}23\text{-}4)$$

Integrating both sides of Eq. (3-23-4) results in a linear relationship between the instantaneous radius of the water drop and evaporation time.

$$R = -Ct + A \qquad (3\text{-}23\text{-}5)$$

where A is the integration constant. A can be determined from the initial condition of $R = R_i \; at \; t = 0$ or $A = R_i$. Eq. (3-23-5) takes the following final linear form.

$$R = R_i - Ct \qquad (3\text{-}23\text{-}6)$$

Under certain atmospheric conditions, it is determined that the proportionality constant $C = 0.1 \; mm/min$. Then a spherical water drop with an initial radius of 2 mm will evaporate entirely in 20 minutes. Drop surface area is a parabolic function of evaporation time as shown in Eq. (3-23-7).

$$Drop\ Surface\ Area = 4\pi(2 - 0.1t)^2 \qquad (3\text{-}23\text{-}7)$$

Drop volume is a cubic function of evaporation time.

$$Drop\ Volume = \frac{4}{3}\pi(2 - 0.1t)^3 \quad (3\text{-}23\text{-}8)$$

Water drop radius, surface area, and volume are shown as a function of evaporation time in Figure 3-23-1.

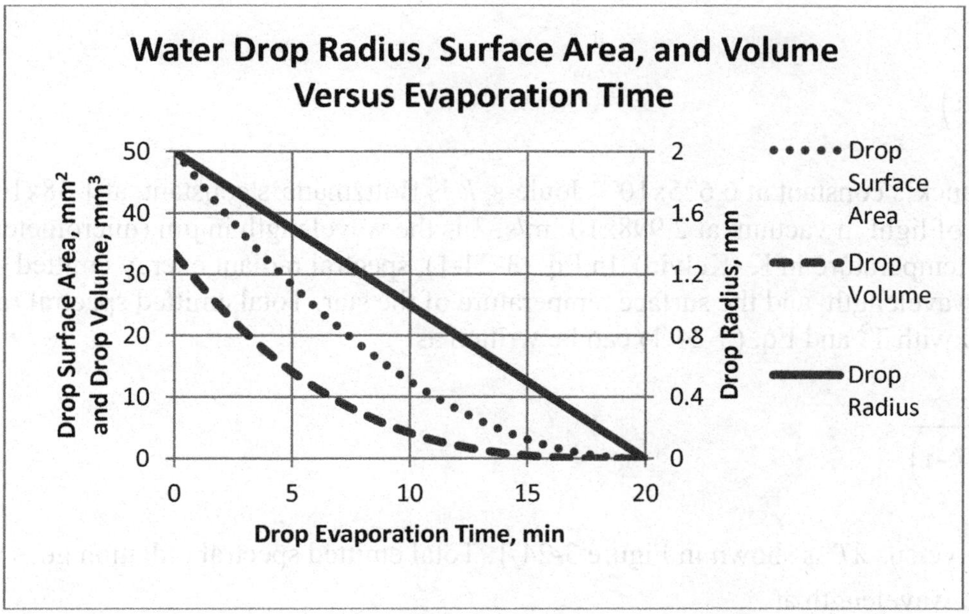

Figure 3-23-1 Water drop radius, surface area, and volume versus evaporation time

Chapter 3-24 Star Temperatures

A star emits radiation as a function of wavelength and its surface temperature similar to a blackbody. Total emitted spectral radiation from a star per unit surface area and per unit wavelength band is estimated by Planck's blackbody radiation formula, see Reference [12], as shown in below.

$$F_\lambda = \frac{2\pi hc^2}{\lambda^5\left(e^{\frac{hc}{k\lambda T}}-1\right)} \tag{3-24-1}$$

where h is Planck's constant at $6.625\text{x}10^{-34}$ Joule-s, k is Boltzmann's constant at $1.38\text{x}10^{-23}$ Joule/K, c is the speed of light in vacuum at $2.998\text{x}10^{8}$ m/s, λ is the wavelength in μm (micrometers), and T is star's surface temperature in K (Kelvin). In Eq. (3-24-1), spectral radiant energy emitted depends on the radiation wavelength and the surface temperature of the star. Total emitted spectral radiation can be normalized with T^5 and Eq. (3-24-1) can be written as

$$\frac{F_\lambda}{T^5} = \frac{2\pi hc^2}{(\lambda T)^5\left(e^{\frac{hc}{k\lambda T}}-1\right)} \tag{3-24-2}$$

The plot of $\frac{F}{T^5}$ versus λT is shown in Figure 3-24-1. Total emitted spectral radiation goes through a maximum at a wavelength at

$$\lambda_{max}T = 2898\ \mu m - K \tag{3-24-3}$$

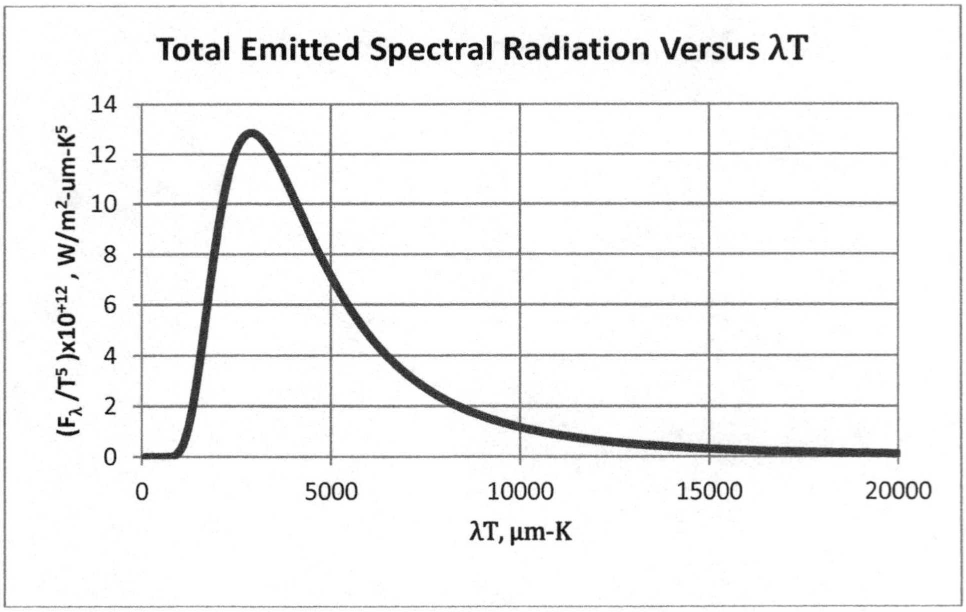

Figure 3-24-1 Total emitted spectral radiation versus λT

If the total emitted spectral radiation from a blackbody is measured at different spectral bands and the wavelength where the maximum radiation is occurring is found, then the surface temperature of that blackbody can be determined from Eq. (3-24-3). A list of blackbody radiation emission sources, their maximum radiation emission wavelengths and their surface temperatures calculated from Eq. (3-24-3) are given in Table 3-24-1.

Blackbody Name	**Maximum Radiation Emission Wavelength, μm**	**Surface Temperature, K**
Regulus In Constellation Leo	0.24	12000
Our Sun	0.50	5800
Oxyacetylene Welding Flame	0.89	3273
Melting Silver	2.35	1235
Living Room	9.66	300

Table 3-24-1 Some blackbodies with their maximum radiation emission wavelengths and surface temperatures

Total emitted radiation from a star per unit surface area can be determined by integrating Eq. (3-24-1) over all wavelengths.

$$F = \int_0^\infty F_\lambda d\lambda \qquad (3\text{-}24\text{-}4)$$

By letting $X = \frac{hc}{k\lambda T}$, Eq. (3-24-4) can be rewritten as

$$F = T^4 \frac{2\pi k^4}{h^3 c^2} \int_0^\infty \frac{X^3}{e^X - 1} dX \qquad (3\text{-}24\text{-}5)$$

The value of the integral in Eq. (3-24-5) is determined by summing all the areas of the rectangles from 0 to 15 with 0.1 thickness . The areal contributions to the integral above $X = 15$ are less than 0.0016% as shown in Figure 3-24-2. The value of the integral is

$$\int_0^\infty \frac{X^3}{e^X - 1} dX = 6.49 \qquad (3\text{-}24\text{-}6)$$

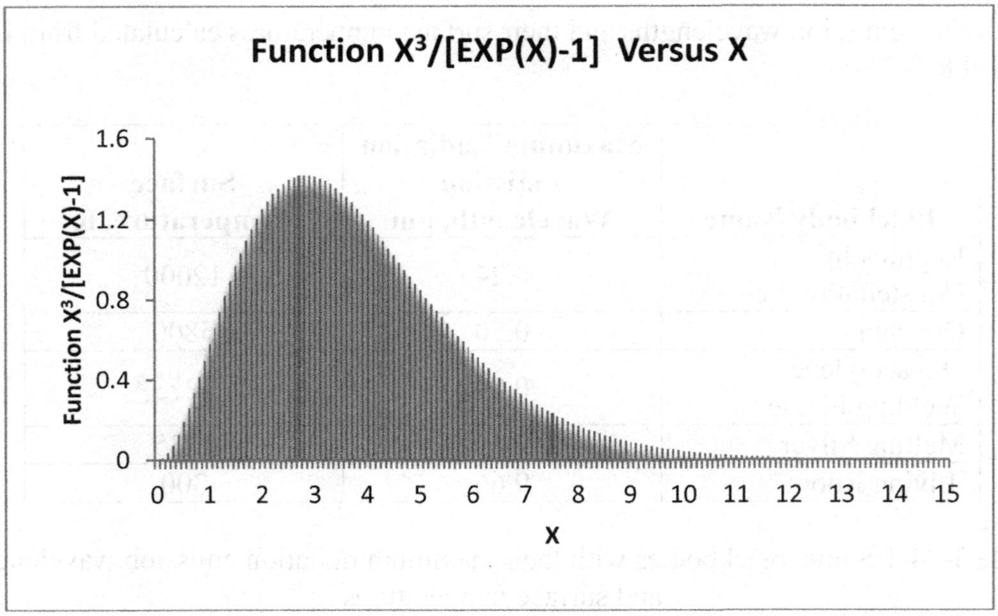

Figure 3-24-2 Sums of areas of rectangles to determine $\int_0^{\infty} \frac{X^3}{e^X - 1} dX$

Total emitted radiation from a star per unit surface area, in Watts/meter^2, from Eq. (3-24-5) is

$$F = \sigma T^4 \tag{3-24-7}$$

where $\sigma = 5.67 \times 10^{-8} \frac{W}{m^2 K^4}$. The total luminosity of a spherical star with a surface area of $4\pi R^2$ becomes $4\pi R^2 \sigma T^4$ Watts. So a small diameter star with a high surface temperature can be as luminous as a big diameter star with a low surface temperature.

Chapter 3-25 String Tension in Musical Instruments

A guitar or a violin string that is under tension can be plucked to generate a wave that travels along the string (X-direction) and pulsates in the Y-direction which is perpendicular to the X-direction. The wave equation for a long string under tension is a second order partial differential equation and its derivation can be found in Reference [13].

$$\frac{\partial^2 Y}{\partial X^2} = \frac{\rho}{T}\frac{\partial^2 Y}{\partial t^2} \qquad (3\text{-}25\text{-}1)$$

where X is the distance along the string in meters, Y is the distance perpendicular to X-direction in meters where XY forms the plane of the string pulse, t is time in seconds, ρ is linear mass density of the string in kg/m, and T is the tension on the string in Newtons (kg-m/s^2). A tensioned string with fixed ends at a guitar bridge and a support post when plucked has a sinusoidal pulse pattern as given in Eq. (3-25-2), see Reference [13].

$$Y(X,t) = A \times sin[\frac{2\pi}{\lambda}(X \pm Vt)] \qquad (3\text{-}25\text{-}2)$$

where A is the amplitude of the pulse in meters, $\frac{2\pi}{\lambda}$ is the period when the sinusoidal pulse motion repeats itself, λ is the wavelength of the sinusoidal pulse in meters per cycle, V is the longitudinal speed, also called phase speed, of the pulse in m/s. In the pulse pattern given in Eq. (3-25-2), X-coordinate is moving with the longitudinal speed of the pulse. Partial derivatives of Eq. (3-25-2) with respect to the independent variables X and t are as follows.

$$\frac{\partial^2 Y}{\partial X^2} = -A \times \left(\frac{4\pi^2}{\lambda^2}\right) sin[\frac{2\pi}{\lambda}(X \pm Vt)] \qquad (3\text{-}25\text{-}3)$$

$$\frac{\partial^2 Y}{\partial t^2} = -A \times \left(\frac{4\pi^2}{\lambda^2}\right) \times V^2 \times sin[\frac{2\pi}{\lambda}(X \pm Vt)] \qquad (3\text{-}25\text{-}4)$$

Eqs. (3-25-3) and (3-25-4) can be substituted into the original wave equation, Eq. (3-25-1) to obtain the relationship between the phase velocity and the tension on the string.

$$V = \sqrt{\frac{T}{\rho}} \qquad (3\text{-}25\text{-}5)$$

From wave mechanics, it is known that a pulse train's wavelength times its frequency is equal to its phase velocity as follows.

$$\lambda F = V \qquad (3\text{-}25\text{-}6)$$

where F is the pulse train frequency in cycles per second or Hertz. Tension on a string that is required for a given frequency is obtained from Eq. (3-25-5)

$$T = \rho\lambda^2 F^2 \qquad (3\text{-}25\text{-}7)$$

Sounds of open strings for a guitar and for a violin are tuned to certain fundamental frequencies called first harmonics as shown in Tables 3-25-1 and 3-25-2, respectively. The wavelength λ at fundamental frequencies is always twice the string length, , between the instrument's bridge and its support post.

$$\lambda = 2L \tag{3-25-8}$$

Then for tuning the strings, Eq. (3-25-7) becomes

$$T = 4L^2 \rho F^2 \tag{3-25-9}$$

Open String Note	**F, Hz**	**L, m**	**ρ, kg/m**
E_4	330	0.65	0.0004
B_3	247	0.65	0.0007
G_3	196	0.65	0.00114
D_3	147	0.65	0.00233
A_2	110	0.65	0.00447
E_2	82	0.65	0.00679

Table 3-25-1 Typical guitar string properties

Open String Note	**F, Hz**	**L, m**	**ρ, kg/m**
E_5	659	0.328	0.0004
A_4	440	0.328	0.0006
D_4	294	0.328	0.0013
G_3	196	0.328	0.0024

Table 3-25-2 Typical violin string properties

Tension required for correct tuning of open strings in a typical guitar versus the string fundamental frequency is shown in Figure 3-25-1. Similar results are shown for a typical violin in Figure 3-25-2. String tension is a balance between the linear mass density of the instrument and square of its open string fundamental frequency.

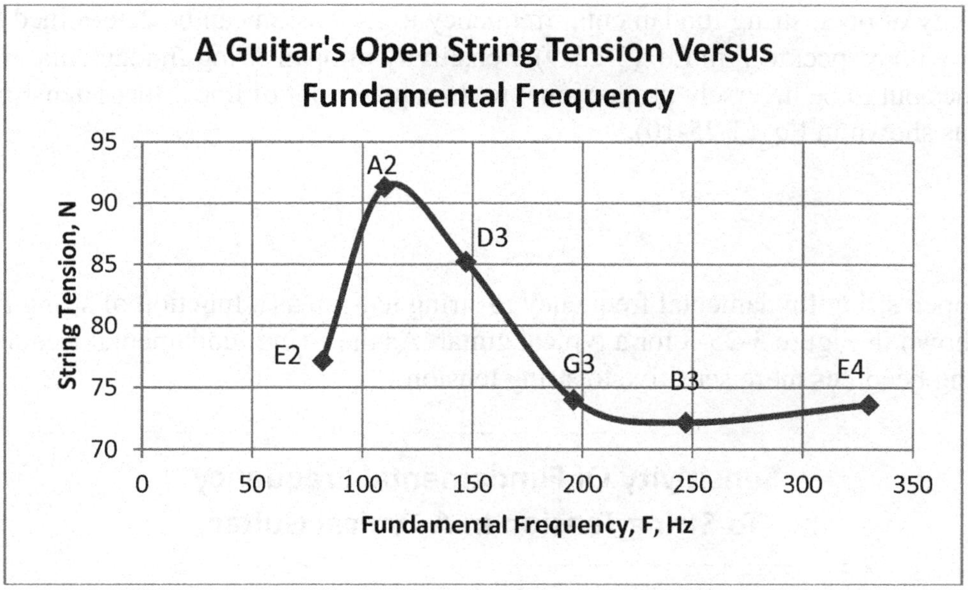

Figure 3-25-1 A guitar's open string tension versus fundamental frequency

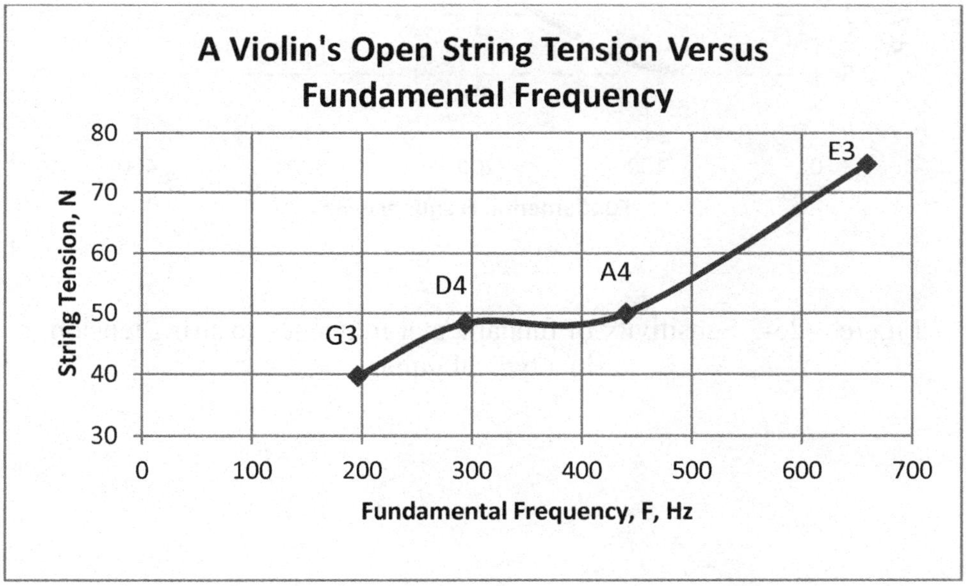

Figure 3-25-2 A violin's open string tension versus fundamental frequency

Sensitivity of open string fundamental frequency to its tension can be determined by taking the derivative of *F* with respect to *T* in Eq. (3-25-9). Sensitivity of open string fundamental frequency to its tension comes out to be inversely proportional to the square root of linear mass density and the string tension as shown in Eq. (3-25-10).

$$\frac{dF}{dT} = \frac{0.5}{\sqrt{4L^2 \rho T}} \qquad (3\text{-}25\text{-}10)$$

Sensitivity of open string fundamental frequency to string tension as a function of string fundamental frequency is shown in Figure 3-25-3 for a typical guitar. As the string fundamental frequency increases, tuning becomes more sensitive to string tension.

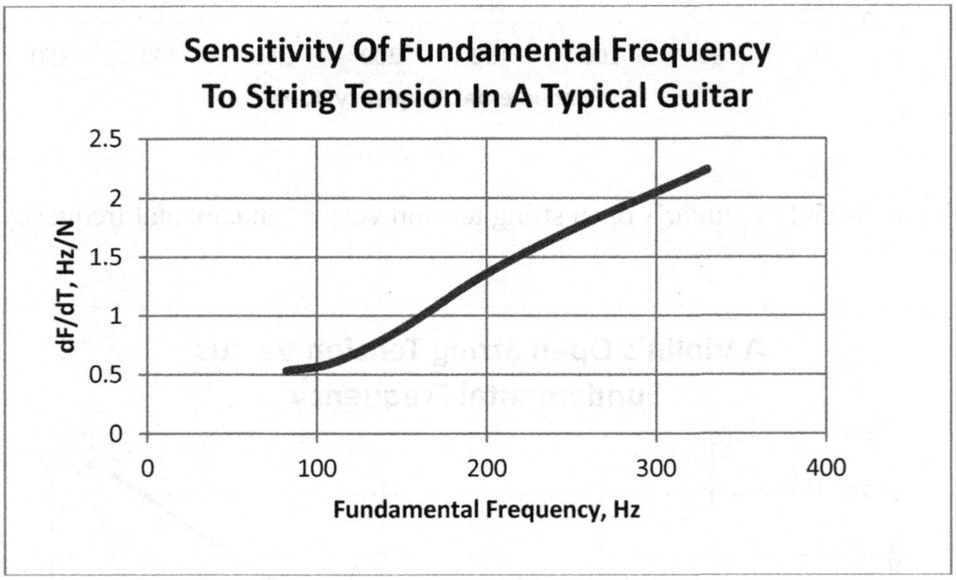

Figure 3-25-3 Sensitivity of fundamental frequency to string tension in a typical guitar

Chapter 3-26 Income-Consumption-Savings

Annual consumption for a family depends on the family's annual net income. As the annual net income increases so does the annual consumption. Annual consumption, C, for families with annual net income, I, between \$50,000 and \$1,000,000 can be approximated by the following relationship.

$$C = I - 30 \times \sqrt{2I} \qquad (3\text{-}26\text{-}1)$$

Annual consumption versus annual net income is shown in Figure 3-26-1. Annual savings, S, is the difference between annual net income and annual consumption.

$$S = I - C = 30 \times \sqrt{2I} \qquad (3\text{-}26\text{-}2)$$

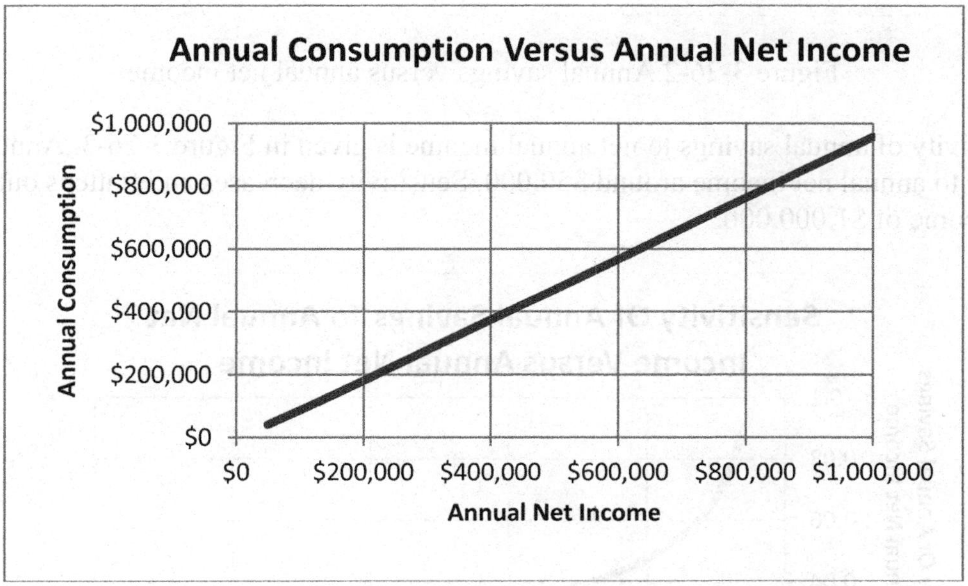

Figure 3-26-1 Annual consumption versus annual net income

Annual savings versus annual net income is shown in Figure 3-26-2. As the annual net income increases so does the annual savings. As the annual net income increases, sensitivity of annual savings to annual net income decreases. This can be shown by taking the derivative of S with respect to I in Eq. (3-26-2).

$$\frac{dS}{dI} = \frac{30}{\sqrt{2I}} \qquad (3\text{-}26\text{-}3)$$

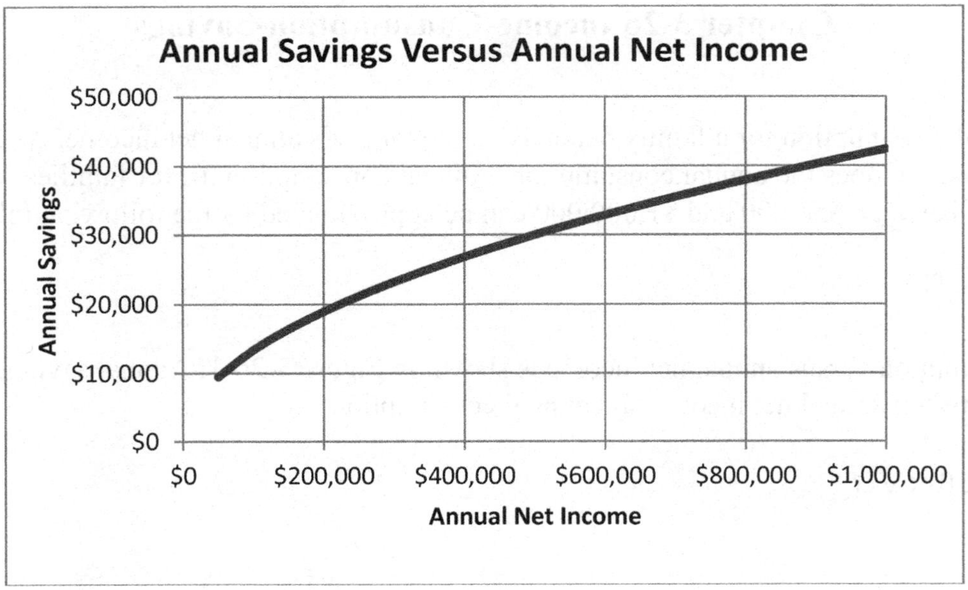

Figure 3-26-2 Annual savings versus annual net income

Sensitivity of annual savings to net annual income is given in Figure 3-26-3. Annual savings is very sensitive to annual net income around $50,000. Sensitivity decreases and flattens out around annual net income of $1,000,000.

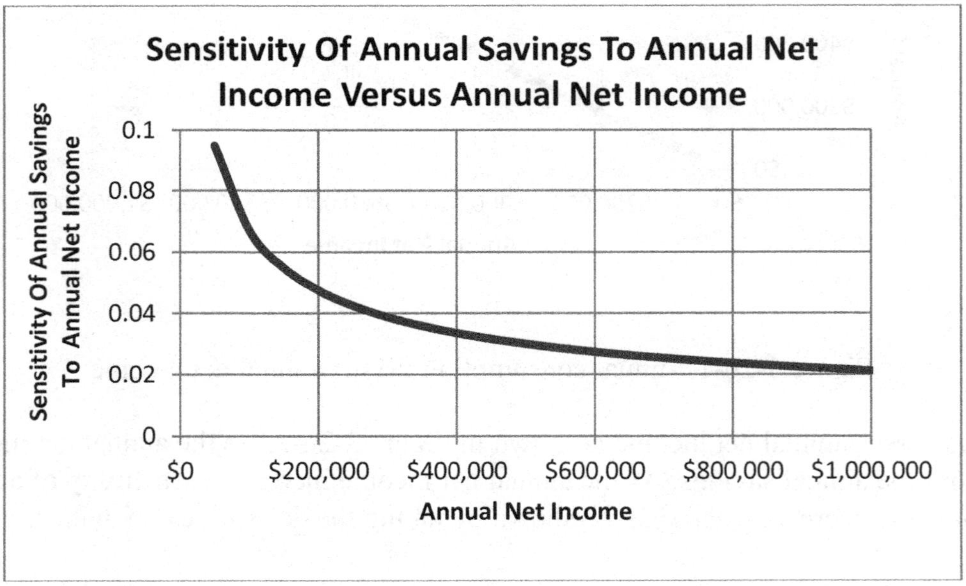

Figure 3-26-3 Sensitivity of savings to annual net income versus annual net income

Chapter 3-27 Maximize Volume of an Open Top Square Box

Let us try to build an open top box from a square cardboard with sides length L. The box will be built by cutting and bending the cardboard ends along the lines shown in Figure 3-27-1. The cardboard is cut at four corners as slits and bent along the dashed lines to form the open top box. The slits and the dashed lines are at X distances from the corners. The object is to maximize the volume of the open top box.

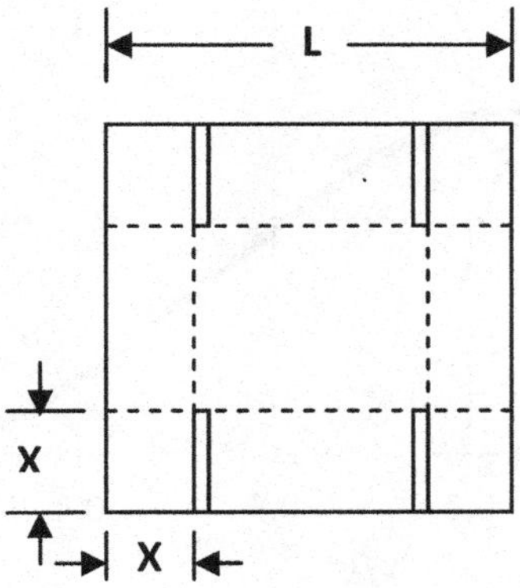

Figure 3-27-1 Square cardboard with corner slits and dashed lines

Volume of the box can be expressed in terms of the variables L and X.

$$V = X(L - 2X)^2 \tag{3-27-1}$$

To maximize the volume of the box, the first derivative of V is taken with respect to X and set it equal to zero.

$$\frac{dV}{dX} = 12X^2 - 8XL + L^2 = 0 \tag{3-27-2}$$

Solutions of the quadratic Eq. (3-27-2) are

$$X = \frac{8L \pm \sqrt{64L^2 - 48L^2}}{24} \quad \text{or} \quad X = \frac{L}{2} \text{ and } X = \frac{L}{6} \tag{3-27-3}$$

From Eq. (3-27-1), for $X = \frac{L}{2}$, $V = 0$ and for $X = \frac{L}{6}$, $V = \frac{2}{27}L^3$. Maximum open top box volume is obtained when the corner slits and bends are at $X = \frac{L}{6}$. This maximum can also be verified by taking the second derivative of volume with respect to X and show that $\frac{d^2V}{dX^2} < 0$ at $X = \frac{L}{6}$.

$$\frac{d^2V}{dX^2} = 24X - 8L < 0 \quad \text{or} \quad X < \frac{L}{3} \tag{3-27-4}$$

Eq. (3-27-1) can be rewritten in a non-dimensional form as follows.

$$\frac{V}{L^3} = \left(\frac{X}{L}\right)\left[1 - 2\left(\frac{X}{L}\right)\right]^2 \qquad (3\text{-}27\text{-}5)$$

Non-dimensional volume versus non-dimensional X is plotted in Figure 3-27-2. The maximum volume occurs at $\frac{X}{L} = \frac{1}{6} = 0.167$.

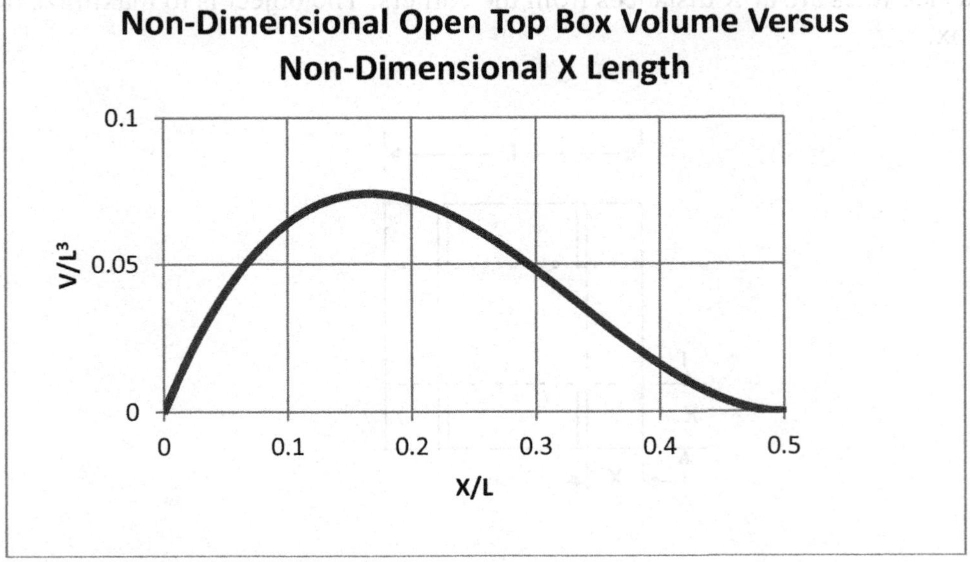

Figure 3-27-2 Non-dimensional open top box volume versus non-dimensional X length

Chapter 3-28 Maximize Volume of an Open Top Rectangular Box

Previous chapter's problem on open top square box can be generalized to rectangular boxes. Let us try to build an open top box from a rectangular cardboard with sides length L and length kL where $k \geq 1$. The box will be built by cutting and bending the cardboard ends along the lines shown in Figure 3-28-1. The cardboard is cut at four corners as slits and bent along the dashed lines to form the open top box. The slits and the dashed lines are at X distances from the corners. The object is to maximize the volume of the open top rectangular box.

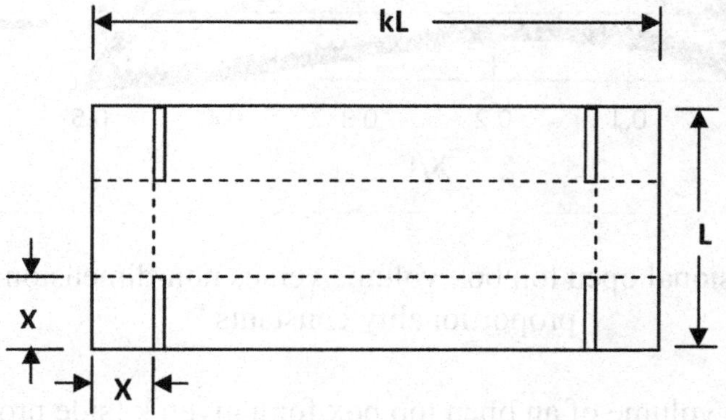

Figure 3-28-1 Rectangular cardboard with corner slits and dashed lines

Volume of the box can be expressed in terms of the variables L, X, and k.

$$V = X(kL - 2X)(L - 2X) \quad (3\text{-}28\text{-}1)$$

To maximize the volume of the open top rectangular box, the derivative of volume with respect to the unknown corner cut length X is taken and set it equal to zero.

$$\frac{dV}{dX} = kL^2 - 4L(k+1)X + 12X^2 = 0 \quad (3\text{-}28\text{-}2)$$

The solution of the quadratic Eq. (3-28-2) which is less than $\frac{L}{2}$ and which maximizes the volume of the open top rectangular box is

$$\frac{X}{L} = \frac{(k+1) - \sqrt{(k+1)^2 - 3k}}{6} \quad (3\text{-}28\text{-}3)$$

Figure 3-28-2 shows how the open top box volume changes with X for different k values. The $\frac{X}{L}$ value that maximizes the open top box volume increases as k increases.

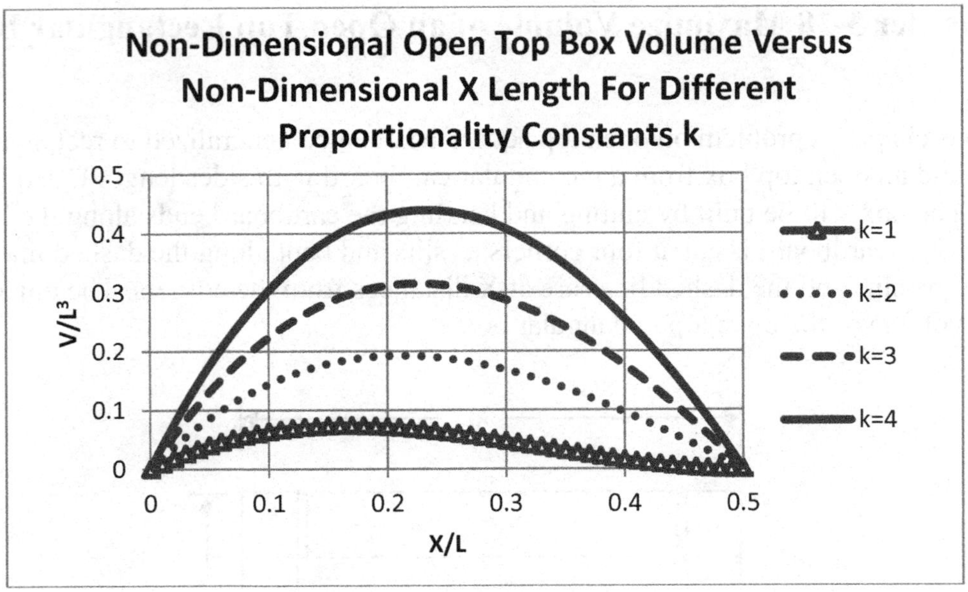

Figure 3-28-2 Non-dimensional open top box volume versus non-dimensional X length for different proportionality constants k

$\frac{X}{L}$ value that maximizes the volume of an open top box for a given k (side proportionality) are shown in Figure 3-28-3. As k gets large, $\frac{X}{L}$ value that maximizes the volume of an open top box approaches 0.25. This can be shown by finding the limiting value of $\frac{X}{L}$ in Eq. (3-28-3) as $k \rightarrow \infty$. Either large values of k can be inserted into Eq. (3-28-3) or L'Hopital's rule, see Reference [1], can be applied to the following indeterminate form of Eq. (3-28-3).

$$\lim_{k\to\infty} \frac{X}{L} = \lim_{k\to\infty} \frac{f(k)}{g(k)} = \left(\frac{1}{6}\right) \frac{\left(1+\frac{1}{k}\right) - \sqrt{1-\frac{1}{k}+\frac{1}{k^2}}}{\left(\frac{1}{k}\right)} = \frac{0}{0} \qquad (3\text{-}28\text{-}4)$$

Differentiating the numerator and the denominator with respect to k gives

$$\lim_{k\to\infty} \frac{X}{L} = \lim_{k\to\infty} \frac{f'(k)}{g'(k)} = \left(\frac{1}{6}\right) \frac{\left(-\frac{1}{k^2}\right) - \left(\frac{1}{2}\right)\left(1-\frac{1}{k}+\frac{1}{k^2}\right)^{-1/2}\left(\frac{1}{k^2}-\frac{2}{k^3}\right)}{\left(-\frac{1}{k^2}\right)} \qquad (3\text{-}28\text{-}5)$$

Dividing the numerator and the denominator into $\left(-\frac{1}{k^2}\right)$ provides

$$\lim_{k\to\infty} \frac{X}{L} = \lim_{k\to\infty} \frac{f'(k)}{g'(k)} = \left(\frac{1}{6}\right)\left[1 + \left(\frac{1}{2}\right)\left(1-\frac{1}{k}+\frac{1}{k^2}\right)^{-1/2}\left(1-\frac{2}{k}\right)\right] = 0.25 \qquad (3\text{-}28\text{-}6)$$

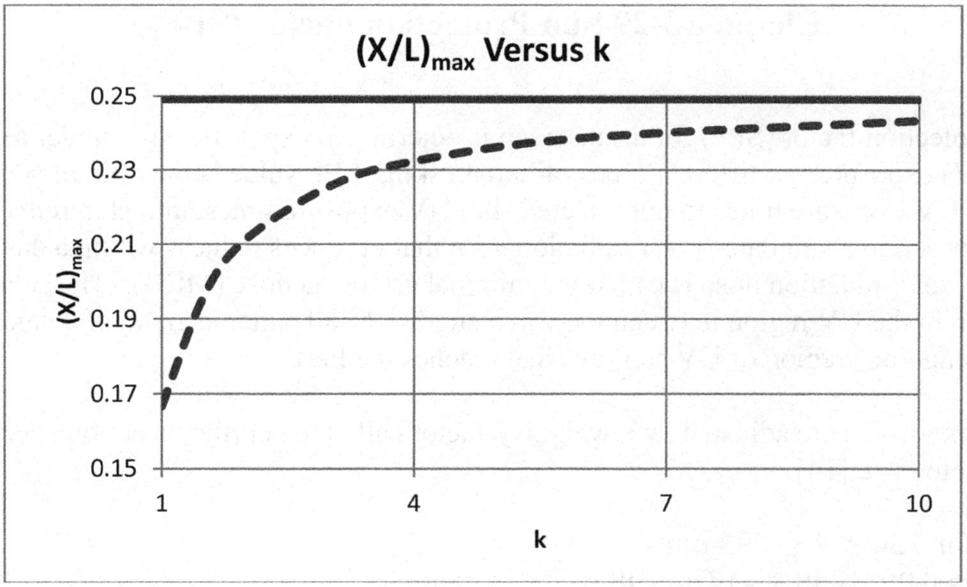

Figure 3-28-3 $(X/L)_{max}$ values that maximizes the volume of an open top box for a given k

Chapter 3-29 Sun Protection Factor (SPF)

Sun protection factor (SPF) for a sunscreen is determined experimentally under a series of ultraviolet (UV) exposures to different types of human skins. SPF value is the ratio of sunscreen protected skin UV exposure times to unprotected skin UV exposure times until skin redness occurs. The skin is exposed to a simulated solar radiation dose that produces redness within a day after exposure. This solar radiation dose is called the minimal erythema dose (MED). The solar radiation doses are given in the UV region between the wavelengths of 280 nanometers to 400 nanometers which is the dangerous region of UV sun rays that reaches the Earth.

UV radiation exposures are adjusted by a weighing factor called the erythema action spectrum as follows, see Reference [14].

$$W(\lambda) = 1.0 \; for \; 250 < \lambda \leq 298 \; nm$$
$$\log_{10}[W(\lambda)] = 0.094(298 - \lambda) \; for \; 298 < \lambda \leq 328 \; nm \qquad (3\text{-}29\text{-}1)$$
$$\log_{10}[W(\lambda)] = 0.015(139 - \lambda) \; for \; 328 < \lambda \leq 400 \; nm$$

Effective exposure on the skin is defined as the sum of all the spectral irradiance weighted by the erythema action spectrum over the UV wavelengths between 280 nm and 400 nm and multiplied by the exposure time.

$$E = T_{exposure}\left(\int_{280}^{400} I(\lambda)W(\lambda)d\lambda\right) \qquad (3\text{-}29\text{-}2)$$

where E is the erythema effective exposure in Joules/m^2, $T_{exposure}$ is the skin exposure time in seconds, $I(\lambda)$ is the spectral irradiance in Watts/(m^2-nm), $W(\lambda)$ is the erythema action spectrum, weighing factor, given by Eq. (3-29-1). The spectral irradiance and the erythema action spectrum are shown as a function of wavelength in Figure 3-29-1.

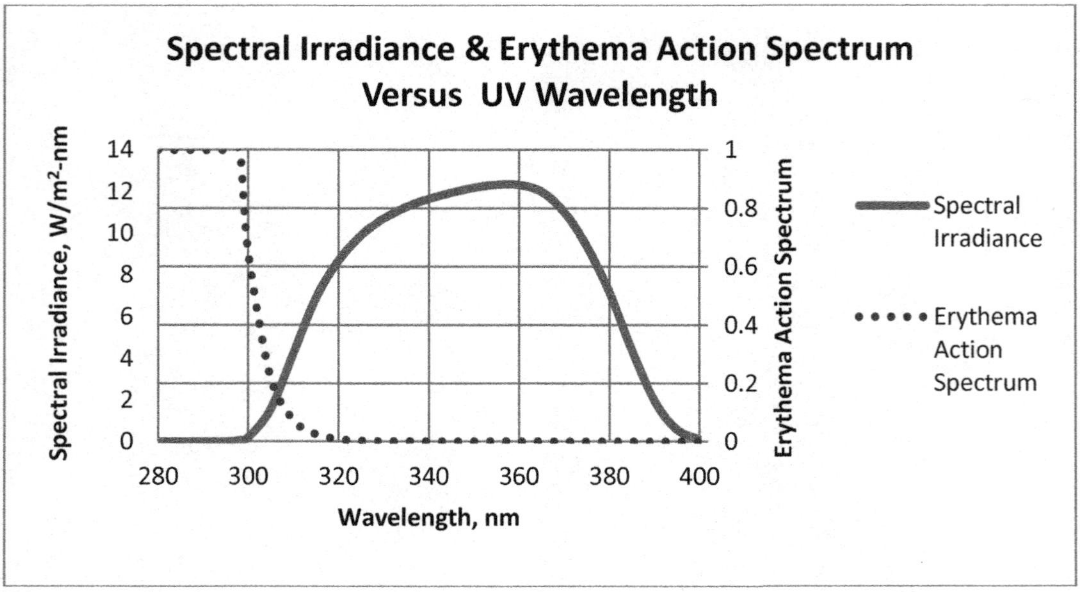

Figure 3-29-1 Spectral irradiance and erythema action spectrum versus UV wavelength

Spectral irradiance multiplied by the erythema action spectrum is shown in Figure 3-29-2 as a function of wavelength.

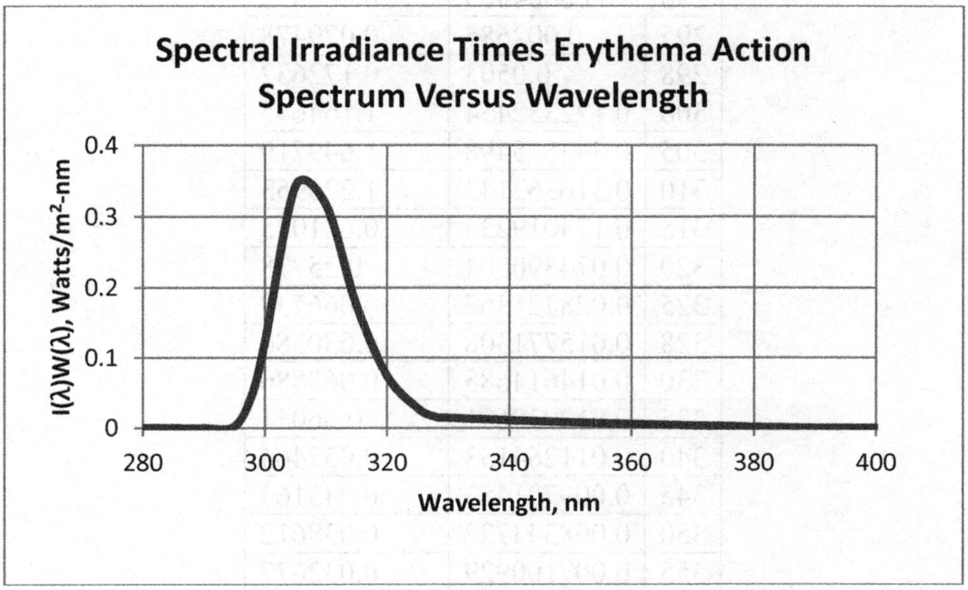

Figure 3-29-2 Spectral irradiance times erythema action spectrum versus wavelength

The area under the curve in Figure 3-29-2 is approximated by numerical methods using the trapezoidal rule, see Reference [10]. The area of each trapezoid is determined by

$$Area_i = 0.5 \times [I(\lambda_i)W(\lambda_i) + I(\lambda_{i+1})W(\lambda_{i+1})] \times \Delta\lambda \qquad (3\text{-}29\text{-}3)$$

where i goes from 1 to N. The area for each trapezoid and for the total function is shown in Table 3-29-1.

λ_i	$I(\lambda_i)W(\lambda_i)$	$Area_i$
280	0.00001523	0.000109
285	0.00002837	0.000181
290	0.00004405	0.006823
295	0.002685	0.079478
298	0.0503	0.172632
300	0.122332454	1.164645
305	0.343525498	1.649719
310	0.316362143	1.225953
315	0.174019234	0.621025
320	0.074390804	0.25778
325	0.028721367	0.066739
328	0.015771308	0.030386
330	0.014614685	0.068886
335	0.012939691	0.06051
340	0.011264153	0.052464
345	0.009721453	0.045163
350	0.008343722	0.038612
355	0.007100929	0.032677
360	0.005969845	0.027146
365	0.004888563	0.021665
370	0.003777301	0.016199
375	0.002702338	0.011109
380	0.001741335	0.006611
385	0.000903061	0.003115
390	0.000342895	0.001071
400	1.30497E-05	
	SUM AREA	**5.660944 Joules/(m^2-s)**

Table 3-29-1 Approximating the integral in Eq. (3-29-2) by trapezoidal rule

Effective exposure on the skin has to be

$$E = 5.66 \times T_{exposure} \quad (3\text{-}29\text{-}4)$$

The solar simulator has to be calibrated to deliver 5.66 Joules per square meter of UV energy every exposed second to the skin. SPF value is determined from the ratio of sunscreen protected skin UV exposure times to unprotected skin UV exposure times until skin redness occurs.

Chapter 3-30 Cooling of an Electronic Chip

Today everyone uses computers or cellular phones that are composed of several electronic chips. Power dissipation and cooling of these chips become a problem as they get smaller and smaller. Temperature rise and power limitations in a chip can be modeled as a first order linear differential equation.

Energy balance for a chip equates the change in internal energy of the chip with respect to time to power generated by the chip minus the energy lost from the chip to its heat sink and to the environment by different heat transfer mechanisms.

$$C\frac{dT_{chip}}{d\Theta} = P - \left(T_{chip} - T_{env}\right)/R \qquad (3\text{-}30\text{-}1)$$

where C is chip's thermal capacitance in W-s/°C, T_{chip} is the average chip temperature in °C, Θ is time in seconds, P is the power generated by the chip in Watts, T_{env} is the chip's environment temperature in °C, and R is the total heat transfer resistance between the chip and its environment in °C/W. Assuming that all the parameters introduced in Eq. (3-30-1) are constants with respect to time and introducing a new variable $T^* = T_{chip} - T_{env}$, Eq. (3-30-1) can be rewritten as

$$RC\frac{dT^*}{d\Theta} = PR - T^* \qquad (3\text{-}30\text{-}2)$$

In order to be able to separate the variables and solve this first order differential equation with ease, a steady-state solution and a transient solution are introduced. Total solution to Eq. (3-30-2) is the sum of a steady-state, T_{ss} , solution and a transient, T_{tr} , solution.

$$T^* = T_{ss} + T_{tr} \qquad (3\text{-}30\text{-}3)$$

We also need one initial condition in order to be able to solve Eq. (3-30-2). The initial condition is

$$T^* = 0 \ at \ \Theta = 0 \qquad (3\text{-}30\text{-}4)$$

It is easy to determine the steady-state solution to first order differential equation, Eq. (3-30-2). Left hand side of this equation goes to zero as time becomes large, namely chip's temperature goes into an equilibrium state and does not change with time as time becomes large enough. Then Eq. (3-30-2) becomes

$$0 = PR - T_{ss} \quad or \quad T_{ss} = PR \qquad (3\text{-}30\text{-}5)$$

To find the transient solution to first order differential equation, Eq. (3-30-2), let us substitute Eqs. (3-30-3) and (3-30-5) into it and the following first order differential equation is obtained.

$$RC\frac{dT_{tr}}{d\Theta} = -T_{tr} \qquad (3\text{-}30\text{-}6)$$

For the transient case, the initial condition, Eq. (3-30-4) becomes

$$T_{tr} = -PR \; at \; \Theta = 0 \tag{3-30-7}$$

A solution to Eq. (3-30-6) can be found by separating the variables and then integrating both sides.

$$\int \frac{dT_{tr}}{T_{tr}} = -\int \frac{d\Theta}{RC} \tag{3-30-8}$$

or

$$\ln T_{tr} = -\frac{\Theta}{RC} + M_1 \tag{3-30-9}$$

where M_1 is the constant of integration. Eq. (3-30-9) can be rewritten by taking the exponential power of both sides.

$$T_{tr} = M \times exp\left(-\frac{\Theta}{RC}\right) \tag{3-30-10}$$

where $M = e^{M_1}$ is now the integration constant and it is determined by using the initial condition Eq. (3-30-7).

$$M = -PR \tag{3-30-11}$$

The transient solution to first order differential equation, Eq. (3-30-6) becomes

$$T_{tr} = -PR \times exp\left(-\frac{\Theta}{RC}\right) \tag{3-30-12}$$

Substituting the steady-state and transient solutions into Eq. (3-30-3) gives the final solution to the first order differential equation, Eq. (3-30-2).

$$T^* = T_{chip} - T_{env} = PR \times \left[1 - exp\left(-\frac{\Theta}{RC}\right)\right] \tag{3-30-13}$$

In Eq. (3-30-13) PR in °C is the temperature potential between the steady-state chip temperature and the environment's temperature. RC is the chip's thermal time constant in seconds, namely the chip climbs to 63% of its temperature potential, PR, in a duration of one time constant. The thermal time constant RC is generally symbolized by the Greek letter tau, τ, in literature. Thermal time constant has the similar notion to the inductive time constant that is defined in Chapter 3-15. As the thermal time constant gets larger – thermal capacitance of the chip times its resistance to transfer heat to the environment gets larger – then the chip temperature reaches its steady-state temperature at a much longer time. Non-dimensional form of chip's temperature versus time, Eq. (3-30-13), is plotted for different thermal time constants, τ's, in Figure 3-30-1.

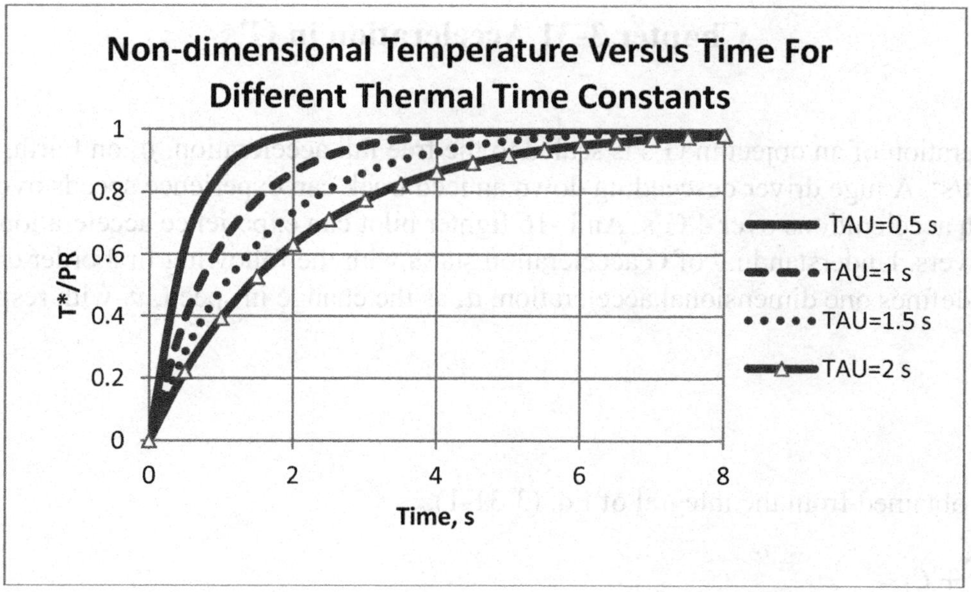

Figure 3-30-1 Non-dimensional temperature versus time for different thermal time constants, τ

As an example, consider a small silicon dioxide chip, 10 mm x 10 mm x 0.1 mm with an internal thermal capacitance, C, of 0.02 W-s/°C. This chip with its thermal sinks has a total thermal resistance of being able to release heat to its environment, R, of 25°C/W. This chip is going to operate in an environment of 30°C temperature and the maximum temperature that the chip can reach is 90°C . What is the maximum power rating of the chip? What is the thermal time constant for this chip?

$$T_{ss} = T_{chip\ max} - T_{env} = 90 - 30 = 60°C$$

or

$$T_{ss} = PR = 60°C \ or \ P = \frac{60\ °C}{25 \frac{°C}{W}} = 2.4\ W$$

The thermal time constant is

$$\tau = RC = 25\ \frac{°C}{W} \times 0.02\ \frac{W-s}{°C} = 0.5\ s$$

This particular chip has a power rating of 2.4 Watts and it reaches 68°C , namely $30 + 60 \times 0.63$, in 0.5 second after it starts to operate.

Chapter 3-31 Acceleration in G's

Acceleration of an object in G's is scaled to the free fall acceleration, g, on Earth, namely 9.8 m/s^2 or 32.2 ft/s^2. A luge driver descending down an iced track can experience speeds over 90 mph or 150 km/h with accelerations over 4 G's. An F-16 fighter pilot can experience accelerations over 8 G's during maneuvers. Understanding of G acceleration starts with the following first order differential equation that defines one dimensional acceleration, a, as the change in speed, v, with respect change in time, t.

$$a = \frac{dv}{dt} \quad (3\text{-}31\text{-}1)$$

Speed can be obtained from the integral of Eq. (3-31-1).

$$v = \int a(t)dt + C_1 \quad (3\text{-}31\text{-}2)$$

where C_1 is the constant of integration. If a constant acceleration in time is assumed and the initial condition of $v = v_0 \; at \; t = t_0$ is given, then

$$G = \frac{a}{g} = \left(\frac{1}{g}\right) \times \frac{v-v_0}{t-t_0} \quad (3\text{-}31\text{-}3)$$

If a Porsche 911 accelerates from 0 to 60 mph in 3.7 s, the driver experiences the following G acceleration.

$$G = \left(\frac{1}{32.2\frac{ft}{s^2}}\right)\frac{(60-0)\frac{mi}{h}\left(5280\frac{ft}{mi}\right)}{(3.7-0)s\left(3600\frac{s}{h}\right)} = 0.74$$

Free fall of acceleration on the surface of the Sun is about 274 m/s^2 or 28 G's. On the other end of the spectrum, free fall acceleration on the surface of the Moon is about 1.67 m/s^2 or about 1/6'th of the free fall acceleration on Earth's surface.

With a special initial condition of $v = 0 \; at \; t = 0$, Eq. (3-31-3) becomes

$$v = at \quad (3\text{-}31\text{-}4)$$

Speed can be integrated with respect to time to get the distance traveled, $s = \int vdt + C_2$. With a special initial of $s = 0 \; at \; t = 0$, Eq. (3-31-4) becomes

$$s = \frac{1}{2}at^2 \quad (3\text{-}31\text{-}5)$$

Combining Eqs. (3-31-4) and (3-31-5), a relationship between acceleration in G's, speed of the object, and the distance traveled.

$$G = \frac{a}{g} = \frac{v^2}{2gs} \quad (3\text{-}31\text{-}6)$$

For example, if a bullet comes out of the muzzle of a rifle at a speed of 900 m/s, average acceleration of the bullet in a 1 m long rifle barrel can be calculated using Eq. (3-31-7).

$$G = \frac{a}{g} = \frac{(900)^2 \frac{m^2}{s^2}}{2\times 9.8 \frac{m}{s^2} \times 1\, m} = 41{,}327$$

Another example of G acceleration usage is in car crash evaluations. For example, if a 3,000 lbf car slams from front into a rigid wall at 35 mph and its frontal crush is 1 foot, what would be the average G acceleration that the vehicle body experiences? Since the net work done by the forces acting on the front of the car equals to the loss of kinetic energy by the car, the relationship between kinetic energy loss and work done on the car can be written as

$$\frac{1}{2}mv^2 = Fs \qquad (3\text{-}31\text{-}7)$$

With $F = ma$, Eq. (3-31-7) can be reduced to Eq. (3-31-6). The average G acceleration that the vehicle body experiences is

$$G = \frac{a}{g} = \frac{(35)^2 \frac{mi^2}{h^2} (5280)^2 \frac{ft^2}{mi^2}}{2\times 32.2 \frac{ft}{s^2} \times 1\, ft \times (3600)^2 \frac{s^2}{h^2}} = 41$$

Of course the driver of this car experiences much less G accelerations due to seat belts, airbags, and other passive restraints.

Chapter 3-32 Surface Area of a Leaf

A plant generates its own food by photosynthesis process that occurs on the surfaces of its leaves. Surface areas of leaves are important for a plant. It is difficult to find the exact surface areas of leaves by integration because of their unconventional geometric shapes. However, the surface area can be estimated by laying the leaf on a graph paper and tracing the leaf boundary on a graph paper and then counting the number of squares on the graph paper that the leaf boundary covers.

Another way of estimating the leaf area is to use the trapezoidal approximation, see Reference [10], for each ΔX element inside the traced boundary. An example is given for an ivy leaf that is traced on a graph paper as shown in Figure 3-32-1.

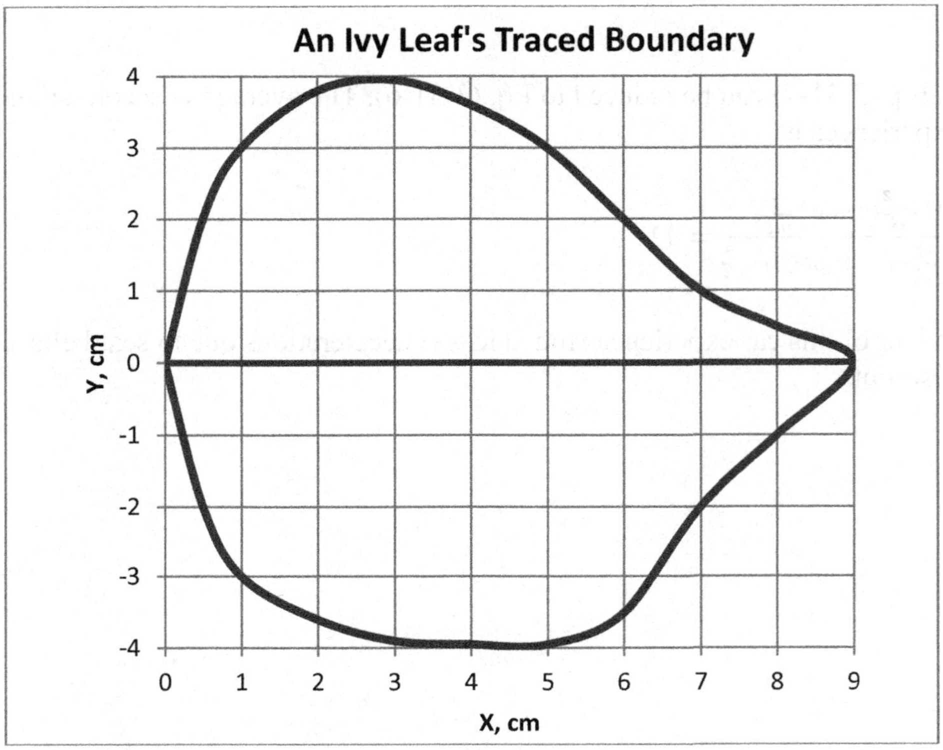

Figure 3-32-1 An ivy leaf's traced boundary

For the trapezoidal approximation, see Reference [10], of ivy leaf's area, ΔX is taken to be 1 cm. Then i'th element's area, A_i , inside the traced boundary is

$$A_i = \frac{1}{2}(Y_i + Y_{i+1}) \times \Delta X \qquad (3\text{-}32\text{-}1)$$

where Y_i is the boundary location at X_i and Y_{i+1} is the boundary location at X_{i+1}.

Each element on positive Y side is summed from 0 to 9 cm and then each element on negative Y side is summed from 0 to 9 cm to obtain the total surface area for one side. The total surface area for one side using the trapezoidal approximation came out to be 46.25 cm^2. For both sides of the ivy leaf, total surface area is then 92.5 cm^2.

By just counting the 1 cm^2 square elements, a total surface area for one side came out to be 46 cm^2. In this example, error introduced by just counting the square elements is -0.54% which is very good.

Chapter 3-33 Volume of a Spherical Candle Holder

A designer is creating a spherical glass candle holder with a cylindrical cavity and a disk shaped base. He wants to determine the volume of this candle holder and therefore its weight. The volume of the holder whose cross-section is shown in Figure 3-33-1 can be determined by splitting the candle holder into slices.

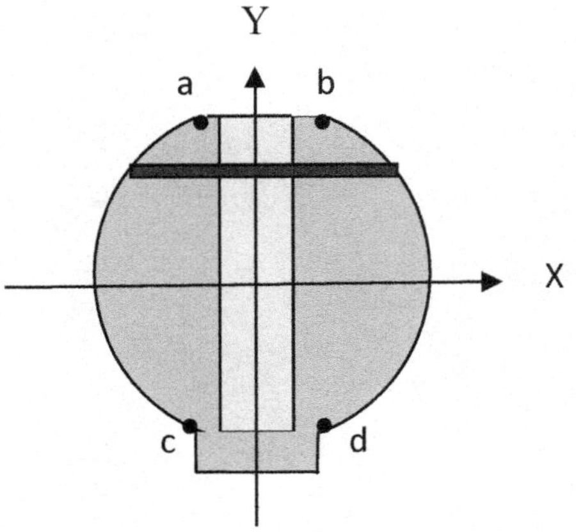

Figure 3-33-1 Cross-section of a spherical candle holder

The radius of the sphere is 10 cm. Therefore a point on the boundaries of the cross-sectional circle can be defined by

$$X^2 + Y^2 = 100 \qquad (3\text{-}33\text{-}1)$$

Since the designer wants to locate the points a, b, c, and d 5 cm away from X=0 in his design, points a, b, c, and d are determined from Eq. (3-33-1) to be (-5, $5\sqrt{3}$), (5, $5\sqrt{3}$), (-5, -$5\sqrt{3}$), and (5, -$5\sqrt{3}$) respectively. From points a, b, c, and d to center cavity the glass thickness is 1 cm. Also from points c and d to the base of the candle holder the thickness of the glass is 1 cm. The designer knows the density of the glass he is going to specify to be 2,000 kg/m^3. To determine the candle holder's weight, he has to find its volume since an object's mass (weight) is its volume times its density. The volume of the candle holder whose cross-section is shown in Figure 3-33-1 can be found by figuring out the incremental volume of the slice shown in blue color in Figure 3-33-1 and summing up or integrating over all the slices. The incremental volume of the slice is

$$dV = \pi(X^2 - 4^2)dY \qquad (3\text{-}33\text{-}2)$$

Combining Eq. (3-33-1) and Eq. (3-33-2) gives

$$dV = \pi(84 - Y^2)dY \qquad (3\text{-}33\text{-}3)$$

where incremental volume slices in Eq. (3-33-3) has to be integrated from -$5\sqrt{3}$ to +$5\sqrt{3}$ as follows.

$$\int_{-5\sqrt{3}}^{+5\sqrt{3}} \pi(84 - Y^2)dY = \pi \times 590\sqrt{3} = 3210.4\ cm^3 \qquad (3\text{-}33\text{-}4)$$

The volume obtained in Eq. (3-33-4) does not include the disk shaped base. The disk shaped base's volume is

$$V_{base} = \pi \times 5^2 \times 1 = 78.5\ cm^3 \qquad (3\text{-}33\text{-}5)$$

The total volume of the glass candle holder is the sum of Eqs. (3-33-4) and (3-33-5) which is 3,289 cm^3.

The mass (weight) of the glass candle holder is

$$M = 2{,}000\ \frac{kg}{m^3} \times 3{,}289\ cm^3 \times \frac{1\ \ m^3}{10^6\ cm^3} = 6.6\ kg$$

Chapter 3-34 Sensitivity of Drag Forces on a Bicycle Cyclist

A bicycle cyclist applies force to the pedals of a bicycle which is transferred to the rear wheel contact point on the ground that generates a reaction force against the rotation of the rear wheel. This reaction force is balanced by a drag force acting on the cyclist and the bicycle and by a rolling resistance force between the ground and the tires, if no acceleration occurs on a flat surface. A force balance along the flat surface with no acceleration can be written as

$$F_{reaction} = F_{drag} + F_{rolling\ resistance} \tag{3-34-1}$$

Reaction force at the real wheel can be related to the normal force applied to the pedals as follows.

$$F_{reaction} = \left(\frac{L \times F_{normal}}{R_{wheel}}\right)\left(\frac{R_{rear\ sprocket}}{R_{front\ sprocket}}\right) \tag{3-34-2}$$

Where L is the length of the pedal crank in feet, F_{normal} is the average normal force applied by the cyclist to the pedals in lbf, R_{wheel} is the radius of the rear wheel in feet, $R_{front\ sprocket}$ is the radius of front sprocket in feet, and $R_{rear\ sprocket}$ is the radius of rear sprocket in feet.

Power input by the cyclist can be expressed as the product of the torque on the front sprocket and the rotational speed of the pedals.

$$P = 2\pi \times L \times F_{normal} \times \Omega_{pedal} \tag{3-34-3}$$

where P is the power input by the cyclist in hp and Ω_{pedal} is the pedal rotational speed in rpm. The velocity of the bicycle can be written as

$$V = 2\pi \times R_{wheel} \times \Omega_{wheel} \tag{3-34-4}$$

where V is the speed of the bicycle in ft/s and Ω_{wheel} is the bicycle's wheel rotational speed in rpm. Also the wheel rotational speed and the pedal rotational speed are related.

$$R_{front\ sprocket} \times \Omega_{pedal} = R_{rear\ sprocket} \times \Omega_{wheel} \tag{3-34-5}$$

If no mechanical losses are assumed, i.e. no bearing losses and no chain losses, the cyclist's power input can be equated to power generated at the rear wheel contact point using Eqs. (3-34-2) through (3-34-5).

$$VF_{reaction} = P = V(F_{drag} + F_{rolling\ resistance}) \tag{3-34-6}$$

Drag force without any wind can be expressed as

$$F_{drag} = 0.5 \times \rho \times C_{drag} \times A \times V^2 \tag{3-34-7}$$

where ρ is the density of air in lbm/ft^3, C_{drag} is the non-dimensional drag coefficient acting on the cyclist and the bicycle , and A is the effective area exposed to the drag force in ft^2.

The rolling resistance force from the contact of the wheels to the ground can be expressed as

$$F_{rolling\ resistance} = C_{rolling\ resistance} \times M \times G \qquad (3\text{-}34\text{-}8)$$

where $C_{rolling\ resistance}$ is the non-dimensional rolling resistance coefficient between the tires and the ground, M is the mass of the cyclist and the bicycle in lbm, and G is the gravitational acceleration, namely 32.2 ft/s^2. Combining Eqs. (3-34-6) through (3-34-8) and rearranging provides the following cubic equation for the maximum velocity a bicyclist can reach on a flat surface.

$$V^3 + A1 \times V = B1 \qquad (3\text{-}34\text{-}9)$$

where

$$A1 = \frac{C_{rolling\ resistance} \times M \times G}{0.5 \times \rho \times C_{drag} \times A} \quad \text{and} \quad B1 = \frac{32.2 \times P}{0.5 \times \rho \times C_{drag} \times A}$$

Eq. (3-34-9) is solved by trial and error using the following parameters.

$F_{normal} = 40\ lbf$

$L = 0.5\ ft$

$\rho = 0.0749\ \frac{lbm}{ft^3}$

$A = 5\ ft^2$

$C_{rolling\ resistance} = 0.005$

$M = 170\ lbm$

Ω_{pedal} is varied between 0 to 120 rpm and C_{drag} has three different values, namely 0.1, 0.3, and 0.5. The velocity values were determined by trial and error as the function $f(V) = V^3 + A1 \times V - B1$ crosses zero as shown in Figure 3-34-1.

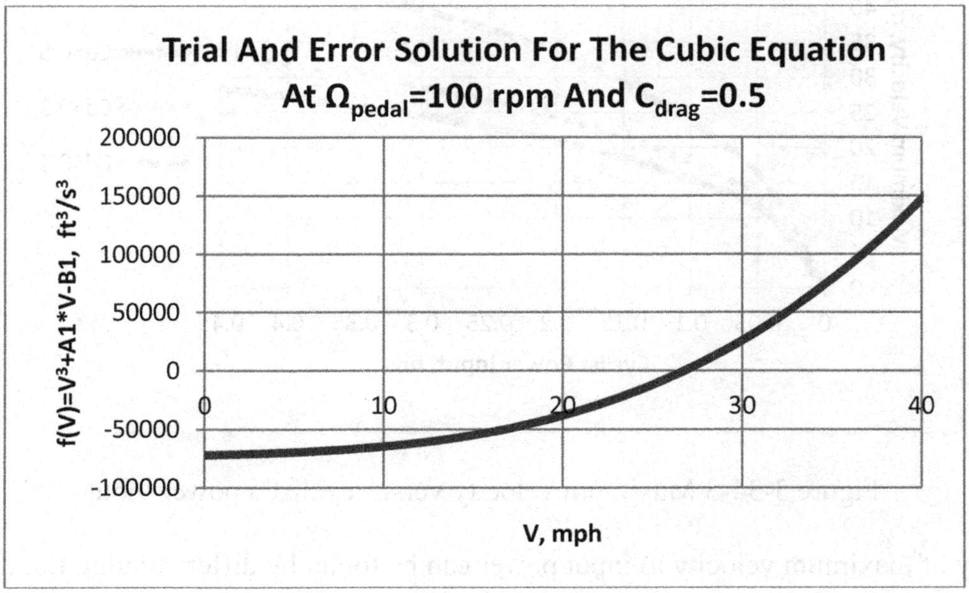

Figure 3-34-1 Trial and error solution for the cubic equation at $\Omega_{pedal} = 100\ rpm$ and $C_{drag} = 0.5$

The curve in Figure 3-34-1 crosses the zero, $f(V) = 0$, at a velocity value of 26.76 mph. The solutions to Eq. (3-34-9) are shown in Figure 3-34-2 for three different drag coefficients. As the drag coefficient decreases maximum velocity of the bicycle increases. Maximum velocity of the bicycle is also shown as a function of input power in Figure 3-34-3.

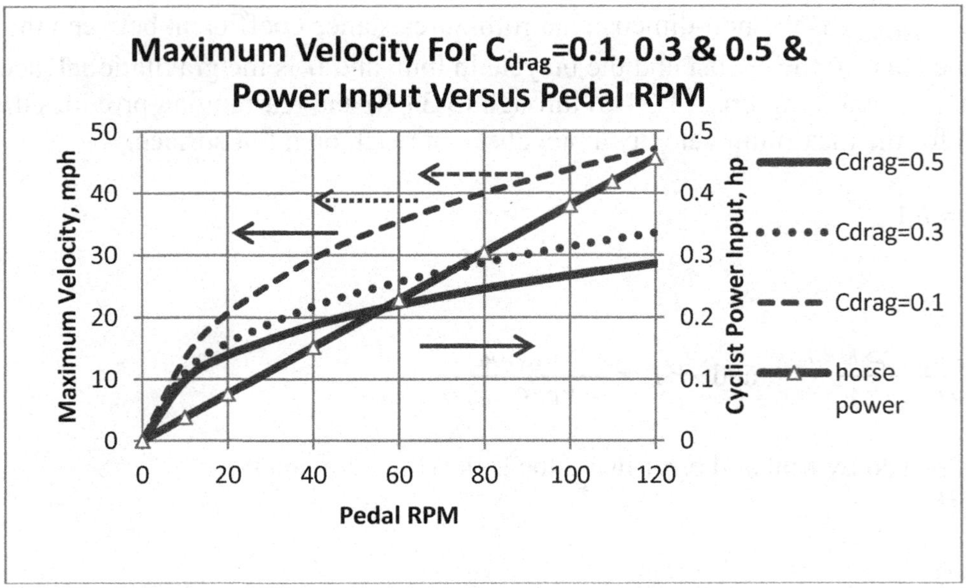

Figure 3-34-2 Maximum velocity for $C_{drag} = 0.1, 0.3, 0.5$ and cyclist power input versus pedal rpm

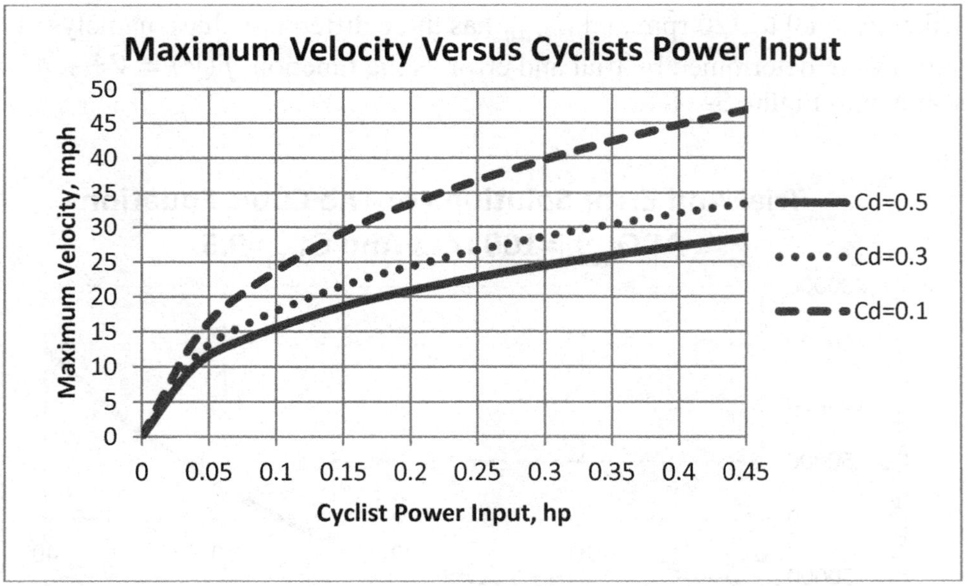

Figure 3-34-3 Maximum velocity versus cyclist's power input

The sensitivity of maximum velocity to input power can be found by differentiating Eq. (3-34-9).

$$\frac{dV}{dP} = \frac{\left(\frac{32.2}{0.5 \times \rho \times C_{drag} \times A}\right)}{3V^2 + B1} \qquad (3\text{-}34\text{-}10)$$

The sensitivities of maximum velocity to input power can be obtained from the slopes of the curves in Figure 3-34-3. The sensitivities of maximum velocity to input power are shown in Figure 3-34-4. As the drag coefficient decreases and at low pedal rotational speeds, maximum bicycle velocity becomes very sensitive to pedal rpm.

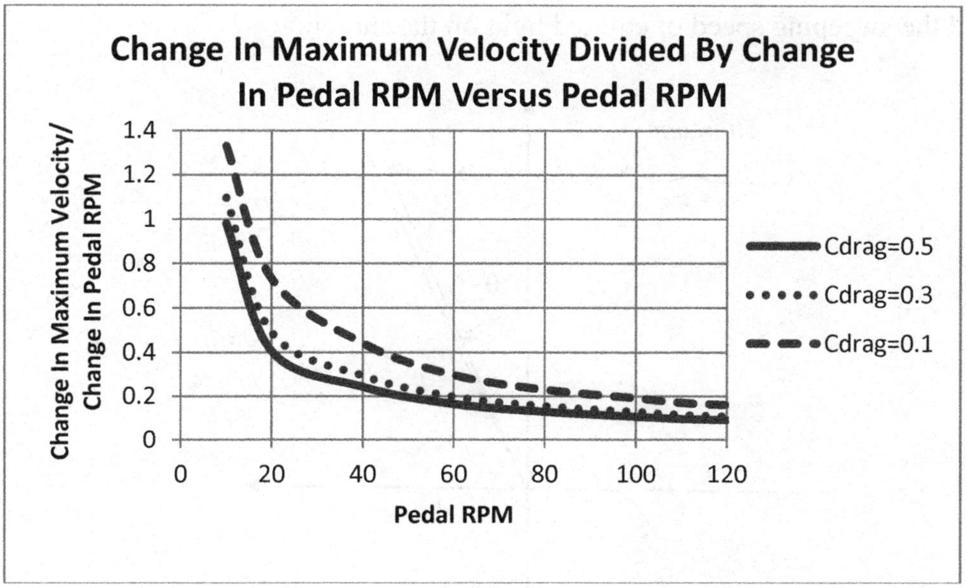

Figure 3-34-4 Change in maximum velocity divided by change in pedal rpm versus pedal rpm

Chapter 3-35 Speed of a Revolving Light

A lighthouse emits a light that rotates at a constant rate of one revolution every four minutes. The lighthouse is 0.5 miles out on the sea from a straight shoreline as shown in Figure 3-35-1. The object is to find the sweeping speed of emitted light on the shoreline.

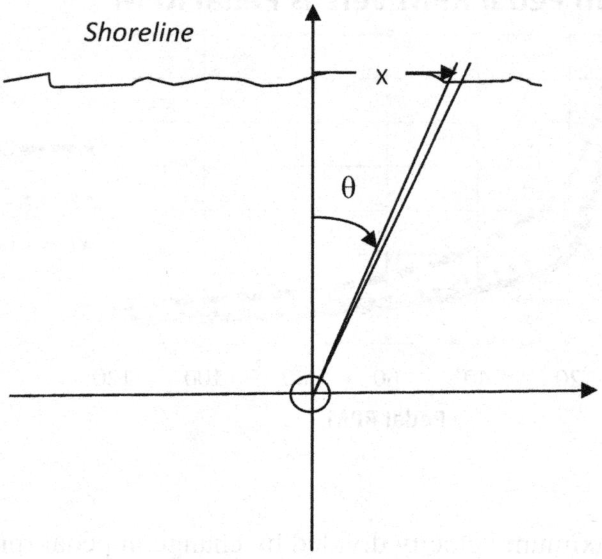

Figure 3-35-1 Lighthouse emitting light to a straight shoreline

The distance X is defined as

$$X = 0.5 \times \tan\theta \quad (3\text{-}35\text{-}1)$$

The sweeping speed of emitted light on shoreline can be obtained from the time derivative X in Eq. (3-35-1).

$$\frac{dX}{dt} = 0.5 \times \frac{d(\tan\theta)}{d\theta} \times \frac{d\theta}{dt} = \frac{0.5}{\cos^2\theta}\frac{d\theta}{dt} \quad (3\text{-}35\text{-}2)$$

In Eq. (3-35-2) $\frac{d\theta}{dt}$ is given as 0.25 revolutions per minute and it has to be converted to radians per minute by multiplying it by 2π radians per revolution.

$$\frac{dX}{dt} = \frac{0.5\ mi}{\cos^2\theta}\left(\frac{d\theta}{dt}\right)\frac{rev}{min} \times (2\pi)\frac{radians}{rev} \times (60)\frac{min}{hr} \quad (3\text{-}35\text{-}3)$$

Sweeping speed of emitted light on shoreline is plotted in Figure 3-35-1 as a function of revolving light angle. The minimum sweeping speed occurs when $\theta = 0$ and the sweeping speed increases to $\pm\infty$ as θ approaches $\pm 90°$.

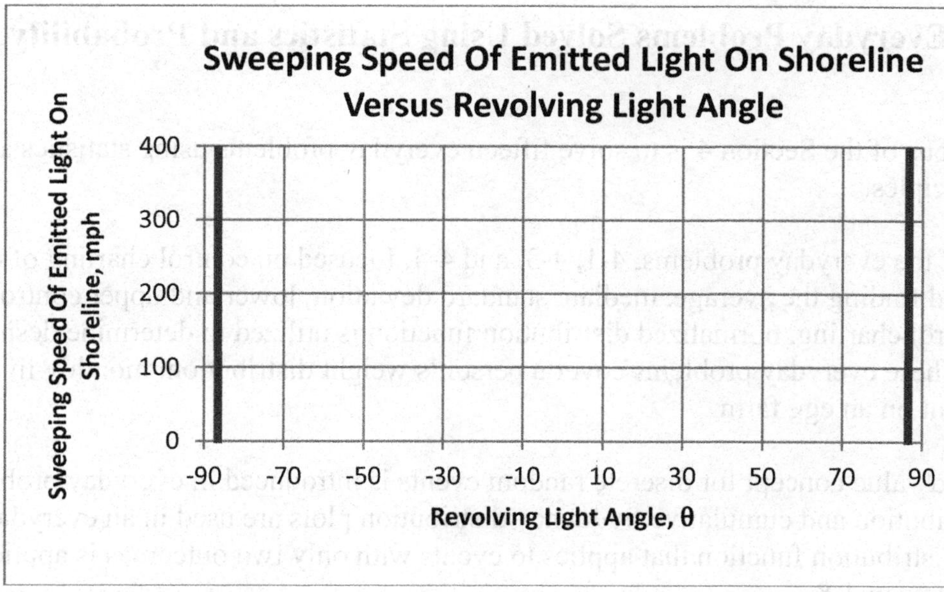

Figure 3-35-1 Sweeping speed of emitted light on shoreline versus revolving light angle

Section 4 – Everyday Problems Solved Using Statistics and Probability Principles

Main focus of the Section 4 is to solve fifteen everyday problems using statistics and probability principles.

Three of the everyday problems, 4-1, 4-3, and 4-4, focused on control charting of different types of data and finding the average, median, standard deviation, lower and upper control limits. In addition to control charting, normalized distribution function is utilized to determine desired probabilities. These everyday problems cover a person's weight distribution, monthly living expenses, and improvement on an egg farm.

Expected value concept for discrete random events is introduced in everyday problems in 4-2. Frequency distribution and cumulative frequency distribution plots are used in an everyday problem in 4-5. Binomial distribution function that applies to events with only two outcomes is applied to everyday problems in 4-8.

Standard deviation of a set of independent events is treated in everyday problems 4-9, 4-10, and 4-12. Standard deviation of a set of independent events is applied to tolerancing and to a sputtering equipment operation. Problem 4-10 also focused on accuracy and precision of a weight scale measurement system along with its repeatability, reproducibility, and gage capability.

A 2x2 full factorial design of experiments is treated in problem 4-11 by introducing the additive and interaction effects of independent variables on the dependent variable in an adhesive application.

Reliable data is key to a statistical analysis. Baseball games have been collecting reliable data over a century. Popular baseball statistics is reviewed in problem 4-15.

Addition rule of probability for mutually exclusive and for mutually non-exclusive events is exercised in everyday problems in 4-6. Joint probability of independent events is used in everyday problems in 4-7. Permutations of possible outcomes of an event are treated in problem 4-13. Percentage combinations are utilized in problem 4-14.

Chapter 4-1 A Person's Weight Distribution

Weight of a person is a very important metric that patterns health and social aspects of one's life. Consider an adult man with a six feet frame. He is overweight, if his weight goes over 184 pounds or 25 body mass index. He is underweight, if his weight goes under 136.5 pounds or 18.5 body mass index.

He goes on his calibrated scale once a week and obtains his weight and plots the data for sixty weeks as shown in Figure 4-1-1.

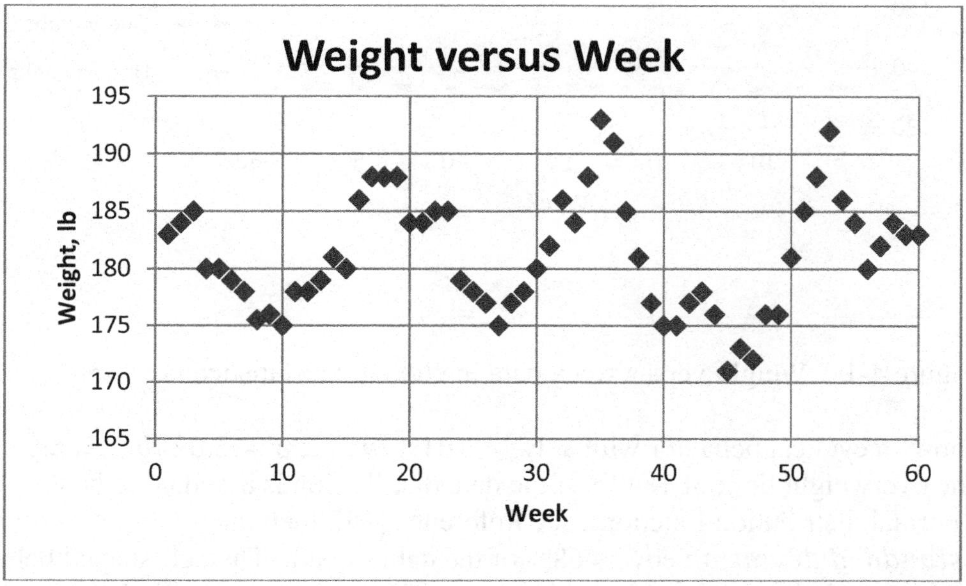

Figure 4-1-1 Data for weight versus week for 60 weeks

This data can be analyzed statistically and the probability of being overweight and being underweight can be determined. On the weight data shown in Figure 4-1-1, the average (mean) of 60 data points from Eq. (4-1-1), the spread (standard deviation) of 60 data points from Eq. (4-1-2), and the overweight and the underweight limits can be plotted to see how the weight behaves through time, Figure 4-1-2.

$$Average\ (mean) weight = \bar{W} = \left(\frac{1}{60}\right) \sum_{i=1}^{60} W_i \qquad (4\text{-}1\text{-}1)$$

$$Spread\ (standard\ deviation) = \sigma = \left(\frac{1}{59}\right) \sum_{i=1}^{60} (W - \bar{W})^2 \qquad (4\text{-}1\text{-}2)$$

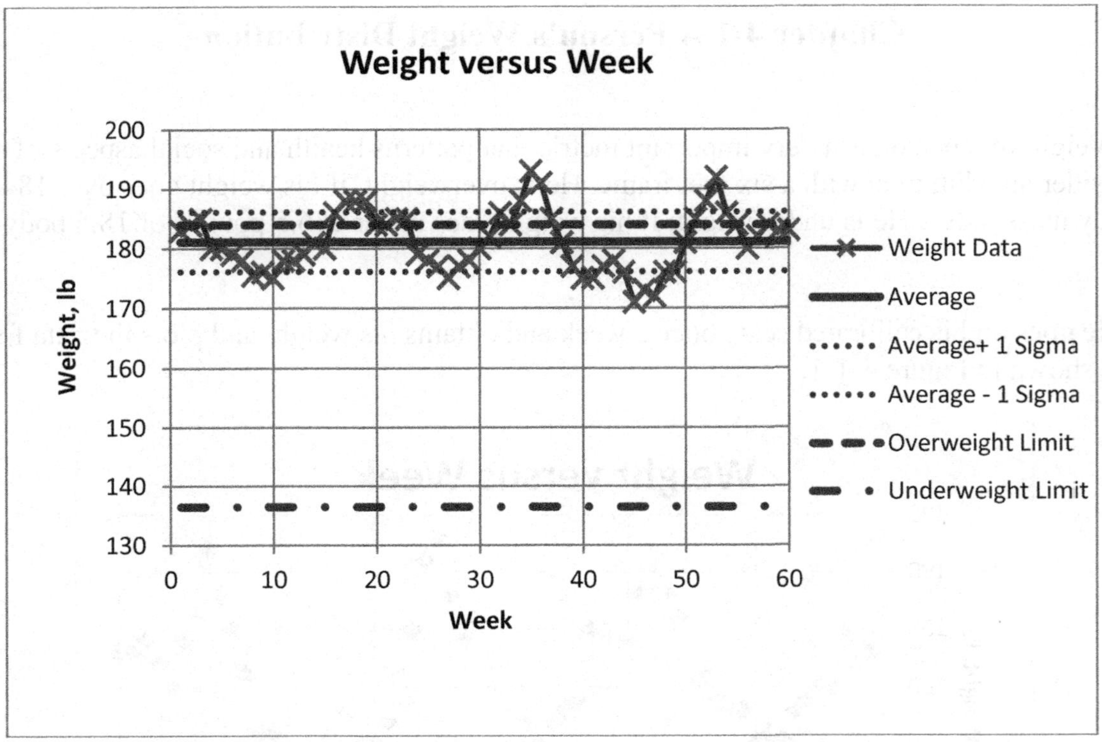

Figure 4-1-2 Weight versus week data including the statistical and limit lines

Figure 4-1-2 shows a cyclical behavior with a $\bar{W} = 181.1\ lb\ and\ \sigma = 5.02\ lb$. Average weight is very close to the overweight limit of 184 lb. If the data distribution is assumed to behave like a bell-shaped curve (normal distribution function), see Reference [15], then the $average \pm 1\ standard\ deviation$ covers 68% of the data spread. The bell-shaped behavior of weight data can be seen in by plotting it as a histogram. The weight data is tabulated in equal intervals as shown in Table 4-1-1 in order to obtain the histogram plot in Figure 4-1-3.

WEIGHT INTERVAL	FREQUENCY
166-169.9	0
170-173.9	3
174-177.9	13
178-181.9	17
182-185.9	16
186-189.9	8
190-193.9	3
194-197.9	0

Table 4-1-1 Weight data frequency distribution

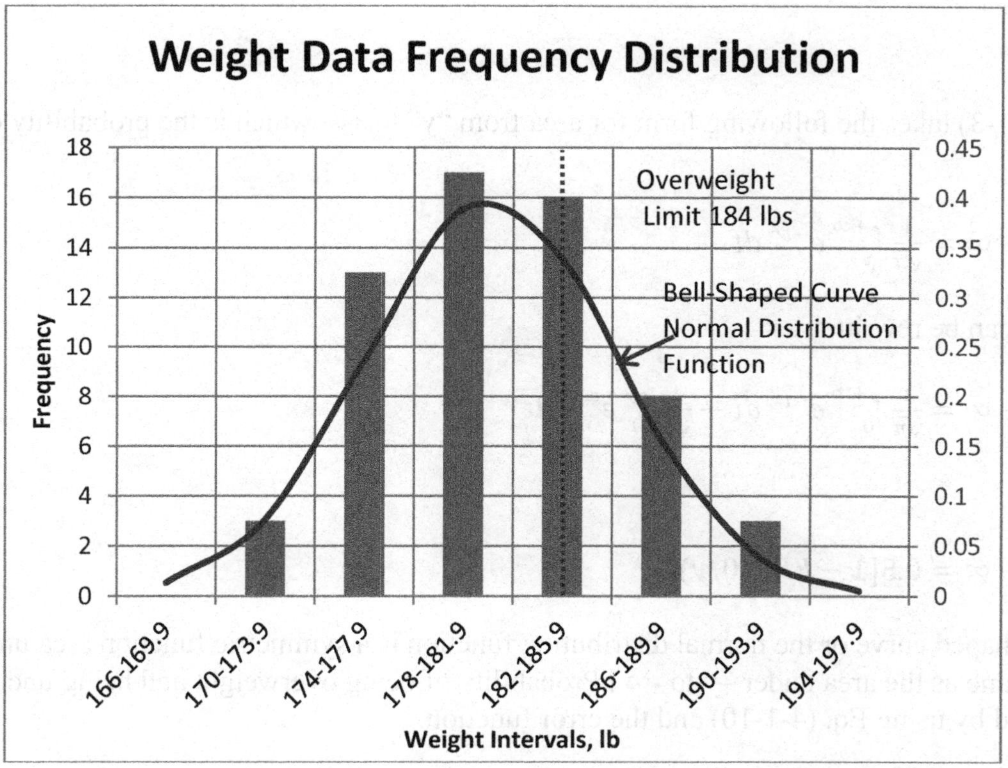

Figure 4-1-3 Weight data frequency distribution and the bell-shaped curve

The bell-shaped curve or the normal distribution function shown in Figure 4-1-3 can be written as follows, see Reference [15].

$$P(W) = \left(\frac{1}{\sigma\sqrt{2\pi}}\right) exp[-\frac{(W-\bar{W})^2}{2\sigma^2}] \quad (4\text{-}1\text{-}3)$$

Eq. (4-1-3) has to be integrated from $-\infty$ to $+\infty$ and normalized to get unity under the bell-shaped curve to be able to assign probabilities from 0 to 1. To achieve this, let

$$z = \frac{(W-\bar{W})}{\sigma} \quad (4\text{-}1\text{-}4)$$

Then the area under the normalized bell-shaped curve is 1.

$$Area = \frac{1}{\sqrt{2\pi}}\int_{-\infty}^{+\infty} e^{\frac{-z^2}{2}}\, dz = 1 \quad (4\text{-}1\text{-}5)$$

The integral in Eq. (4-1-5) can be identified by the well known error function, ERF(0,y), see MS Excel functions.

$$ERF(0,y) = \frac{2}{\sqrt{\pi}}\int_0^y e^{-t^2}\, dt \quad (4\text{-}1\text{-}6)$$

If a new variable "y" is introduced, then values of the error function, Eq. (4-1-6), can be used to evaluate the normalized form of Eq. (4-1-3).

$$y = \frac{z}{\sqrt{2}} \quad (4\text{-}1\text{-}7)$$

then Eq. (4-1-3) takes the following form for area from "y" to +∞ which is the probability of being overweight.

$$Area\ y\ to + \infty = \frac{1}{\sqrt{\pi}}\int_{y}^{+\infty} e^{-t^2}\,dt \quad (4\text{-}1\text{-}8)$$

Eq. (4-1-8) can be rewritten as

$$Area\ y\ to + \infty = \frac{1}{\sqrt{\pi}}\int_{0}^{+\infty} e^{-t^2}\,dt - \frac{1}{\sqrt{\pi}}\int_{0}^{y} e^{-t^2}\,dt \quad (4\text{-}1\text{-}9)$$

or

$$Area\ y\ to + \infty = 0.5[1 - ERF(0, y)] \quad (4\text{-}1\text{-}10)$$

Since bell-shaped curve or the normal distribution function is a symmetric function area under +y to +∞ is the same as the area under –y to -∞ . Probability of being overweight and being underweight can be calculated by using Eq. (4-1-10) and the error function.

In the overweight case,

$$z = \frac{(184-181.1)}{5.02} = 0.573 \quad (4\text{-}1\text{-}11)$$

and

$$y = \frac{0.573}{\sqrt{2}} = 0.405 \quad (4\text{-}1\text{-}12)$$

From error function in MS Excel,

$$ERF(0{,}0.405) = 0.433 \quad (4\text{-}1\text{-}13)$$

Using Eq. (4-1-13) and multiplying the result by 100 give the probability of being overweight which is 28.3%. This result is not surprising because all the weight data is crowded around the overweight limit and even one standard deviation spread from the average goes over the overweight limit. The normalized bell-shaped curve or the normal distribution function with the probability of being overweight is shown in Figure 4-1-4.

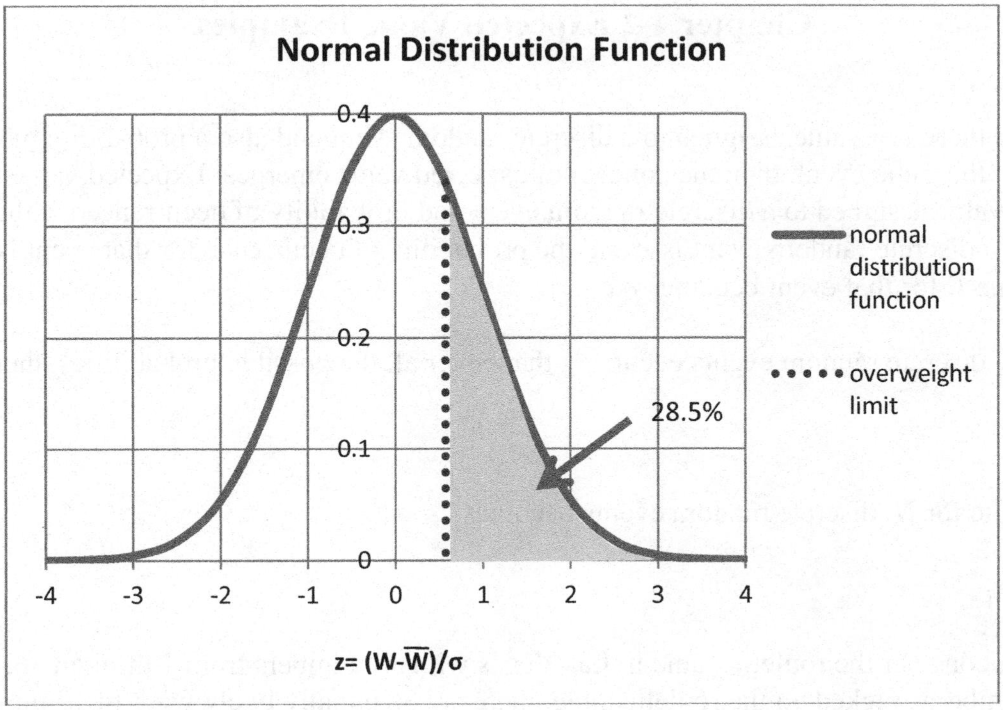

Figure 4-1-4 Normal distribution function indicating 28.5% probability of being overweight

Similar analysis can be performed for the underweight limit of 136.5 pounds. It can be shown that the probability of going below the underweight limit is zero with $z = -8.9$. This can clearly be seen from Figure 4-1-4 that even four times the standard deviation spread below the average gives zero probability of being underweight.

In normalized probability distribution functions $\overline{W} \pm \sigma$ covers 68.3% percent of the data spread, $\overline{W} \pm 2\sigma$ covers 95.4% percent of the data spread, and $\overline{W} \pm 3\sigma$ covers 99.7% percent of the data spread.

Chapter 4-2 Expected Value Examples

When there is a value assigned to a discrete random event and also a probability of occurrence is assigned to the same event, then the concept of expected value emerges. Expected value is the product of a value assigned to a discrete random event and probability of occurrence for that event. If the value for a discrete random event is x and the probability of occurrence for that event is p, then the expected value E for that event becomes px.

If there are N discrete random events occurring that cover all the possible probabilities, then

$$\sum_{i=1}^{N} p_i = 1 \tag{4-2-1}$$

Expected value for N discrete random events becomes

$$E = \sum_{i=1}^{N} p_i x_i \tag{4-2-2}$$

For example, consider the roulette game in Las Vegas with 38 numbers from 1 through 36, a 0, and a 00. If any number is picked on the roulette table, there are 38 equally likely ways of winning and the probability of winning is $1/38$. A $1 bet on any one number pays $35, if your number comes. The probability of losing the $1 bet is $37/38$. The expected value for this single number bet becomes

$$E = \left(\frac{1}{38}\right)\$35 - \left(\frac{37}{38}\right)\$1 = -\$\frac{2}{38} = -\$0.0526 \tag{4-2-3}$$

Half of the 1 through 36 numbers on the roulette table are red and the other half is black. Also half of the 36 numbers are even and the other half is odd. If all red colored numbers or all black colored numbers or all even numbers or all odd numbers is picked, the probability of winning is $18/38$. A $1 bet pays $1, if your pick comes. The probability of losing the $1 bet is $20/38$. The expected value for this bet becomes

$$E = \left(\frac{18}{38}\right)\$1 - \left(\frac{20}{38}\right)\$1 = -\$\frac{2}{38} = -\$0.0526 \tag{4-2-4}$$

If you evaluate the expected values of all possible bets on a roulette table, you will see that the house has an edge of $-\$0.0526$ on every $1 bet. A caution should be given to the effect that roulette rules and payoffs change from place to place.

Next let us look at the expected value of rolling two dice. When two dice are rolled, possible outcomes are 2, 3, 4, 5, 6, 7, 8, 9, 10, 11, and 12. There are 36 possible equally likely results that can occur after the roll. For example the outcome 6 can be obtained 5 equally likely ways resulting from (1,5), (5,1), (2,4), (4,2), and (3,3), namely with a probability of $\frac{5}{36} = 0.1389$. A $1 bet pays $6, if your pick comes. The probability of losing the $1 bet is 31/36. The expected value for this bet becomes

$$E = \left(\frac{5}{36}\right)\$6 - \left(\frac{31}{36}\right)\$1 = -\$\frac{1}{36} = -\$0.028 \tag{4-2-5}$$

All the expected values resulting from rolling two dice are given in Table 4-2-1.

Sum of two dice	Possible number of occurrences	Probability of getting the sum	Payment per $1 bet	Expected winnings, $	Expected losses, $	Net expected value, $
2	1	0.0278	$34	0.944	0.972	-0.028
3	2	0.0556	$16	0.888	0.944	-0.056
4	3	0.0833	$10	0.833	0.916	-0.083
5	4	0.1111	$7	0.777	0.888	-0.111
6	5	0.1389	$6	0.833	0.861	-0.028
7	6	0.1667	$5	0.833	0.833	0
8	5	0.1389	$6	0.833	0.861	-0.028
9	4	0.1111	$7	0.777	0.888	-0.111
10	3	0.0833	$10	0.833	0.916	-0.083
11	2	0.0556	$16	0.888	0.944	-0.056
12	1	0.0278	$34	0.944	0.972	-0.028

Table 4-2-1 Expected values resulting from rolling two dice

Payments for a successful $1 bet are also shown in Table 4-2-1. The house always has an edge on the bettor except when a 7 is picked. In the case of picking 7, the house and the bettor have equal expected values.

Expected values are also used in determining life insurance premiums and in calculating trust fund annuities.

As an example, let us consider a 25 years old male who wants to leave one million dollars to his children when he passes away. The question is how much monthly premium he should pay in order to expect $1,000,000 in his trust account at age 82?

First thing is to find out the yearly probability of mortality for a US male from age 26 on. This can be found in life tables issued by Centers for Disease Control. A typical probability of a US male mortality by age is shown in Figure 4-2-1. In Figure 4-2-1, the age axis goes up to 82 years.

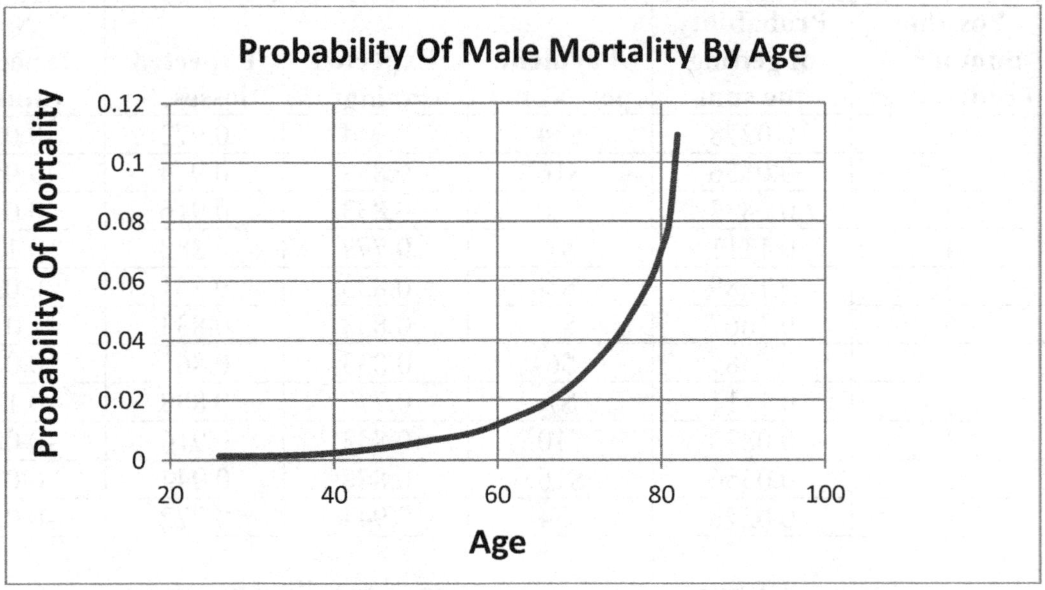

Figure 4-2-1 Probability of a US male mortality by age

Second thing that has to be figured out is the future value, V , of monthly premiums, M , at a constant monthly interest rate i . After N months the future value V_N becomes

$$V_N = M\sum_{j=1}^{N-1}(1+i)^{j-1} \tag{4-2-6}$$

The series in Eq. (4-2-6) looks like a geometric progression with a common ratio of (1+i). The sum of this geometric progression series can be obtained by

$$V_N - (1+i)V_N = M[1-(1+i)^N] \tag{4-2-7}$$

or

$$V_N = M[(1+i)^N - 1]/i \tag{4-2-8}$$

In order to have a future expected value of $1,000,000 at the age of 82, this man has to perform the following summation from age 26 through age 82 for 57 years and determine his monthly premiums, M.

$$\$1{,}000{,}000 = M\sum_{j=1}^{57} p_j\left[(1+i)^{12j} - 1\right]/i \tag{4-2-9}$$

The yearly probability of mortality, p_j, is obtained from Figure 4-2-1 where p_1 is the probability of mortality at age 26, where p_2 is the probability of mortality at age 27, and p_{57} is the probability of mortality at age 82. If an annual interest rate of 6% is assumed, i.e. a monthly interest rate of $i = (0.06/12) = 0.005$, then the monthly premiums come out to be $280.36.

If he does not consider the probability of mortality and keep depositing $280.36 premiums monthly, he would reach $1,000,000 at age 74. This result can be obtained from Eq. (4-2-8) as follows:

$$\$1{,}000{,}000 = \$280.36[(1+0.005)^N - 1]/0.005 \tag{4-2-10}$$

or

$$N = \frac{\log_{10}(\frac{\$1{,}000{,}000 \times 0.005}{\$280.36}+1)}{\log_{10}(1+0.005)} = 589\ months \qquad (4\text{-}2\text{-}11)$$

which is 49 years after the age of 25. By the time he completes the age of 82, $(82-25) \times 12 = 684\ months$, he would have $1,643,465.29 in his trust fund.

$$V_{684} = \frac{\$280.36[(1+0.005)^{684}-1]}{0.005} = \$1{,}643{,}465.29 \qquad (4\text{-}2\text{-}12)$$

Chapter 4-3 Monthly Living Expenses

One way to control our monthly expense is to look at how we have spent the allocated funds historically. After deducting the fixed monthly expenses such as house loan payments, house insurance, house property taxes, car loan payments, car insurance, and car registration fees, hopefully there is money left from our income to live on.

Consider a family with $2,500 per month budget for living expenses. This family records their expenses and categorizes them in food, utilities, cars, general, and outing buckets. They have the expense data for one year as shown in Table 4-3-1.

Month	Food	Utilities	Cars	General	Outing	Vacation	Total Monthly Expenses	Savings
January-09	410	461	303	510	345	333	2362	138
February-09	380	476	310	503	340	333	2342	158
March-09	465	350	530	545	245	333	2468	32
April-09	420	340	315	531	255	333	2194	306
May-09	390	310	334	610	410	333	2387	113
June-09	450	290	341	540	429	333	2383	117
July-09	445	320	353	585	331	333	2367	133
August-09	460	320	361	504	300	333	2278	222
September-09	407	331	630	405	325	333	2431	69
October-09	430	321	326	550	340	333	2300	200
November-09	440	342	354	456	470	333	2395	105
December-09	450	447	343	510	443	333	2526	-26
12 MONTH TOTALS=	5147	4308	4500	6249	4233	4000	28437	1563
MONTHLY AVERAGES=	429	359	375	521	353	333	2370	130

Table 4-3-1 Monthly expense data for one year

Total annual vacation expense of $4,000 is divided equally into 12 months. General category includes things like clothing, presents, sports equipment, etc. Savings is what is left from the monthly budget of $2,500.

Best way to analyze this data is to look at each expense bucket and the monthly total expenses month by month. This is called control charting the expenses month by month. Here only the total monthly expenses will be control charted as shown in Figure 4-3-1.

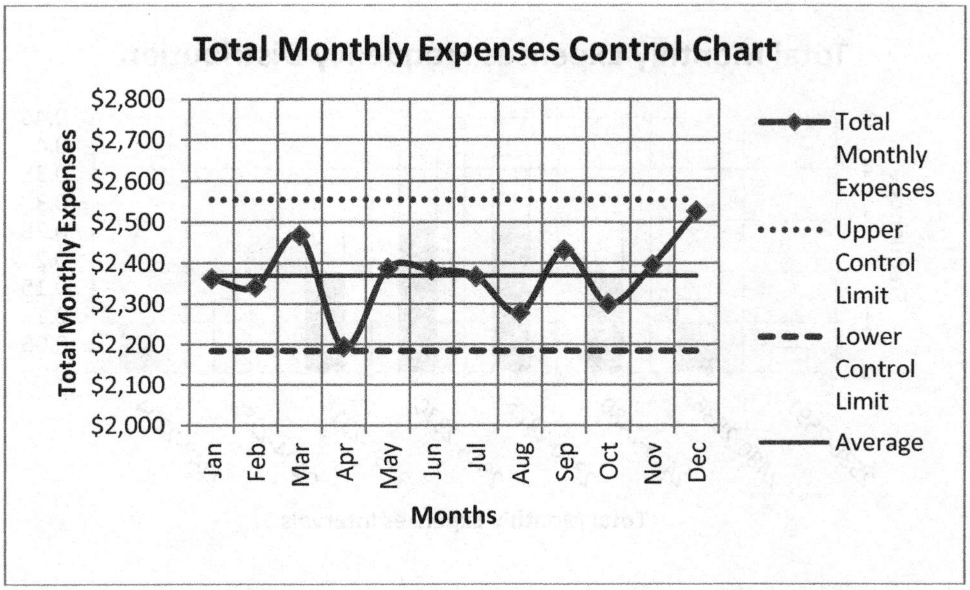

Figure 4-3-1 Total monthly expenses control chart

Each total monthly expense has one data point, X , in Figure 4-3-1. The average, $\bar{X}$, is for 12 months.

$$\bar{X} = \frac{1}{12}\sum_1^{12} X = \$2{,}369.75 \qquad (4\text{-}3\text{-}1)$$

In order to determine the upper and lower control limits in Figure 4-3-1, dispersion of total monthly expenses is needed. Assuming normal distribution for the total monthly expense data, standard deviation which is a metric for dispersion, can be calculated as follows.

$$S = \sqrt{\frac{\sum_1^{12}(X-\bar{X})^2}{N-1}} = \$92.63 \qquad (4\text{-}3\text{-}2)$$

Then upper and lower control limits become

$$UPPER\ CONTROL\ LIMIT = \bar{X} + 2S = \$2{,}555.01 \qquad (4\text{-}3\text{-}3)$$

$$LOWER\ CONTROL\ LIMIT = \bar{X} - 2S = \$2{,}184.49 \qquad (4\text{-}3\text{-}4)$$

Normality assumption for this limited data is a reasonable measure of dispersion for total monthly expenses. The normal (bell-shaped) behavior of total monthly expenses can be seen in Figure 4-3-2 where data is plotted as a histogram of equal intervals. White columns in Figure 4-3-2 are the normal distribution columns and dark columns represent the actual data. Visual histogram picture of data as shown in Figure 4-3-3 is a simple way of checking for normality of data.

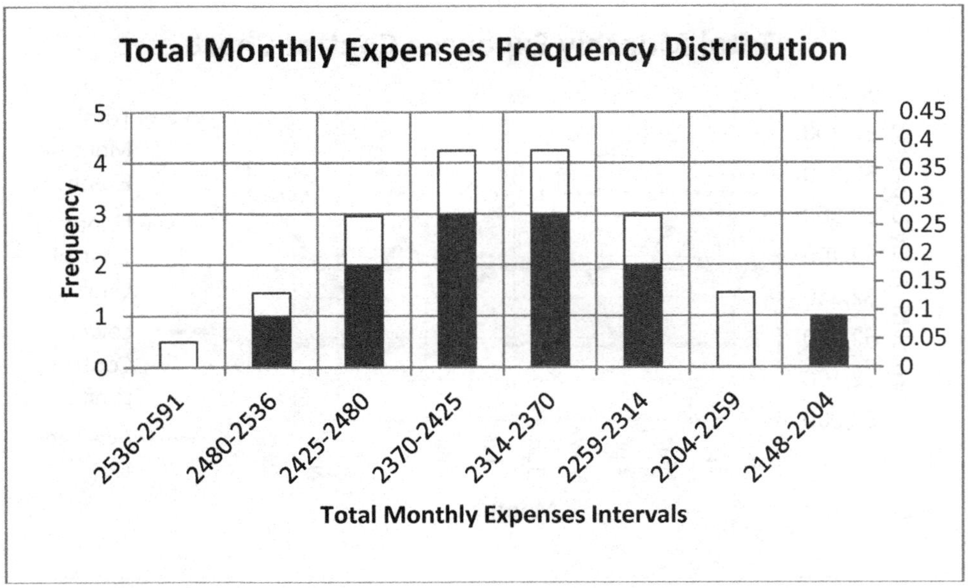

Figure 4-3-2 Total monthly expenses frequency distribution compared to a normal distribution

As the number of data points increases, i.e. over 25, the standard deviation should be recalculated and the upper and lower control limits should be adjusted with the new dispersion values. There are other more involved ways of checking normality of data. One such method is Anderson-Darling normality test which is detailed Reference [16].

Another chart that can be generated from the total monthly expenses is the moving range chart. The moving range chart provides the upper boundary for monthly expense swings from month to month. Previous month's total monthly expense is deducted from the present month's total expense and moving range, R , is determined. Moving ranges are plotted in Figure 4-3-3.

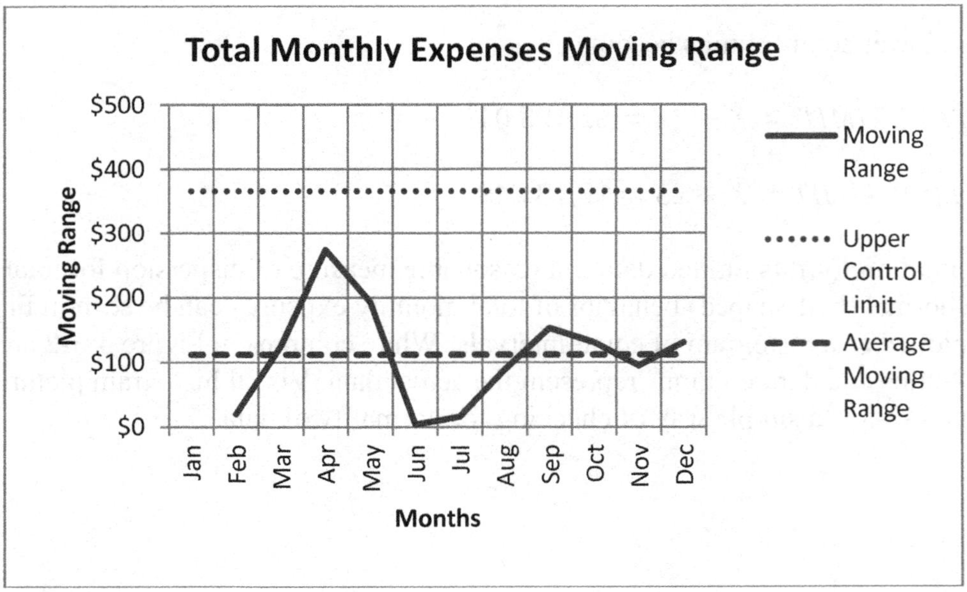

Figure 4-3-3 Total monthly expenses moving range

There is no moving range data point for January since it is the first data point. Since moving ranges are determined from two data points, there is no lower limit to moving range. Average moving range is obtained in a standard averaging method.

$$\bar{R} = \frac{1}{11}\sum_{1}^{11} R = \$112.00 \qquad (4\text{-}3\text{-}5)$$

The upper control limit to moving range is obtained from

$$Moving\ Range\ upper\ Control\ Limit = 3.267\bar{R} = \$365.90 \qquad (4\text{-}3\text{-}6)$$

The constant 3.267 is for N=2 and it varies for different sample size. For more insight into how these constants are obtained see Reference [17].

Upper control limit for total monthly expenses is $2,555 which is above the monthly allocation of $2,500. Also the upper control limit for the moving range data is $366. With these swings in expenses, there can be out of control months when the total monthly expenses can exceed the budgeted $2,500. This can happen 8.1% of the time as shown in Figure 4-3-4. 8.1% is the area under the normal distribution that is over $2,500 or area over $Z = \frac{(X-\bar{X})}{S} = 1.4$.

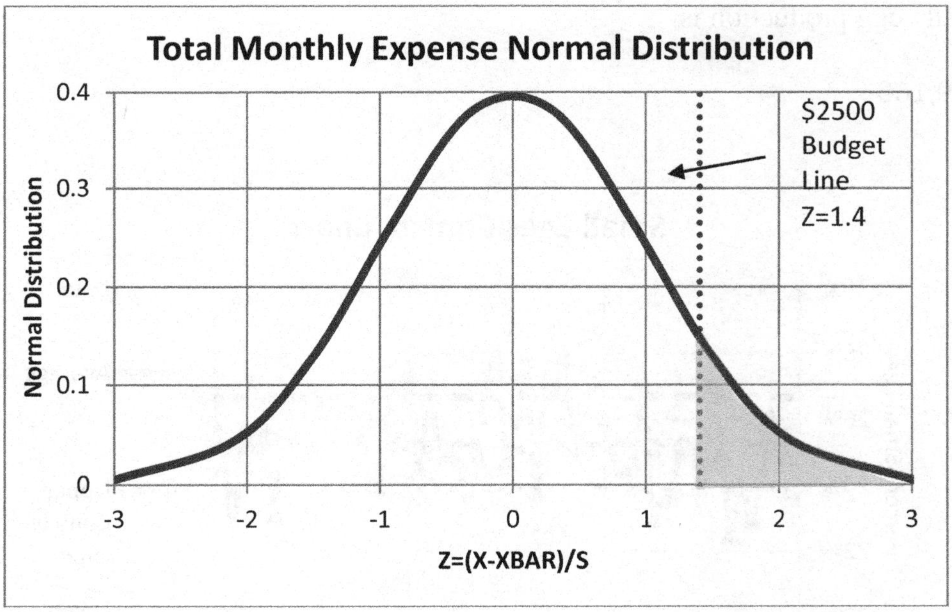

Figure 4-3-4 Total monthly expense normal distribution and 8.1% area over $2,500 budget

Chapter 4-4 Egg Farm

Process control charts can be a very useful tool for attribute data as well as to variable data. Visual defects on a product or open circuits on a printed circuit board are some examples of attribute data that can be controlled by using process control charts called P-charts. In this example, an egg farm is considered. Eggs below a certain size are categorized as small and cannot be sold to the customer. Egg farm manager tracks the number of small eggs every day. The data for 60 days is plotted as a process control P-charts as shown in Figure 4-4-1.

Figure 4-4-1 is based on the percentage of small eggs identified in daily production.

$$P = \frac{\#\ of\ small\ eggs\ in\ one\ day's\ production}{one\ day's\ production} \times 100 \qquad (4\text{-}4\text{-}1)$$

P is plotted for each day for 60 days. The average value is obtained from

$$\bar{P} = \frac{\sum_1^{60} P}{\sum_1^{60} N} = 25.3\% \qquad (4\text{-}4\text{-}2)$$

The average daily egg production is

$$\bar{N} = \sum_1^{60} N = 9{,}163 \qquad (4\text{-}4\text{-}3)$$

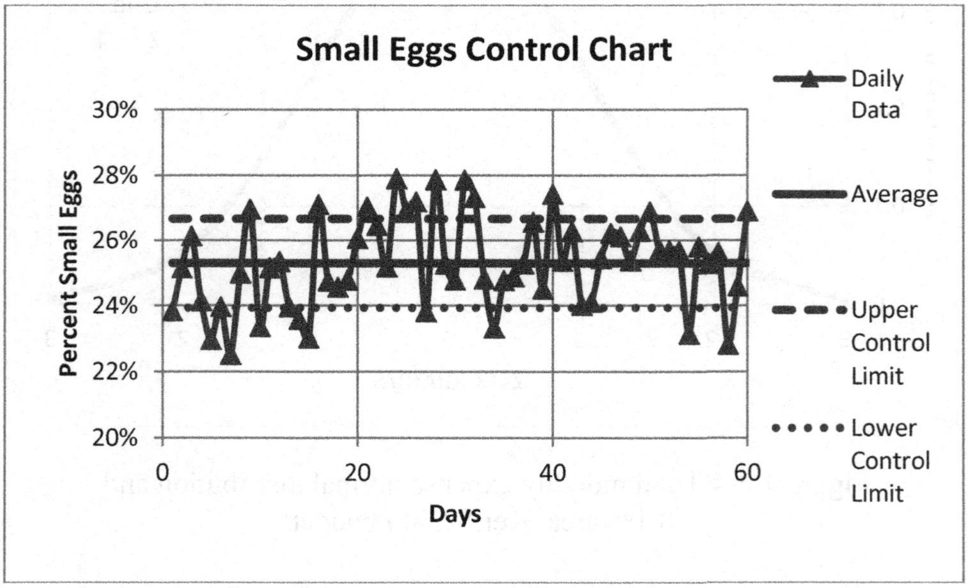

Figure 4-4-1 Small eggs control P-chart

Upper and lower control limits for the P-chart are calculated assuming a binomial distribution, see Reference [8].

$$Upper\ Control\ Limit = \bar{P} + 3\sqrt{\frac{\bar{P}(1-\bar{P})}{\bar{N}}} = 26.7\% \qquad (4\text{-}4\text{-}4)$$

$$Lower\ Control\ Limit = \bar{P} - 3\sqrt{\frac{\bar{P}(1-\bar{P})}{\bar{N}}} = 23.9\% \quad (4\text{-}4\text{-}5)$$

In Figure 4-4-1, data for rejected small eggs show that quite a few points are outside the upper and lower control limits and the average loss is high. The farmer decides to try an enhanced diet for his chicken to lower the small egg losses and obtains the following data shown in Figure 4-4-2 for sixty days after the new diet has settled down.

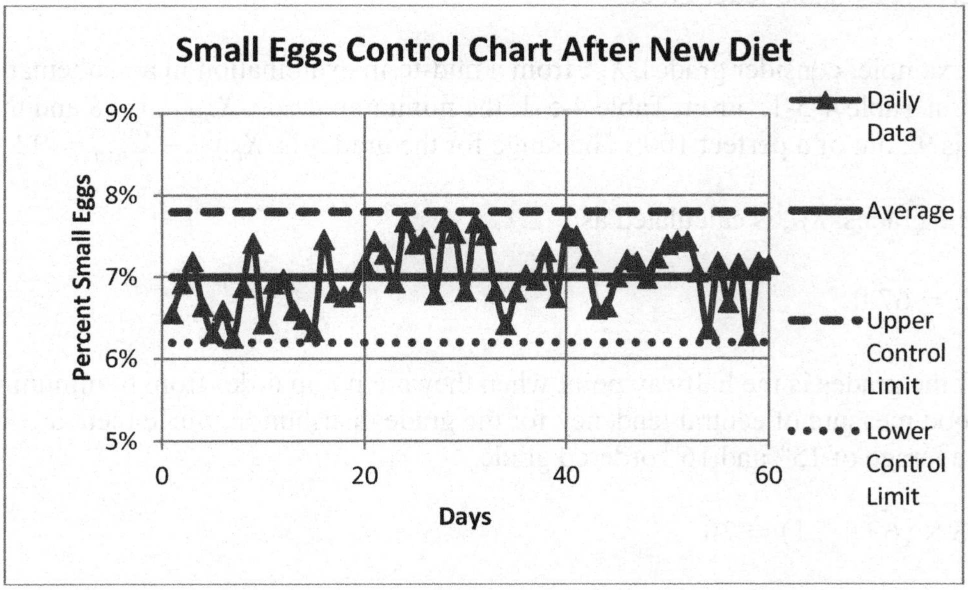

Figure 4-4-2 Small eggs control P-chart after new diet

The average value of rejected eggs drops to 7%. Also the dispersion of the rejected egg distribution shrinks by 1.7 fold. Every data point is within the upper and lower control limits.

Standard deviation, σ , for a binomial distribution, see Reference [8], is defined as

$$\sigma = \sqrt{\bar{N}\bar{P}(1-\bar{P})} \quad (4\text{-}4\text{-}6)$$

Before the diet implementation, dispersion of rejected egg distribution is

$$\sigma_{before\ diet} = \sqrt{9{,}163 \times \left(\frac{25.3}{100}\right) \times \left[1 - \left(\frac{25.3}{100}\right)\right]} = 41.6\ eggs \quad (4\text{-}4\text{-}7)$$

After the diet implementation, dispersion of rejected egg distribution is reduced to 24.4 eggs.

$$\sigma_{after\ diet} = \sqrt{9{,}156 \times \left(\frac{7}{100}\right) \times \left[1 - \left(\frac{7}{100}\right)\right]} = 24.4\ eggs \quad (4\text{-}4\text{-}8)$$

Chapter 4-5 Statistical Analysis of Mid-Term Grades

Grade analysis for a class can start with simple statistical measures such as minimum, maximum, range, mean, median, and standard deviation for grades. The grade analysis can get more involved by looking into grade frequency distribution and grade cumulative frequency distribution. Sometimes numerical grade mean is below the desired letter grade mean and the numerical grades are shifted up to give reasonable letter grades.

As an example, consider grades, X_i , from a mid-term examination in a mathematics class of 30 students given in Table 4-5-1. From Table 4-5-1, the minimum grade, X_{min} , is 28 and the maximum grade, X_{max} , is 92 out of a perfect 100. The range for the grades is $X_{max} - X_{min} = 92 - 28 = 64$.

The mean of the grades, X_i , is calculated as

$$\overline{X} = \frac{1}{30}\sum_{i=1}^{30} X_i = 67.8 \qquad (4\text{-}5\text{-}1)$$

The median of the grades is the halfway point when they are put in order from minimum to maximum. Median is a good measure of central tendency for the grade distribution. Since there are 30 students, median is the average of 15th and 16th ordered grade.

$$Median = 0.5 \times (69 + 71) = 70 \qquad (4\text{-}5\text{-}2)$$

Since the mean and the median of the grades are close to each other, the grade frequency distribution is close to being symmetrical as shown in Figure 4-5-1.

Student	Mid-Term Numerical Grade
1	61
2	75
3	56
4	87
5	74
6	66
7	92
8	28
9	69
10	75
11	85
12	77
13	63
14	52
15	61
16	73
17	82
18	71
19	73
20	67
21	65
22	58
23	56
24	79
25	76
26	78
27	75
28	45
29	68
30	47

Table 4-5-1 Mid-term grades for each student

Student grades given in Table 4-5-1 can be put into grade interval bins and a frequency distribution of grades can be formed as shown in Figure 4-5-1. Figure 4-5-1 also shows the cumulative frequency distribution of 30 students. The cumulative frequency distribution shows how many students are above or below a certain grade. In this case half the students got grades above the mean grade of 67.8 and half of them got below.

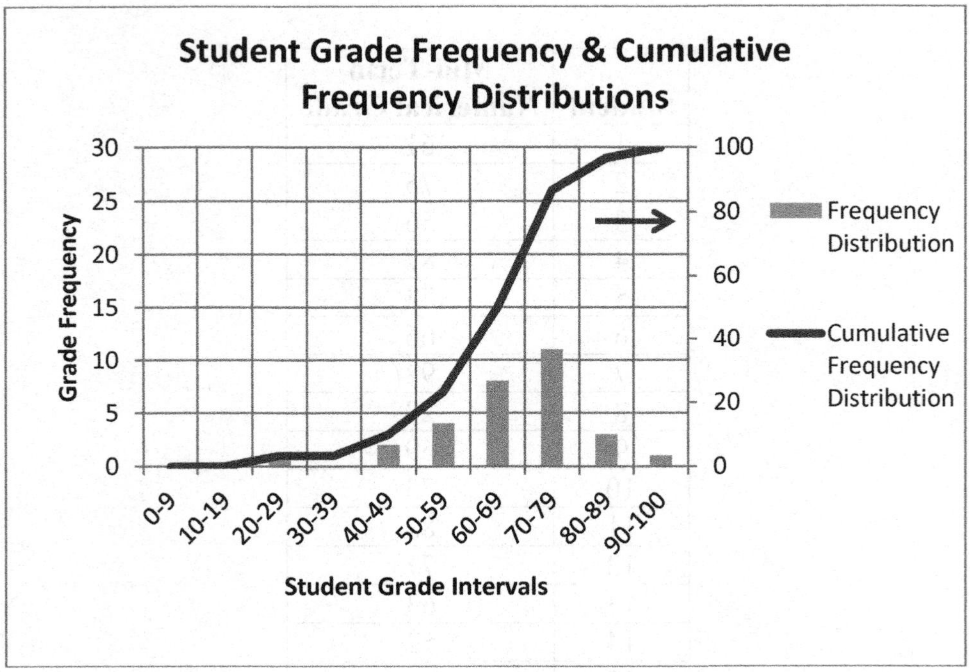

Figure 4-5-1 Student grade frequency and cumulative frequency distributions

The frequency distribution of student grades looks similar to a normal probability distribution function given as follows.

$$F(X) = \frac{1}{\sigma\sqrt{2\pi}} exp\left[-\frac{(X-\bar{X})^2}{2\sigma^2}\right] \quad (4\text{-}5\text{-}3)$$

where σ is the standard deviation of the grade distribution given by

$$\sigma = \sqrt{\frac{\sum_{i=1}^{30}(X_i-\bar{X})^2}{30}} = 13.4 \quad (4\text{-}5\text{-}4)$$

The normal probability distribution given in Eq. (4-5-3) and frequency distribution for the grades are shown in Figure 4-5-2. The normal probability distribution function in Eq. (4-5-3) represents the grade frequency distribution quite well with a standard deviation of 13.4 which represents the spread of the distribution.

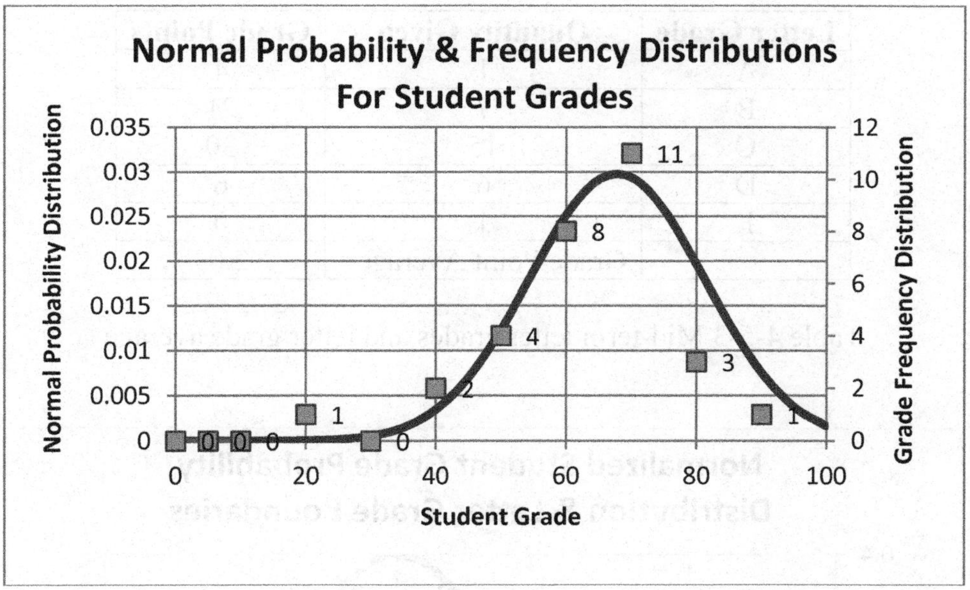

Figure 4-5-2 Normal probability and frequency distributions for student grades

It is customary to normalize the probability function given by Eq. (4-5-3) so that the area from $-\infty\ to + \infty$ is unity. To normalize the probability function a new variable is introduced.

$$Z = \frac{X - \bar{X}}{\sigma} \tag{4-5-5}$$

The normalized probability distribution function becomes

$$P(Z) = \frac{1}{\sqrt{2\pi}} exp\left(-\frac{Z^2}{2}\right) \tag{4-5-6}$$

The professor has a standard way of converting numerical grades to the required letter grades as given in Table 4-5-2. In this case the maximum grade of 100 is bounded by a Z value of 2.4 and the minimum grade of 0 is bounded by a Z value of -5.06.

Letter Grade	Numerical Grade Minimum X Value	Numerical Grade Maximum X Value
A	$\bar{X} + 1.8 \times \sigma$	$\bar{X} + 2.4 \times \sigma$
B	$\bar{X} + 0.6 \times \sigma$	$\bar{X} + 1.8 \times \sigma$
C	$\bar{X} - 0.6 \times \sigma$	$\bar{X} + 0.6 \times \sigma$
D	$\bar{X} - 1.8 \times \sigma$	$\bar{X} - 0.6 \times \sigma$
F	$\bar{X} - 5.06 \times \sigma$	$\bar{X} - 1.8 \times \sigma$

Table 4-5-2 Standard way of converting numerical grades to letter grades

Normalized student grade probability distribution, Eq. (4-5-6), and letter grade boundaries are shown in Figure 4-5-3. With the conditions given in Table 4-5-2, letter grades and letter grade point average are given in Table 4-5-3.

Letter Grade	Quantity Given	Grade Points
A	1	4
B	7	21
C	15	30
D	6	6
F	1	0
	Grade Point Average=	2.03

Table 4-5-3 Mid-term letter grades and letter grade average

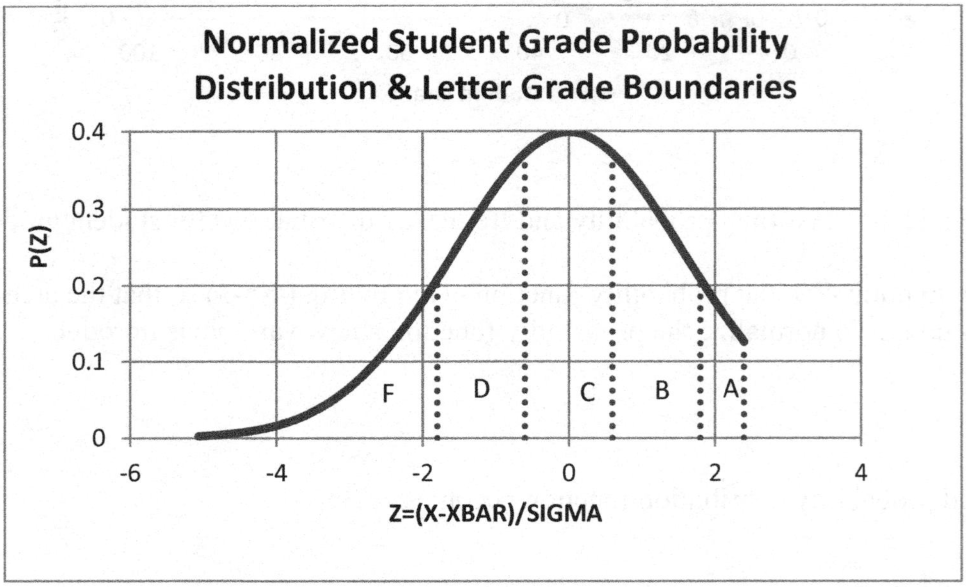

Figure 4-5-3 Normalized student grade probability distribution and letter grade boundaries

The professor did not like the low letter grade average and decided to shift the letter grade boundaries downwards by 10 points as shown in Table 4-5-4.

Letter Grade	Numerical Grade Minimum X Value	Numerical Grade Maximum X Value
A	$\bar{X} + 1.05 \times \sigma$	$\bar{X} + 2.4 \times \sigma$
B	$\bar{X} - 0.15 \times \sigma$	$\bar{X} + 1.05 \times \sigma$
C	$\bar{X} - 1.35 \times \sigma$	$\bar{X} - 0.15 \times \sigma$
D	$\bar{X} - 2.55 \times \sigma$	$\bar{X} - 1.35 \times \sigma$
F	$\bar{X} - 5.06 \times \sigma$	$\bar{X} - 2.55 \times \sigma$

Table 4-5-4 10 point downward shifted letter grade boundaries

By shifting the letter grade boundaries downwards by 10 points, the professor obtained the letter grades given in Table 4-5-5. Normalized student grade probability distribution, Eq. (4-5-6), and letter grade boundaries are shown in Figure 4-5-4.

Letter Grade	Quantity Given	Grade Points
A	4	16
B	15	45
C	8	16
D	2	2
F	1	0
	Grade Point Average=	2.63

Table 4-5-5 Mid-term letter grades and letter grade average after shifting the letter grade boundaries downward by 10 points

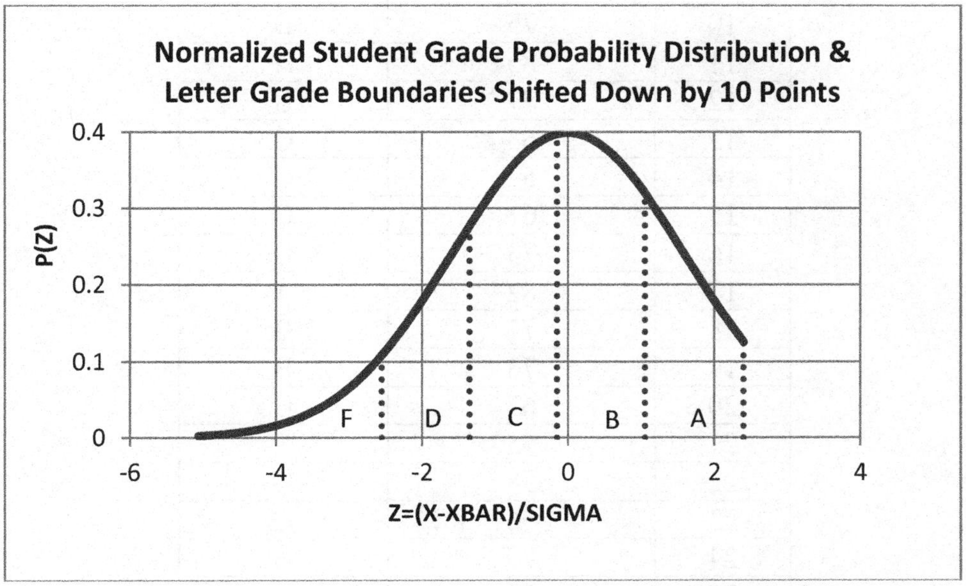

Figure 4-5-4 Normalized student grade probability distribution and letter grade boundaries shifted downward by 10 points

The professor was satisfied with the final letter grade point average of 2.63 and provided the following mid-term results to his students given in Table 4-5-6.

Student	Mid-Term Numerical Grade	Mid-Term Letter Grade
1	61	C
2	75	B
3	56	C
4	87	A
5	74	B
6	66	B
7	92	A
8	28	F
9	69	B
10	75	B
11	85	A
12	77	B
13	63	C
14	52	C
15	61	C
16	73	B
17	82	A
18	71	B
19	73	B
20	67	B
21	65	C
22	58	C
23	56	C
24	79	B
25	76	B
26	78	B
27	75	B
28	45	D
29	68	B
30	47	D

Table 4-5-6 Finalized mid-term letter grades for each student

Chapter 4-6 Addition Rule of Probability Examples

The probability of one event can be added to the probability of another event, if the events are mutually exclusive and if the total probability of both events occurring is desired. For example a deck of 52 cards have four 2's and twelve faces in it. If we are interested in the probability of drawing a 2 card or a face card from the full deck, the probabilities of these two mutually exclusive events can be added.

$$P(2\ or\ face) = P(2) + P(face) \tag{4-6-1}$$

However, the probability of pulling a 2 card is

$$P(2) = \frac{4}{52} = 0.08 \tag{4-6-2}$$

and the probability of pulling a face card is

$$P(face) = \frac{12}{52} = 0.23 \tag{4-6-3}$$

Then the probability of pulling a 2 card or a face card is

$$P(2\ or\ face) = 0.08 + 0.23 = 0.31 \tag{4-6-4}$$

The probability of one event can also be added to the probability of another event, if the events are mutually non-exclusive. If the total probability for two mutually non-exclusive events is desired, then Eq. (4-6-1) has to be updated as follows.

$$P(event\ 1\ or\ event\ 2) = \\ P(event\ 1) + P(event\ 2) - P(event\ 1\ and\ event\ 2\ occurring\ together) \tag{4-6-5}$$

For example, if the probability of pulling a heart or a face card is desired from a deck of 52 cards, these two events are not mutually exclusive because there are three face cards that are hearts at the same time.

The probability of pulling a heart is

$$P(heart) = \frac{13}{52} = 0.25 \tag{4-6-6}$$

The probability of pulling a face card is

$$P(face) = \frac{12}{52} = 0.23 \tag{4-6-7}$$

The probability of pulling a face card that is also a heart is

$$P(face\ and\ heart) = \frac{3}{52} = 0.06 \tag{4-6-8}$$

Then using Eq. (4-6-5), the probability of pulling a face card or a heart is

$$P(face\ or\ heart) = 0.23 + 0.25 - 0.06 = 0.42 \quad (4\text{-}6\text{-}9)$$

In another mutually non-exclusive example, consider blood tests for 420 patients at a health clinic. 41 of the blood test results came out with high glucose levels and 36 of the blood test results came out with high cholesterol levels. However 16 of them had both high glucose and cholesterol levels. What percentage of patients had no problem with high glucose and cholesterol levels?

$$P(high\ glucose) = \frac{41}{420} = 0.098 \quad (4\text{-}6\text{-}10)$$

and

$$P(high\ cholestrol) = \frac{36}{420} = 0.086 \quad (4\text{-}6\text{-}11)$$

The probability of having both high glucose and high cholesterol levels is

$$P(high\ glucose\ and\ cholestrol) = \frac{16}{420} = 0.038 \quad (4\text{-}6\text{-}12)$$

Then using Eq. (4-6-5), the percentage of patients having either high glucose or high cholesterol levels is 14.6% as shown below.

$$P(high\ glucose\ or\ high\ cholestrol) = 0.098 + 0.086 - 0.038 = 0.146 \quad (4\text{-}6\text{-}13)$$

The percentage of healthy patients is

$$100\% - 14.6\% = 85.4\% \quad (4\text{-}6\text{-}14)$$

Chapter 4-7 Joint Probability Examples

If there are two or more events that are independent, then the joint probability of all the events results from the multiplication of individual probability of each event.

$$P(event_1, event_2, \ldots event_N) = P(event_1) \times P(event_2) \times \ldots P(event_N) \quad (4\text{-}7\text{-}1)$$

For example, the joint probability of pulling three aces in succession out of a 52-card deck can be calculated by applying Eq. (4-7-1). Each pull is an independent event and the joint probability is calculated as follows.

$$P(Ace_1, Ace_2, Ace_3) = \left(\frac{4}{52}\right) \times \left(\frac{3}{51}\right) \times \left(\frac{2}{50}\right) = 0.00018 \quad (4\text{-}7\text{-}2)$$

The odds of achieving such a feat are $\frac{1}{0.00018} = 5{,}525$ to 1.

As a second example, let us try to find out the probability of getting exactly two tails in four tosses of a coin. There are four independent events and the probabilities of each event's outcome are shown in Table 4-7-1.

Event 1 Possible Results	Event 1 Probability	Event 2 Possible Results	Event 2 Probability	Event 3 Possible Results	Event 3 Probability	Event 4 possible Results	Event 4 Probability
Head$_1$	0.5	Head$_1$ and Head$_2$	0.25	Head$_1$ and Head$_2$ and Head$_3$	0.125	Head$_1$ and Head$_2$ and Head$_3$ and Head$_4$	0.0625
Tail$_1$	0.5	Head$_1$ and Tail$_2$	0.25	Head$_1$ and Head$_2$ and Tail$_3$	0.125	Head$_1$ and Head$_2$ and Head$_3$ and Tail$_4$	0.0625
		Tail$_1$ and Head$_2$	0.25	Head$_1$ and Tail$_2$ and Head$_3$	0.125	Head$_1$ and Head$_2$ and Tail$_3$ and Head$_4$	0.0625
		Tail$_1$ and Tail$_2$	0.25	Head$_1$ and Tail$_2$ and Tail$_3$	0.125	***Head$_1$ and Head$_2$ and Tail$_3$ and Tail$_4$***	***0.0625***
				Tail$_1$ and Head$_2$ and Head$_3$	0.125	Head$_1$ and Tail$_2$ and Head3 and Head$_4$	0.0625
				Tail$_1$ and Head$_2$ and Tail$_3$	0.125	***Head$_1$ and Tail$_2$ and Head$_3$ and Tail$_4$***	***0.0625***
				Tail$_1$ and Tail$_2$ and Head$_3$	0.125	***Head$_1$ and Tail$_2$ and Tail$_3$ and Head$_4$***	***0.0625***
				Tail$_1$ and Tail$_2$ and Tail$_3$	0.125	Head$_1$ and Tail$_2$ and Tail$_3$and Tail$_4$	0.0625
						Tail$_1$ and Head$_2$ and Head$_3$ and Head$_4$	0.0625
						Tail$_1$ and Head$_2$ and Head$_3$ and Tail$_4$	***0.0625***
						Tail$_1$ and Head$_2$ and Tail$_3$ and Head$_4$	***0.0625***
						Tail$_1$ and Head$_2$ and Tail$_3$ and Tail$_4$	0.0625
						Tail$_1$ and Tail$_2$ and Head$_3$ and Head$_4$	***0.0625***
						Tail$_1$ and Tail$_2$ and Head$_3$ and Tail$_4$	0.0625
						Tail$_1$ and Tail$_2$ and Tail$_3$ and Head$_4$	0.0625
						Tail$_1$ and Tail$_2$ and Tail$_3$ and Tail$_4$	0.0625

Table 4-7-1 Four coin toss events - possible results and probabilities

Since every coin toss is independent, a typical joint probability, event 4 in Table 4-7-1, is obtained as follows.

$$P(Tail_1, Tail_2, Head_3, Tail_4) = 0.5 \times 0.5 \times 0.5 \times 0.5 = 0.0625 \qquad (4\text{-}7\text{-}3)$$

Exactly two tails in tosses occurs six times as shown in Table 4-7-1. These results are highlighted. Therefore the probability of exactly getting two tails in four tosses is $6 \times 0.0625 = 0.375$

Chapter 4-8 Binomial Distribution Examples

If there are independent events that have only two possible outcomes such as "good" or "bad", "success" or "failure", "heads" or "tails", and the probability of each outcome is a constant, then the binomial distribution can be used to define the probability of an outcome after a certain number of repeated independent events, see Reference [8].

$$P(X\ good\ outcomes\ in\ N\ events) = \frac{N!}{X!(N-X)!} PG^X PB^{N-X} \qquad (4\text{-}8\text{-}1)$$

where PG is the probability of good outcomes and is a constant and PB is the probability of bad outcomes and is a constant.

For example, if a coin is tossed 4 times what is the probability of getting 2 tails? Let us assume that $PG = P(tails) = 0.5$. Then from Eq. (4-8-1)

$$P(2\ tails\ in\ 4\ tosses) = \frac{4!}{2!(4-2)!} 0.5^2 0.5^{4-2} = \frac{4\times3\times2\times1}{2\times1\times2\times1} \times 0.25 \times 0.25 = 0.375 \qquad (4\text{-}8\text{-}2)$$

Probability of every possible outcome, namely getting 1 tail, 2 tails, 3 tails, and 4 tails can be calculated using Eq. (4-8-1). Remember that 0!=1 by definition. Since $PG = PB$ in this case, the probabilities of getting tails are equal to the probabilities of getting heads. Results for this case are plotted in Figure 4-8-1.

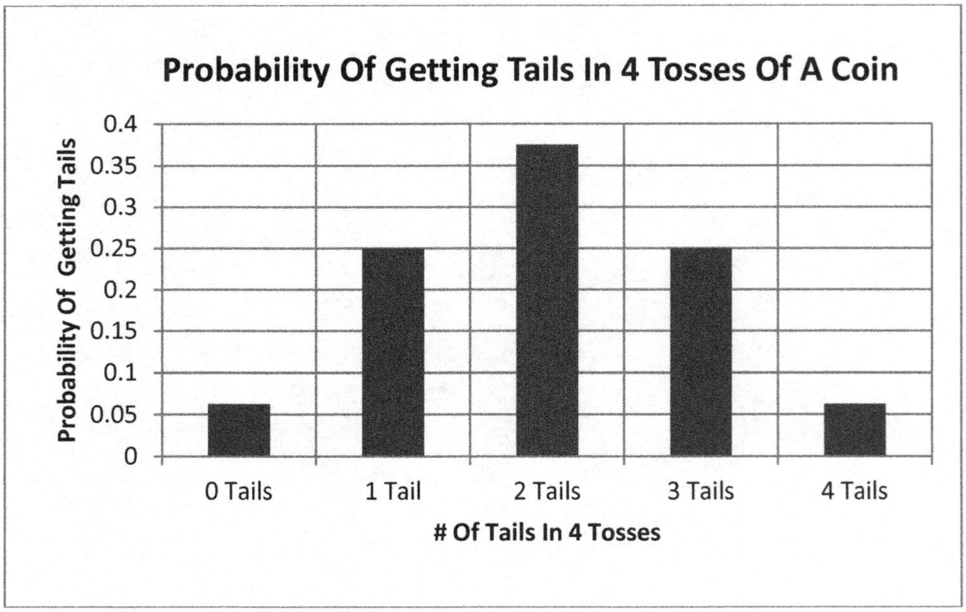

Figure 4-8-1 Probability of getting tails in 4 tosses of a coin

The mean of a binomial distribution is given by

$$Mean = PG \times N \qquad (4\text{-}8\text{-}3)$$

which is $0.5 \times 4 = 2$ in this case. The standard deviation of a binomial distribution is given by

$$\sigma = \sqrt{PG \times (1 - PG) \times N} \tag{4-8-4}$$

which is $\sqrt{0.5 \times 0.5 \times 4} = 1$ in this case.

Let us consider another example where a committee of 10 has to be formed in a city from its voting population. The city is composed of 40% conservatives, 25% moderates, and 35% liberals. If the committee members are selected at random, what is the probability that none are conservative? Using Eq. (4-8-1), the following result is obtained.

$$P(0\ Conservatives\ out\ of\ 10) = \frac{10!}{0!(10-0)!} 0.4^0 0.6^{10} = 0.006 \tag{4-8-5}$$

Probability of selecting different number of conservatives to the committee of 10 can be calculated by using Eq. (4-8-1). Probability of selecting different number of conservatives, C's, to the committee is shown in Figure 4-8-2. Highest probability is 25% and it is to select 4 conservative members to the committee.

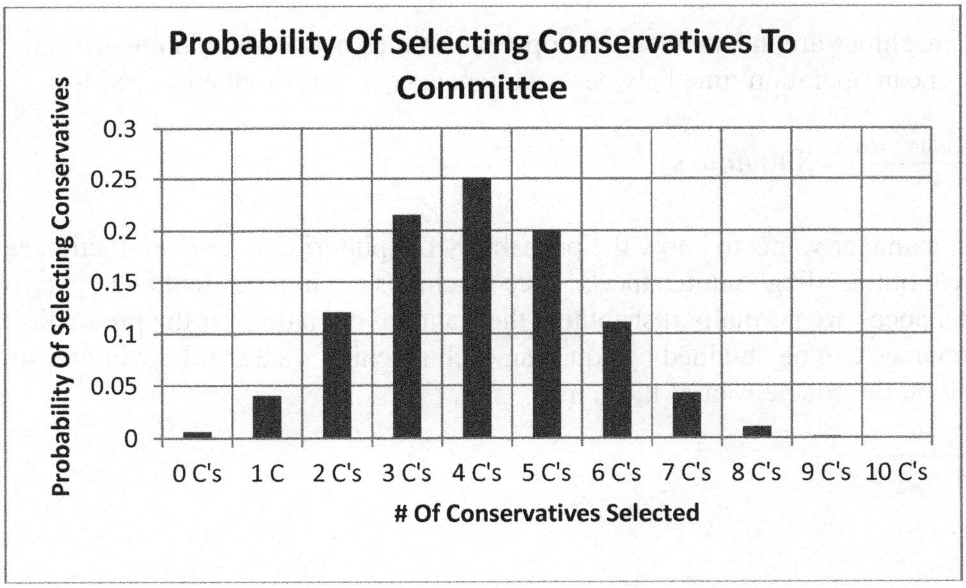

Figure 4-8-2 Probability of selecting different number of conservatives to a committee of 10

Chapter 4-9 Sputtering Equipment

Let us consider a production process in which one of the production cells have three sputtering equipment that perform the same process. Some of the equipment is old and some are new. All three sputtering equipment are sensitive to their maintenance schedules. Machine operation times between maintenances are recorded and statistical results are given in Table 4-9-1.

Sputtering Equipment	**Mean machine operation times between maintenances, hrs**	**Standard deviation of machine operation times between maintenances, hrs**
# 1	300	30
# 2	360	42
# 3	510	20

Table 4-9-1 Statistics for machine operation times between maintenances for three sputtering equipment

Assuming that machines are independent and operation times between maintenances are normally distributed, the mean operation time between maintenances in this production cell is

$$Mean = \frac{(300+360+510)}{3} = 390\ hours \quad (4\text{-}9\text{-}1)$$

The production manager wants to know the probability of sputtering in this production cell 400 hours continuously without needing maintenance? Since machines are independent and operation times between maintenances are normally distributed, the standard deviation for the production cell time between maintenance can be obtained by squaring each machine's standard deviation, summing the squares, and taking the square root of the sum.

$$\sigma_{cell} = \sqrt{\sigma_{\#1}^2 + \sigma_{\#2}^2 + \sigma_{\#3}^2} \quad (4\text{-}9\text{-}2)$$

or

$$\sigma_{cell} = \sqrt{30^2 + 42^2 + 20^2} = 55.35 \quad (4\text{-}9\text{-}3)$$

The probability of sputtering in this production cell 400 hours continuously without needing maintenance is the area when $Z = \frac{X-\bar{X}}{\sigma_{cell}} > \frac{400-390}{55.35} = 0.18$ on a normalized probability distribution function as shown in Figure 4-9-1. The probability can be found by determining the area under the normalized probability distribution function between $Z = 0.18$ and $Z \to \infty$. Values of the integral for the normalized probability distribution function can be found in any statistics textbook, see Reference [15].

$$P(Z > 0.18) = \frac{1}{\sqrt{2\pi}} \int_{0.18}^{\infty} \exp(-0.5 * Z^2)\, dZ = 0.43 \quad (4\text{-}9\text{-}4)$$

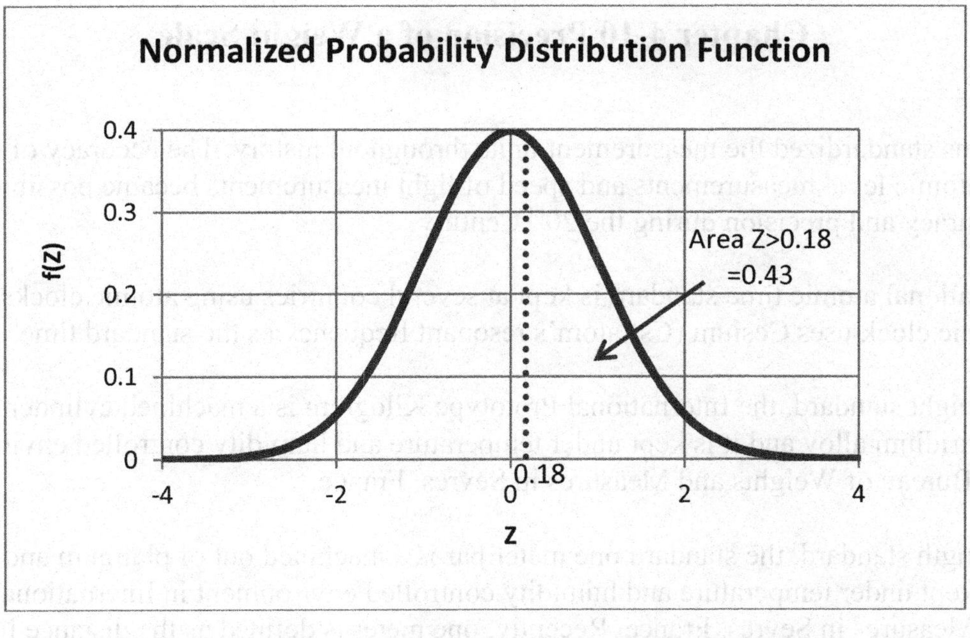

Figure 4-9-1 Normalized probability distribution function

Chapter 4-10 Precision of a Weight Scale

Humans standardized the measurement units throughout history. The accuracy of the standards increased as atomic level measurements and speed of light measurements became possible with high degree of accuracy and precision during the 20th Century.

International atomic time standard is kept at several countries using atomic clocks. The most common atomic clock uses Cesium (Cs) atom's resonant frequency as the standard time.

For weight standard, the International Prototype Kilogram is a machined cylinder made out of platinum and iridium alloy and it is kept under temperature and humidity controlled environment in International Bureau of Weights and Measures in Sevres, France.

For length standard, the standard one meter bar is a machined out of platinum and iridium alloy and it is also kept under temperature and humidity controlled environment in International Bureau of Weights and Measures in Sevres, France. Recently, one meter is defined as the distance light travels in vacuum in $\frac{1}{299{,}792{,}458}$ seconds.

The standard value of a measurement is called the "true value" of that measurement. However, in our daily lives our watches, scales, and measuring tapes deviate from true values.

The differences between average measurements of time, weight, length, etc. and their true values are phrased as "accuracy" of the watch, scale, measuring tape, etc. in question. A very accurate watch tracks the time same as the standard atomic clock and always provides the correct time to us.

There is another phrase that is important in measurement and it is "precision". Precision of a measurement gives an indication of repeatability and reproducibility of that measurement.

If you weigh yourself on a scale and read 161 pounds and weigh again immediately and read 163 pounds, there is a repeatability issue with the scale or with your readings and the scale is not precise.

If you weigh yourself on a scale and read 161 pounds and weigh again immediately on another similar scale and read 162.5 pounds, there is a reproducibility issue with these scales or with your readings and the scales are not precise.

Precision of a measurement is defined by the standard deviation of measurement errors as follows.

$$\sigma_{measurement\ error} = \sqrt{\sigma^2_{repeatability} + \sigma^2_{reproducibility}} \qquad (4\text{-}10\text{-}1)$$

If only one gauge and one person taking the measurements are considered than Eq. (4-10-1) becomes

$$\sigma_{measurement\ error} = \sigma_{repeatability} \qquad (4\text{-}10\text{-}2)$$

Consider a person who has a 110 pounds true weight. This person measures himself 28 times repeatedly on the same scale and gets the results given in Table 4-10-1.

Measurement #	1	2	3	4	5	6	7
Weight, lbs	109	108	108.5	108	109.5	109	108
Measurement #	**8**	**9**	**10**	**11**	**12**	**13**	**14**
Weight, lbs	108.5	107	107	108	109	108.5	108
Measurement #	**15**	**16**	**17**	**18**	**19**	**20**	**21**
Weight, lbs	107	109	107.5	107.5	107	107.5	107
Measurement #	**22**	**23**	**24**	**25**	**26**	**27**	**28**
Weight, lbs	109	107.5	107.5	108	107.5	108	108

Table 4-10-1 Weight measurements by one person on the same scale

The statistics for weight measurements given in Table 4-10-1 are given below.

$Average = 108\ lbs$
$Minimum = 107\ lbs$
$Maximum = 109.5\ lbs$
$Range = 2.5\ lbs$
$Standard\ Deviation = 0.73\ lbs$

The average measurement is 2 pounds less than the true value. The precision of the measurements is

$$\bar{X} \pm 3\sigma = 108 \pm 2.2\ lbs \quad (4\text{-}10\text{-}3)$$

If there is a requirement for the weight tolerance of $\pm 5\ lbs$, then measurement precision, $6\sigma_{measurement\ error}$, divided by total tolerance should be less than 10% in order to be able to use this scale and it's reader for this purpose.

$$\frac{6\sigma_{measurement\ error}}{Total\ Tolerance} \times 100 < 10\% \quad (4\text{-}10\text{-}4)$$

In this case $\frac{6\sigma_{measurement\ error}}{Total\ Tolerance} \times 100 = 44\%$, so this measurement system cannot be used with a precision satisfying $\pm 5\ lbs$ tolerance. We have to find a weight measurement system, scale plus the reader, which will have measurement error standard deviation, using Eq. (4-10-4), of $\sigma_{measurement\ error} \leq 0.167$.

Chapter 4-11 Example of a 2x2 Design of Experiments

Let us perform a simple experiment with two factors (variables) and each factor having two levels. This is called a 2x2 full factorial design of experiment, see Reference [18].

We would like to investigate the effects of an additive (factor #1) and cure temperature (factor #2) to an adhesive that is going to be applied to a joint. The dependent variable of our tests is the pull strength of two components that is being joined together with this adhesive. There can be other factors that can affect this adhesion process such as cure time, uniformity of mixing the additive to adhesive, amount of adhesive applied to the joint, adhesive shelf life, additive shelf life, etc. In this experiment only two of the significant factors mentioned above are going to be considered.

Each factor has two levels. Tests involving factor #1 has either no additive or the same amount of additive. Tests involving factor #2 is being cured either at room temperature (20 C) for an hour or at 100 C for an hour. 2x2 full factorial experiment matrix is given in Table 4-11-1.

Experiment #	Additive Factor	Cure Temperature Factor
1	NO	20 C
2	NO	100 C
3	YES	20 C
4	YES	100 C

Table 4-11-1 2x2 full factorial experiment matrix

Five tests are performed for each experiment that is defined in Table 4-11-1 and results of pull strength for each test are given in Table 4-11-2.

Experiment #	Test #1 Pull Strength, kg	Test #2 Pull Strength, kg	Test #3 Pull Strength, kg	Test #4 Pull Strength, kg	Test #5 Pull Strength, kg
1	10.1	11.2	10.4	9.5	10.3
2	25.8	25.1	26.1	26.2	26.9
3	23.2	24	23.8	23.7	24.1
4	15.1	14.6	14.8	14.6	15.3

Table 4-11-2 Results of pull strength for each test

Let us analyze the results of these tests by taking the averages of the test results for each experiment. The test averages are shown in Table 4-11-3.

Experiment #	Additive Factor	Cure Temperature Factor	Test Averages, kg
1	NO	20 C	10.30
2	NO	100 C	26.02
3	YES	20 C	23.76
4	YES	100 C	14.88

Table 4-11-3 Test averages

Test averages for each experiment can be plotted, Figure 4-11-1, to see how the results are behaving. When an additive is added to the adhesive the pull strength decreases with increasing cure temperature. On the other hand, with no additive, the pull strength increases with increasing cure temperature. The "NO Additive" and the "YES Additive" curves cross each other as the cure temperature increases from 20 C to 100 C. This means that there is an interaction between the additive and the cure temperature.

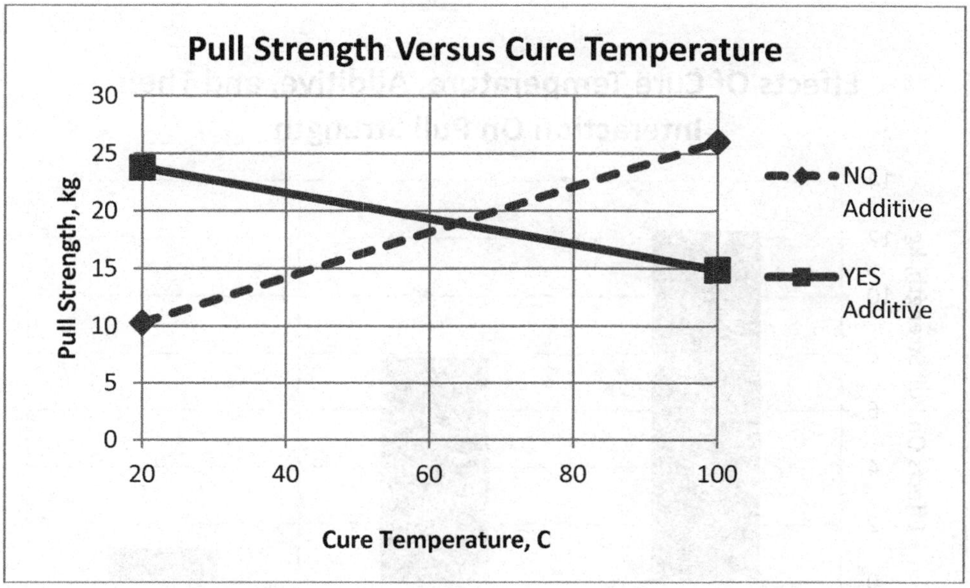

Figure 4-11-1 Pull strength versus cure temperature with no additive and with additive

Next we would like to find out the effects of each independent variable and also their interaction on the pull strength.

For the additive effect: The averages of Experiment #1, 10.30 kg, and Experiment #2, 26.02 kg, are averaged to give 18.16 kg and the averages of Experiment #3, 23.76 kg, and Experiment #4, 14.88 kg, are averaged to give 19.32 kg. The additive effect is calculated as the difference between averages of "YES Additive" and "NO Additive" experiments, namely 19.32 kg – 18.16 kg = 1.16 kg. For the cure temperature effect: The averages of Experiment #1, 10.30 kg, and Experiment #3, 23.76 kg, are averaged to give 17.03 kg and the averages of Experiment #2, 26.02 kg, and Experiment #4, 14.88 kg, are averaged to give 24.89 kg. The cure temperature effect is calculated as the difference between averages of "20 C Cure Temperature" and "100 C Cure Temperature" experiments, namely 24.89 kg – 17.03 kg = 7.86 kg.

For the interaction effect between having either no additive or additive and the cure temperature: The averages of Experiment #1, 10.30 kg, and Experiment #4, 14.88 kg, are averaged to give 12.59 kg and the averages of Experiment #2, 26.02 kg, and Experiment #3, 23.76 kg, are averaged to give 24.89 kg. The interaction effect is calculated as the difference between averages of "NO Additive and 20 C Cure Temperature (Experiment #1) & YES Additive and 100 C Cure Temperature (Experiment #4)" and "NO Additive and 100 C Cure Temperature (Experiment #2) & YES Additive and 20 C Cure Temperature (Experiment #3)" experiments, namely 12.59 kg – 24.89 kg =-12.30 kg.

Absolute values of all the effects are plotted in Figure 4-11-2. Interaction has the highest effect on pull strength. Cure temperature has the second highest effect on pull strength, being 64% of interaction effect. Additive has the least effect on pull strength, being 9.4% of interaction effect. If we want the highest possible strength from the joint, we can either go with NO additive and 100 C cure temperature or go with YES additive and 20 C cure temperature. Always confirmation experiments should be performed after the best levels for factors are chosen.

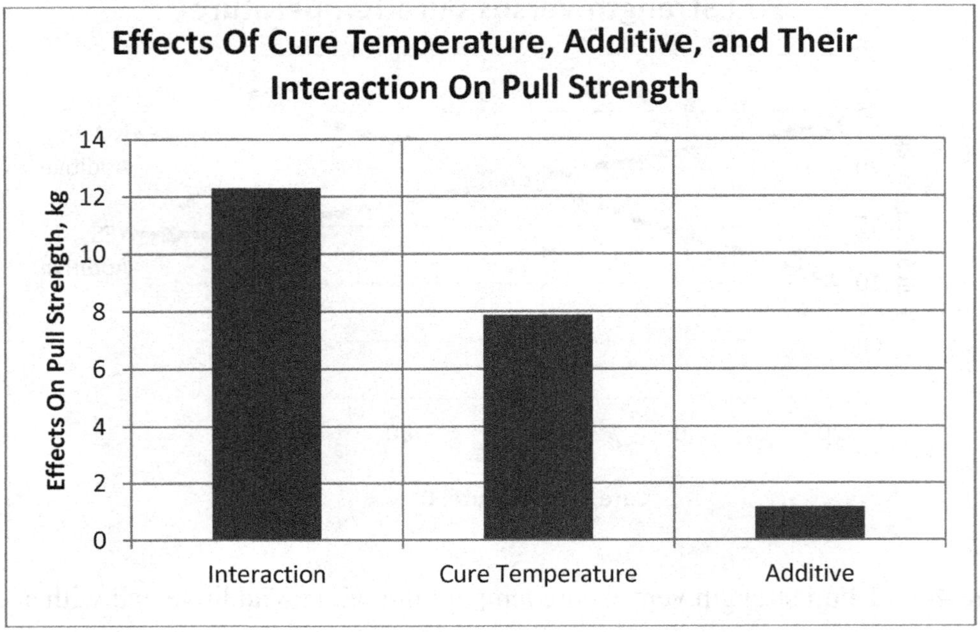

Figure 4-11-2 Effects of cure temperature, additive, and their interaction on pull strength

Chapter 4-12 Tolerancing

When two or more components are assembled together lengthwise, average length of the assembled system will be the sum of average length of each component.

Let us consider an opening between two walls which is 96.5 inches wide. We would like to install two beams end to end into this opening. The lengths of the beams come as 48 +- 1 inches. The lengths are distributed normally within the +-1 inch tolerance. Therefore with the normal distribution assumption, the standard deviation of the beam length is

$$\sigma_{beam} = \left(\frac{1}{3}\right) = 0.333\ inches \qquad (4\text{-}12\text{-}1)$$

What is the probability that the carpenter does not have to cut the end of a beam in order to be able to fit two of them end-to-end into the 96.5 inches opening between the walls?
The mean of the two assembled beams will be sum of the average length of each beam.

$$\bar{X}_{assembly} = \bar{X}_{beam\ \#1} + \bar{X}_{beam\ \#2} = 48 + 48 = 96\ inches \qquad (4\text{-}12\text{-}2)$$

The standard deviation of the assembled two beams will be obtained by squaring each beam's standard deviation, summing the results, and taking the square root of the sum.

$$\sigma_{assembly} = \sqrt{\sigma^2_{beam\ \#1} + \sigma^2_{beam\ \#2}} = \sqrt{0.333^2 + 0.333^2} = 0.471\ inches \qquad (4\text{-}12\text{-}3)$$

Area under the normal probability distribution function can be obtained from any standard statistics book, see Reference [8]. The normalized area is commonly given as a function of *Z* variable defined as follows.

$$Z = \frac{X_{wall\ opening} - \bar{X}_{assembly}}{\sigma_{assembly}} = \frac{96.5 - 96.0}{0.471} = 1.061 \qquad (4\text{-}12\text{-}4)$$

The probability that the two beams assembled end-to-end is going to have a length greater than 96.5 inches is 0.144 or 14.4% as shown in Figure 4-12-1. The odds are for the carpenter to shave the end of a beam in order to fit them into the opening is 1 out of 7 such assemblies, namely (1/0.144=6.94).

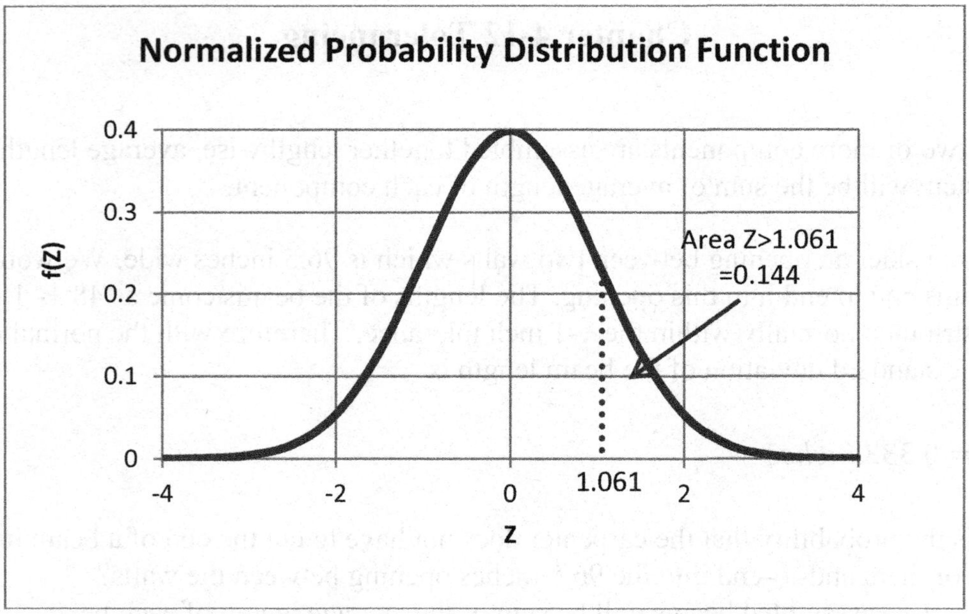

Figure 4-12-1 Area for Z>1.061 for normal probability distribution function

Chapter 4-13 Betting on Soccer and Baseball Games

If you try to know the outcomes of ten soccer games in a betting pool, how many different predictions are possible? A soccer game has three different outcomes, namely a team can win, lose or tie a game.

One single soccer game has three different predictions. Let us see how many different predictions are possible, if two soccer games are considered? Calling win, lose, or tie W_1, L_1, T_1 for game #1 and W_2, L_2, T_2 for game #2, the following nine different predictions or permutations are possible.

$W_1W_2, W_1L_2, W_1T_2, L_1W_2, L_1L_2, L_1T_2, T_1W_2, T_1L_2, T_1T_2$

If you bet one soccer game you have 3^1 different predictions or permutations. If you bet two soccer games you have 3^2 different predictions or permutations. If you bet three soccer games you have 3^3 different predictions or permutations and so on. So for ten soccer games the following number of different predictions or permutations is possible.

$3^{10} = 1{,}594{,}323$

In a baseball game the number of different predictions or permutations changes because there are two different predictions, win or lose, per game. For ten baseball games the following number of different predictions or permutations is possible.

$2^{10} = 1{,}024$

If one event can result in X_1 ways, a second event can result in X_2 ways, and a third event can result in X_3 ways and the N'th event can result in X_N ways, then the possible number of predictions or permutations for the N events is

$X_1X_2X_3 \ldots X_N$

For example, if you bet on five soccer games and five baseball games the possible number of predictions or permutations is

$3^5 \times 2^5 = 7{,}776$

Chapter 4-14 Percentage Combinations

A large department store wants to order all possible combinations of a teen fashion dress to a manufacturer. 70% of the dresses will be polyester and the rest cotton. Each material type will have four different colors, namely 40% black, 30% orange, 15% green, and 15% blue. Also each material type and color combination will have six different sizes with percentages given in Table 4-14-1.

Dress Size	Percentage Of Each Material Type And Color Combination
0	5
2	10
4	20
6	30
8	25
10	10

Table 4-14-1 Dress size percentages for each material type and color combination

What is the minimum order size that will include every possible combination as an integer?

Percentages of each component can be analyzed separately and then combined to find the minimum order quantity. For material type polyester is 70% and cotton is 30%. 10 is the maximum integer common factor for material type percentages.

For four different colors, namely 40% black, 30% orange, 15% green, and 15% blue, 5 is the maximum integer common factor for color percentages. For six different dress size percentages in Table 4-14-1, 5 is the maximum integer common factor.

Combination of material type, dress color, and dress size for minimum common factors gives

$$\frac{10}{100} \times \frac{5}{100} \times \frac{5}{100} = \frac{250}{1{,}000{,}000} = \frac{1}{4{,}000}$$

So the minimum order quantity for this teen fashion dress that will provide integer values for all combinations is 4,000.

For polyester material, dress quantities for every combination are given in Table 4-14-2.

Dress Quantities For Polyester Material, 70%					
Size, %	**Black, 40%**	**Orange, 30%**	**Green, 15%**	**Blue, 15%**	**Total**
0, 5%	56	42	21	21	140
2, 10%	112	84	42	42	280
4, 20%	224	168	84	84	560
6, 30%	336	252	126	126	840
8, 25%	280	210	105	105	700
10, 10%	112	84	42	42	280
			Total # Of Polyester Dresses		2,800

Table 4-14-2 Size and color combinations for polyester dresses

For example, in Table 4-14-2, size 0 black dress quantity in polyester material is obtained as follows:

$$\%\, polyester \times \%\, size \times \%\, color \times Total\ qty = specified\ dress\ qty$$
$$0.7 \times 0.05 \times 0.4 \times 4{,}000 = 56$$

For cotton material, dress quantities for every combination are given in Table 4-14-3.

Dress Quantities For Cotton Material, 30%					
Size, %	**Black, 40%**	**Orange, 30%**	**Green, 15%**	**Blue, 15%**	**Total**
0, 5%	24	18	9	9	60
2, 10%	48	36	18	18	120
4, 20%	96	72	36	36	240
6, 30%	144	108	54	54	360
8, 25%	120	90	45	45	300
10, 10%	48	36	18	18	120
			Total # Of Cotton Dresses		1,200

Table 4-14-3 Size and color combinations for cotton dresses

For example, in Table 4-14-3, size 6 orange dress quantity in cotton material is obtained as follows:

$$\%\, cotton \times \%\, size \times \%\, color \times Total\ qty = specified\ dress\ qty$$
$$0.3 \times 0.3 \times 0.3 \times 4{,}000 = 108$$

Chapter 4-15 Baseball Statistics

We start to learn statistics at a very early age by going to baseball games or by playing baseball. Every kid knows the batting average of his favorite players. The most popular offensive statistic is the batting average of a hitter.

$$Batting\ Average = \frac{\sum Hits}{\sum At\ Bat} \tag{4-15-1}$$

Sum of hits does not differentiate between singles, doubles, triples, and homeruns. When the at bat numbers are summed up, base on balls, hit by a pitch, sacrifice fly, etc. are excluded. Players with batting averages above 0.3 are considered to be good hitters. All time high for career batting average is Ty Cobb's at 0.366.

Data obtained for batting average of a player can be sorted out to provide subset statistics such as batting average against left-handed pitchers, batting average against a team, batting average every season, life-time batting average, etc.

Another important offensive statistic is the strikeout ratio. For a player, it is not good to have a high strikeout ratio. Strikeout ratio for a player is defined as

$$Strikeout\ Ratio = \frac{\sum Strikeout}{\sum At\ Bat} \tag{4-15-2}$$

Power of a hitter is measured by his slugging average. Slugging average weighs single hits by a factor of one, double hits by a factor of two, triple hits by a factor of three, and homeruns by a factor of four.

$$Slugging\ Average = \frac{1\times\sum Single + 2\times\sum Double + 3\times\sum Triple + 4\times\sum Homerun}{\sum At\ Bat} \tag{4-15-3}$$

Players with high slugging averages tend to have high strikeout ratios. All time high for career slugging average is Babe Ruth's at 0.690. There are many more offensive statistics in baseball. Only three most important ones are presented here.

Pitching statistics also leave an impression in the minds of many baseball fans. Everyone knows his or her favorite pitcher's earned run average (ERA).

$$Earned\ Run\ Average = \frac{9\times\sum Earned\ Runs}{\sum Innings\ Pitched} \tag{4-15-4}$$

Earned run is a run that the pitcher is accountable for. If a runner reaches home plate by safe hits, sacrifice hits, stolen bases, hit batters, wild pitches, on walks, etc. then the run is charged to the pitcher. An excellent career earned run average for a pitcher is around 2.

Defense statistics play a crucial role in deciding who plays in what defensive position. Range factor (RF) is a good indicator for a player's defensive position.

$$Range\ Factor = \frac{\sum Assists + \sum Putouts}{\sum Defensive\ Innings\ Played} \tag{4-15-5}$$

Higher the range factor, better the fielding for a player's defensive position. It is common to see some first basemen accumulate a range factor above 10 during a season.

There are many other offensive, defensive, and pitching statistics in baseball. Data have been kept for every inning of every game for more than a century.

Section 5 – Everyday Problems Solved Using Spreadsheets

Main focus of the Section 5 is to solve six everyday problems using inputs, calculations, and outputs in a spreadsheet format. What if scenarios can be applied with ease by changing inputs and seeing their effects on outputs. Everyday problems that are treated using spreadsheets are in chapters 5-1 through 5-6. Spreadsheets are applied to determining amount of party drinks, levels of home energy usage, levels of carbon footprint, nutrition facts, house remodeling, and to foreign currency exchange.

Chapter 5-1 Party Drinks

When there are a lot of inputs, calculations, and steps to be taken to reach results, spreadsheets such as Microsoft Excel is the best way to attack the problem. As an example, drink requirements for a party for 400 people will be considered.

Invitations were sent out with RSVP "**R**epondez **S**'il **V**ous **P**lait" or "Respond Please" request. Also each guest's choices for drinks and main courses were requested. Drinks choices were white wine, red wine, martini, tequila punch, diet coke, and diet seven-up. 400 guests responded the ways shown in Table 5-1-1.

Drink Type	# Of Guests Chose By RSVP
White Wine	42
Red Wine	61
Martini	76
Tequila Punch	53
Diet Coke	75
Diet Seven-Up	25
Undecided	68
TOTAL	400

Table 5-1-1 Guests choices for drinks

It was agreed that undecided guests' drinks will be decided by drink choice percentages of decided guests. Also a 10% buffer will be added to the final tally. Since Table 5-1-1 represents the inputs from guests, final drink demand can be determined from the spreadsheet as shown in Table 5-1-2.

A	B	C	D	E
Drink Type	# Of Guests Chose (Input From RSVP)	% Of Total Decided Guests (Column B x 100 / 332)	Drink Distribution For 68 Undecided Guests (Column C x 68 / 100)	Drink Distribution For 68 Undecided Guests Rounded Up Column D
White Wine	42	12.7	8.6	9
Red Wine	61	18.4	12.5	13
Martini	76	22.9	15.6	16
Tequila Punch	53	16.0	10.9	11
Diet Coke	75	22.6	15.4	16
Diet Seven-Up	25	7.5	5.1	6
TOTAL	332	100.0	68.1	

A	F	G	H
Drink Type	Total Drink Demand (Column B + Column E)	Add 10% Buffer To Total Drink Demand (1.1 x Column F)	Final Drink Demand Round Up Column G
White Wine	51	56.1	57
Red Wine	74	81.4	82
Martini	92	101.2	102
Tequila Punch	64	70.4	71
Diet Coke	91	100.1	101
Diet Seven-Up	31	34.1	35
TOTAL	403		448

Table 5-1-2 Final demand for drinks

Number of glasses of drinks allocated to a guest's drink choice and the liquid volume for each drink glass are given as inputs in Table 5-1-3. Also each guest is allocated two glasses of water which is also added on as shown in Table 5-1-3.

A	I	J
Drink Type	# Of Glasses Allocated For A Drink Choice	Liquid Volume In Each Drink Glass, mL (milliliters)
White Wine	3	200
Red Wine	3	200
Martini	3	150
Tequila Punch	3	150
Diet Coke	2	250
Diet Seven-Up	2	250
Water for 400 guests	2	250

Table 5-1-3 Number of glasses of drinks allocated to a guest's drink choice and liquid volume in each glass

In this example liquid volume is in metric units. However, there are a lot of drink containers that use US volume units. Volume relationships between metric units and US units are given in Table 5-1-4.

1 L (liter)= 1000 mL (milliliter)
1 US Fluid Gallon=4 US Fluid Quarts=3.786 L
1 US Fluid Quart=2 US Fluid Pints=0.946 L
1 US Fluid Pint=16 US Fluid Ounces=0.473 L
1 US Fluid Ounce=29.6 mL

Table 5-1-4 Volume relationships between metric and US units

Now the volumetric demand for each drink type can be calculated. The volumetric demand for each drink type is given in Table 5-1-5.

A	K
Drink Type	Volumetric Demand For Each Drink Type, L (Column H x Column I x Column J)
White Wine	34.2
Red Wine	49.2
Martini	45.9
Tequila Punch	31.95
Diet Coke	50.5
Diet Seven-Up	17.5
Water (400 x Column I x Column J)	200

Table 5-1-5 Volumetric demand for each type of drink in liters

Martini is a mixture of 90% vodka and 10% vermouth by volume. Tequila punch is a mixture of 80% tequila, 10% orange juice, and 10% granadine by volume. Amount of liquid in each drink container (bottle or can) and container's unit price are given as inputs in Table 5-1-6.

L	M	N
Drink Type	Volume Per Container, mL	Price Per Container, $
White Wine (bottle)	750	15.00
Red Wine (bottle)	750	20.00
Martini		
Vodka (bottle)	1750	20.00
Vermouth (bottle)	750	25.00
Tequila Punch		
Tequila (bottle)	1500	18.00
Orange Juice (bottle)	2000	6.00
Granadine (bottle)	500	25.00
Diet Coke (can)	355	0.50
Diet Seven-Up (can)	355	0.50
Water (bottle)	1500	2.00

Table 5-1-6 Volume and price per container

Number of containers for each drink type needed and the total cost of the drinks can be calculated from the information provided in Table 5-1-6. The results are shown in Table 5-1-7.

L	O	P	Q	R
Drink Type	Volumetric Demand, mL	# Of Bottles Or Cans Needed	# Of Bottles Or Cans Needed Round Up Column P	Total Cost, $
White Wine (bottle)	34200 (Column K x 1000)	45.6	**46**	**690.00**
Red Wine (bottle)	49200 (Column K x 1000)	65.6	**66**	**1320.00**
Martini				
Vodka (bottle)	41310 (0.9 x Column K x 1000)	23.6	**24**	**480.00**
Vermouth (bottle)	4590 (0.1 x Column K x 1000)	6.1	7	**175.00**
Tequila Punch				
Tequila (bottle)	25560 (0.8 x Column K x 1000)	17.04	**18**	**324.00**
Orange Juice (bottle)	3195 (0.1 x Column K x 1000)	1.6	2	**12.00**
Granadine (bottle)	3195 (0.1 x Column K x 1000)	6.4	7	**175.00**
Diet Coke (can)	50500 (Column K x 1000)	142.3	**143**	**71.50**
Diet Seven-Up (can)	17500 (Column K x 1000)	49.3	**50**	**25.00**
Water (bottle)	200000 (Column K x 1000)	133.3	**134**	**268.00**
GRAND TOTAL				**$3,540.50**

Table 5-1-7 Number of bottles or cans needed and total costs

The beauty of a spreadsheet approach to a problem like the one presented here is that any change in the inputs can be put into the appropriate cells and the final results come out with ease. Also different scenarios can be investigated with ease. For example, number of glasses allocated for a drink choice, guest's choices for drinks, or any other inputs can be altered and results of each scenario can be on your finger tips.

Chapter 5-2 Carbon Footprint

Carbon footprint is the sum of all emissions of carbon dioxide, CO_2, released to atmosphere by a person's activities and living style such as usage of electricity in our home, heating our home, driving our car(s), taking an airplane trip, a bus trip or a train trip, eating meat or vegetables, eating at home or at restaurants, recycling or no recycling, walking for exercise or riding a jet ski for recreation. As the Earth's standard of living and population increase so does her carbon footprint. In order to prevent global warming, we have to stabilize the increase in our carbon footprint.

This chapter provides a simple way of calculating our personal carbon footprint in a spreadsheet format. Personal carbon footprint is generally determined by kilograms of CO_2 released to atmosphere in a full year. As the CO_2 release amounts get larger, tons of CO_2 released per year per person is utilized where a metric ton equals to 1,000 kilograms. Top ranked nations for CO_2 release are given in Table 5-2-1 for year 2006, see Union of Concerned Scientists at ucsusa.org.

Country	Billion Tons Of CO_2 Released Per Year	Population, Million People	Average Tons Of CO_2 Released Per Year Per Person
1. CHINA	6.018	1,340	4.49
2. USA	5.903	308	19.17
3. RUSSIA	1.704	142	12.00
4. INDIA	1.293	1,170	1.11
5. JAPAN	1.247	128	9.74
6. GERMANY	0.858	82	10.46
7. CANADA	0.614	33	18.61

Table 5-2-1 Average tons of CO_2 released per year per person by highest carbon footprint countries

Next let us analyze significant carbon footprint contributors such as electric power, home heating, cars, airplane trips, bus or train trips, and living styles.

Electric power can come from a mix of different power sources such as coal, petroleum, natural gas, nuclear, hydroelectric, solar, wind, and geothermal. Highest CO_2 release comes from coal powered sources at 1 kg of CO_2 per kWh (kilo-Watt-hour) of electricity. None or little CO_2 release comes from nuclear, hydroelectric, solar, wind, and geothermal powered sources. Mix of electric power sources and therefore CO_2 releases from electric power plants can vary a lot from state to state. For example, Washington State releases about 0.1 kg of CO_2/kWh as compared to West Virginia State of 0.9 kg of CO_2/kWh. Instead of analyzing the electric power carbon footprint state by state, a weighted average for the whole United States can be determined by using the electric power source mix in Table 5-2-2.

Column 1	Column 2	Column 3	Column 4
Electric Power Source	Carbon Footprint kg CO_2/kWh	% Of Electric Power Sources In USA	Column 2 x Column 3 kg CO_2/kWh
COAL	1.0	52	52
NUCLEAR	0	21	0
NATURAL GAS	0.59	16	9.44
HYDROELECTRIC	0	7	0
PETROLEUM	0.89	3	2.67
OTHER	0	1	0
		SUM=100	SUM=64.11
Weighted average CO_2 release for electric power=64.11/100=0.64 kg CO_2/kWh			

Table 5-2-2 Weighted average CO_2 release for electric power in USA

A weighted average CO_2 release of 0.64 kg CO_2/kWh will be used in the present personal carbon footprint calculations.
Home heating can come from different energy sources such as natural gas, heating oil, coal, or wood. Standard units and CO_2 release for each home heating energy source is given in Table 5-2-3. Conversion from BTU (British Thermal Unit) to kilo-Watt-hour is: 3,413 BTU=1 kWh.

Energy Source	Standard Consumption Units	Conversion To kWh	kg CO_2 Released Per kWh
Natural Gas	1 Therm=10^5 BTU	1 Therm=29.3 kWh	0.19
Heating Oil	1 Gallon Heating Oil=138,000 BTU	1 Gallon Heating Oil=40.43 kWh	0.28
Coal	kg	6 kWh per kg of Coal	0.33
Wood	kg	4 kWh per kg of Wood	0.175

Table 5-2-3 Home heating energy sources kilogram CO_2 releases per kilo-Watt-hour

CO_2 release values given in Table 5-2-3 can be used to determine the home heating carbon footprint. We should not forget to divide the total CO_2 release from home heating per year into the number of persons living in that house in order to obtain the home heating carbon footprint per year per person. For cars or motorcycles, the carbon footprint can be determined from the fuel economy of the vehicle and the miles driven with that vehicle in a year. For motor vehicles running on gasoline, 8.7 kg of CO_2 release per gallon of gasoline consumed can be used. For motor vehicles running on diesel fuel, 9.95 kg of CO_2 release per gallon of diesel fuel consumed can be used.

Airplane trip carbon footprint can be tricky to determine, since there are many variables that affect the airplane travel carbon footprint. Some of these variables are aircraft type, engine type, passenger load, baggage load, cargo load, short haul, long haul, cruising altitude, head or tail wind, congestion during takeoff and landing, etc. If a person knows the total air miles he or she has flown in a year, then the airplane travel carbon footprint can be calculated by assuming 4 air miles of travel per kilogram of CO_2 released.

Similar variability arguments can be made for bus or train travel. If a person knows the total miles he or she has traveled by bus or train in a year, then the bus or train travel carbon footprint can be calculated by assuming 7 miles of travel per kilogram of CO_2 released.

Living style also affects your carbon footprint. For example, if you are a meat lover and a vegetable hater, you will release 1,000 kilograms more CO_2 in a year for food because of the methane gas release from animals. If you recycle as compared to no recycling, you will release 1,000 kilograms less CO_2 in a year. For your exercise, if you walk or play tennis instead of jet skiing or motorcycle racing, you will release 1,000 kilograms less CO_2 in a year. If you eat at the restaurants a lot, you will release 2,000 kilograms more CO_2 in a year. We can put a person's life style into three categories: A person with a high carbon footprint life style will release about 14,500 kg CO_2 a year, i.e. a meat eater, no recycling, jet skier, and a constant restaurant eater. A person with a low carbon footprint life style will release about 9,500 kg CO_2 a year, i.e. a vegetarian, recycling, walking, and once a month restaurant eater. A person with a moderate carbon footprint life style will release about 12,000 kg CO_2 a year.

After determining the major contributors to a person's carbon footprint above, total carbon footprint of a person can now be calculated as shown in Table 5-2-4. Then a person's carbon footprint can be compared to the average tons of CO_2 released per year per person in USA as shown in Table 5-2-1.

CO_2 Source	**Energy Source**	**CO_2 Release**	**Energy Consumption /Year/Person**	**kg CO_2 Release/ Year/Person**
Home Electricity	Weighted Average Mix	0.64 kg CO_2/kWh	3300 kWh	2,112
Home Heating	Natural Gas	0.19 kg CO_2/kWh	150 therms =4,395 kWh	835
Cars	Gasoline	8.7 kg CO_2/gallon	20,000 miles/year ÷ 30 miles/gallon= 666.7 gallons	5,800
Airplane Trips	Jet Fuel	0.25 kg CO_2/mile	25,000 miles/year	6,250
Bus Or Train Trips		0.14 kg CO_2/mile	450 miles/year	63
Living Style	Moderate Carbon Footprint			12,000
	TOTAL CARBON FOOTPRINT =			27,060

Table 5-2-4 A person's approximate carbon footprint in a year

$$\frac{\textit{Total Carbon Footprint}}{\textit{Average Carbon Footprint In USA}} = \frac{27{,}060}{19{,}170} \times 100 = 1.412 \qquad (5\text{-}2\text{-}1)$$

So this person's carbon footprint is 41.2% over the national average.

Chapter 5-3 Home Energy Usage

Energy that we consume at home can be broken down to its components and total energy consumption per year can be determined. Once the baseline energy consumption is calculated, improvements such as insulating the doors and windows, using thermostats with time control, using compact fluorescent light bulbs (CFL), solar or wind electric energy generation, and solar water heating can bring down the energy consumed at home tremendously.

Assuming space heating and water heating are achieved by natural gas, energy expended in a year can be found in the gas bill statements. For natural gas, the gas bill statements provide the energy expenditure in therms which is 10^5 BTU's (British Thermal Unit). Also 1 therm equals 29.3 kWh (kilo-Watt-hour). For space heating, for the present calculations, assume that 300 therms per year is expended which equals to 8,790 kWh/yr. For water heating, for the present calculations, assume that 100 therms per year is expended which equals to 2,930 kWh/yr.

Assume that space cooling is achieved with electrical energy. By knowing the cooling capacity of the air conditioners in kW and by estimating the number of hours they operate during the summer, kWh electrical energy that is consumed per year for space cooling can be determined. For the present calculations, a 5 kW cooling capacity air conditioner which operates about 400 hours per year is assumed, namely for space cooling 2,000 kWh/yr electrical energy is consumed.

Electrical energy consumption for home lighting is determined by finding the power rating of each bulb in every location of the house and estimating the number of hours per year each light bulb is used. If CFL bulbs are used consumes 75% less energy. Home lighting energy consumption spreadsheet is presented in Table 5-3-1.

Location	Lighting Power, W	Usage, hours/day	Usage hours/year	kWh/year
TV Room	660	4	1320	871.2
Living Room	900	2	660	594
Kitchen	300	4	1320	396
3 Bedrooms	600	2	660	396
3 Bathrooms	600	2	660	396
Garage	100	1	330	33
Outside	600	1	330	198
Total Yearly Electrical Energy Consumption For Lighting = 2,884 kWh				

Table 5-3-1 Total yearly electrical energy consumption for home lighting

In Table 5-3-1, 330 days per year of home lighting is assumed. Rest of the days during the year, home lighting is not used due to vacations, eating outside, etc. Home computers, TV's, and other electronics electrical energy consumption per year is presented in a spreadsheet format in Table 5-3-2. Also in Table 5-3-2, 330 days of computer, TV, and other electronics usage per year is assumed.

Equipment	Equipment Power, W	Usage, hours/day	Usage hours/year	kWh/year
2 Laptops	240	5	1650	396
2 24" Monitors	60	5	1650	99
1 46" LCD Flat TV	250	6	1980	495
Sound System	90	3	990	89.1
3 Cellular Phones	1.77	5	1650	2.9
Total Yearly Electrical Energy Consumption For Computers, TV's, and Other Electronics = 1,082 kWh				

Table 5-3-2 Total yearly electrical energy consumption for computers, TV's, and other electronics

Appliance energy consumptions are presented in Table 5-3-3 in a spreadsheet format. All appliances are assumed to be energy star compliant, namely they comply with energy efficiency guidelines of EPA (Environmental Protection Agency) and US Department of Energy. All appliances are assumed to operate with electrical energy.

Appliance	Electricity Consumption	Appliance Usage	Usage per year	kWh/year
2 Refrigerators	550 kWh/ refrigerator/yr			1,100
1 Freezer	500 kWh/ freezer/yr			500
1 Dishwasher	2 kWh/run	1 run/two nights	150 runs/yr	300
1 Oven	2 kW for bake & 3.6 kW for broil	1 bake & 1 broil for 1 hr/wk	48 wks/yr	(2+3.6)x48= 268.8
1 Microwave	1.7 kW	10 min/day	330 days/yr	93.5
1 Dryer	3 kWh/load	3 loads/wk	48 wks/yr	432
1 Washer	6 kWh/load	3 loads/wk	48 wks/yr	864
Other Appliances: Coffee Maker, Toaster, Vacuum Cleaner, Blender, Garage Opener, Garbage Disposal, Telephones, Clocks	3,308 W	5 min/day	330 days/yr	90.97
Total Yearly Electrical Energy Consumption For Appliances = 3,649 kWh				

Table 5-3-3 Total yearly electrical energy consumption for appliances

Another energy consumption come from hot tubs and pools. We will only consider a hot tub here. A hot tub with two pumps with each having a 2.5 HP electric motor has 2x2.5x745.7=3728.5 W power. If it is used 2 hrs/day and 60 days/yr, then the annual electrical energy consumption for this hot tub is 447 kWh.

We are ready to add all energy consumptions in our home and see how the annual energy consumptions are distributed. Summary of all energy consumptions in our home are shown in Table 5-3-4.

Energy Consumer	**Annual Energy Consumption kWh/yr**	**% Of Grand Total**
Space Heating	8,790	40.4
Water Heating	2,930	13.4
Natural Gas Energy Consumption Total=11,720		53.8
Space Cooling	2,000	9.2
Lighting	2,884	13.2
Computer, TV's, and Other Electronics	1,082	5.0
Appliances	3,649	16.7
Hot Tub	447	2.1
Electrical Energy Consumption Total=10,062		46.2
Grand Total Of Annual Energy Consumption=21,782 kWh/yr		

Table 5-3-4 Home energy consumers and grand total of annual energy consumption

For natural gas energy consumption, this house releases

11,720 kWh x 0.19 kg- CO_2/kWh = 2,227 kg CO_2

and for electrical energy consumption, this house releases

10,062 kWh x 0.64 kg-CO_2/kWh = 6,440 kg CO_2

a year into the atmosphere. If 4 people live in this house, the average annual release of CO_2 into the atmosphere per person is (2,227 + 6,440)/4 = 2,167 kg. For CO_2 release per kWh rate estimates see Chapter 5-2.

Chapter 5-4 Nutrition Facts

Exercise and the food that goes down our throats determine the health and quality of our lives. Every package of food has a label explaining its nutrition facts. These nutrition facts have to match your daily nutrition requirements determined by your age, gender, height, weight, and daily activity level. In this example, daily diet requirements will be divided into breakfast, lunch, and dinner diet requirements. These diet requirements will be compared to a set of food serving nutrition facts in a spreadsheet format.

This example is based on a 2,000 Calorie daily diet given in Table 5-4-1. Vitamins are excluded from this example.

Daily Values Based On A 2,000 Calorie Diet			
Calories	2,000	Calories	Maximum
Total Fat	65	grams	Maximum
Saturated Fat	20	grams	Maximum
Trans Fat	-	-	Minimize Intake
Cholesterol	300	milligrams	Maximum
Sodium	2,400	milligrams	Maximum
Total Carbohydrates	300	grams	At Least
Dietary Fiber	25	grams	At Least
Sugar	-	-	Minimize Intake
Protein	50	grams	At Least

Table 5-4-1 Dietary goals for a 2,000 Calorie Daily Diet Limiting nutrients are highlighted in darker shaded cells and at least nutrients are highlighted in non-shaded cells

Let us divide the daily values to three equal dietary goals for breakfast, lunch, and dinner. For breakfast two servings of cereal (2 cups or 98 grams) and one serving of mixed fruit (1 cup or 8 fluid ounces) are served. Nutrition facts for breakfast are shown in Table 5-4-2.

BREAKFAST ALLOWANCE			**Units**	**Cereal Servings**	**% allowed**	**Fruit Serving**	**% allowed**	**TOTAL**	**% allowed**
Calories	Max	**666.7**	Calories	360	***54.0***	130	***19.5***	490	***73.5***
Total Fat	Max	**21.7**	g	2	***9.2***	0.2	***0.9***	2.2	***10.2***
Saturated Fat	Max	**6.7**	g	0	***0.0***	0.2	***3.0***	0.2	***3.0***
Trans Fat	Min	-	g	0	-	0.1	-	0.1	-
Cholesterol	Max	**100.0**	mg	0	***0.0***	0	***0.0***	0	***0.0***
Sodium	Max	**800.0**	mg	0	***0.0***	11	***1.4***	11	***1.4***
Total Carbohydrates	At Least	**100.0**	g	76	***76.0***	34	***34.0***	110	***110.0***
Dietary Fiber	At Least	**8.3**	g	10	***120.0***	4	***48.0***	14	***168.0***
Sugar	Min	-	g	0	-	25.8	-	25.8	-
Protein	At Least	**16.7**	g	10	***60.0***	1.1	***6.6***	11.1	***66.6***

Table 5-4-2 Breakfast nutrition facts

For breakfast, the calorie intake is 26.5% less than the allowed. Fat, cholesterol, and sodium levels are in excellent shape. Total carbohydrates and dietary fiber intakes are also good. 25.8 grams of sugar can be okay, but keep daily sugar levels to a minimum. Protein is below the breakfast allowance. However, these nutrition excesses and deficits at breakfast can be carried over to the lunch allowance and a modified lunch allowance is established.

For lunch, the food is one serving of Caesar salad with grilled chicken and with Caesar sauce where one serving is four cups. Nutrition facts for lunch are shown in Table 5-4-3.

LUNCH ALLOWANCE			Units	CARRY OVER FROM BREAKFAST	TOTAL LUNCH ALLOWANCE	Caesar Salad With Grilled Chicken	% allowed
Calories	Max	**666.7**	Calories	176.7	**843.3**	930	*110.3*
Total Fat	Max	**21.7**	g	19.5	**41.1**	71	*172.6*
Saturated Fat	Max	**6.7**	g	6.5	**13.1**	13	*99.0*
Trans Fat	Min	-	g	0.1	-	0.2	-
Cholesterol	Max	**100.0**	mg	100.0	**200.0**	60	*30.0*
Sodium	Max	**800.0**	mg	789.0	**1589.0**	1840	*115.8*
Total Carbohydrates	At Least	**100.0**	g	-10.0	**90.0**	28	*31.1*
Dietary Fiber	At Least	**8.3**	g	-5.7	**2.7**	6	*225.0*
Sugar	Min	-	g	25.8	-	4	-
Protein	At Least	**16.7**	g	5.6	**22.2**	43	*193.4*

Table 5-4-3 Lunch nutrition facts

Caesar salad with grilled chicken and with Caesar sauce exceeds the total lunch allowance for calories, fat, and sodium. After lunch, the nutrition facts are at the limit of saturated fats, good at cholesterol, dietary fiber, and protein levels, but short on total carbohydrates. After lunch daily nutrition excesses and deficits can be carried over to the dinner allowance and a modified dinner allowance is established.

For dinner, the food is one serving of green salad (3 cups) with one tablespoon of Italian salad dressing, one serving of grilled salmon at 240 grams, and two servings (2 slices=100 grams) of bread. Nutrition facts for dinner are shown in Table 5-4-4.

DINNER ALLOWANCE			Units	CARRY OVER FROM LUNCH	TOTAL DINNER ALLOWANCE	Green Salad	Salmon	Bread	TOTAL DINNER	% OF TOTAL DINNER ALLOWANCE
Calories	Max	**666.7**	Calories	-86.7	**580.0**	63	240	220	523	90.2
Total Fat	Max	**21.7**	g	-29.9	**NONE**	4.2	6	1	11.2	NA
Saturated Fat	Max	**6.7**	g	0.1	**6.8**	0.7	0	0	0.7	10.3
Trans Fat	Min	**-**	g	0.3	**-**	0	0	0	0	-
Cholesterol	Max	**100.0**	mg	140.0	**240.0**	0	260	0	260	108.3
Sodium	Max	**800.0**	mg	-251.0	**549.0**	210	320	150	680	123.9
Total Carbohydrates	At Least	**100.0**	g	62.0	**162.0**	4.5	0	46	50.5	31.2
Dietary Fiber	At Least	**8.3**	g	-3.3	**5.0**	2	0	8	10	200.0
Sugar	Min	**-**	g	29.8	**-**	2.2	0	2	4.2	-
Protein	At Least	**16.7**	g	-20.8	**CAN INTAKE MORE**	1.1	48	8	57.1	NA

Table 5-4-4 Dinner nutrition facts

After dinner, the calories are 9.8% lower than the daily value which is good. If one serving of mousse au chocolat at 400 Calories is offered for dessert, we should pass it. There is no allowance for total fat after over running it by 8.2 grams after breakfast and lunch, but still 11.2 grams are added at dinner. Saturated fat allowance is in good shape. Cholesterol and sodium allowances are exceeded. After dinner carbohydrate allowance is short, but dietary fiber and protein allowances are exceeding the requirements.

After the breakfast, lunch, and dinner intakes, nutrition totals and daily values are compared in Table 5-4-5.

DAILY VALUE ALLOWANCE			**Units**	**Total Of Breakfast+ Lunch+ Dinner**	**% OF DAILY VALUE**
Calories	Max	2,000	Calories	1943.0	97.2
Total Fat	Max	65	g	84.4	129.8
Saturated Fat	Max	20	g	13.9	69.5
Trans Fat	Min	-	g	0.3	-
Cholesterol	Max	300	mg	320.0	106.7
Sodium	Max	2,400	mg	2531.0	105.5
Total Carbohydrates	At Least	300	g	188.5	62.8
Dietary Fiber	At Least	25	g	30.0	120.0
Sugar	Min	-	g	34.0	-
Protein	At Least	50	g	111.2	222.4

Table 5-4-5 Total daily nutrition facts – In the final column, excess daily values are highlighted in darker shaded cells and good daily values are highlighted in non-shaded cells

The results of this day's eating adventure show that the daily value nutrition facts are good in calories, dietary fiber, and protein. Today's food intake exceeds the total fat, cholesterol, and sodium daily allowances. The saturated fat level is lower than its daily value. Total carbohydrates are deficient with respect to its daily value. Trans fat (unsaturated fat) and sugar levels are kept to a minimum.

Chapter 5-5 House Remodeling

When planning a sizeable project such as remodeling a house, usage of spreadsheets can help in detailing every task of the project. The estimated cost or the cost allowance for each task, the actual cost when the job is completed, and the differential cost can be calculated by ease. Also comments can be added for each task for reference and to ease stress on your memory.

A typical house remodeling spreadsheet is presented in Figure 5-5-1. The remodeling project has three major task groups. Each task group is divided into its tasks. Each task group's budgeted cost, actual cost and the cost differential are summed under "SUBTOTAL". Total project's budgeted cost, actual cost and the cost differential are summed under "TOTAL".

The object is to keep the spreadsheet simple and manageable. It is important to fill in cost details of every task before the project starts and after the project ends. If the project is too large, say over 200 tasks and ten task groups, you can create a separate spreadsheet for each project task group and then create a master spreadsheet for the whole project that summarizes the results of each task group.

REMODELING A'S HOUSE 10-Feb-2010						
TASK GROUP	**TASK**	**ESTIMATE OR ALLOWANCE**	**BUDGETED COST, $**	**ACTUAL COST, $**	**COST DIFFERENTIAL, $**	**COMMENTS**
GENERAL REQUIREMENTS	Permits	Allowance	500	472	-28	Roof and Electrical
	Supervision	Allowance	7500	6750	-750	10 hrs/wk for 10 wks @ $75/hr
	Fire Extinguishers	Allowance	80	78	-2	Two
	Temporary Sanitation	Estimate	380	391	11	
	Job Site Cleanup	Allowance	4880	4280	-600	117 hrs @ $40/hr
	Dump Fees & Trash Removal	Allowance	600	594	-6	
	Temporary Storage For Owner	Allowance	400	412	12	
	SUBTOTAL		**14340**	**12977**	**-1363**	
ROOF	Removal Of Present Roof	Estimate	2400	2400	0	48 hrs @ $50/hr
	Repair Of Rotten Sections	Allowance	1400	1680	280	20 hrs @ $70/hr
	Undercoat Installation	Estimate	1400	1400	0	
	Tiles	Estimate	11000	11450	450	4,000 ft2
	Tile Installation	Estimate	10500	10710	210	150 hrs @ $70/hr
	Gutter Installation	Estimate	2000	2155	155	
	SUBTOTAL		**28700**	**29795**	**1095**	
3 BATHROOMS AND SHOWERS	3 Counter Tops Marble Slab Material & Installation	Estimate	7350	8120	770	$4000 Allowance for Marble Material
	Finish Carpentry	Allowance	1050	1200	150	21 hrs @ $50/hr
	Cabinet Knobs	Allowance	375	395	20	
	Tile Material	Allowance	5250	5670	420	350 Tiles at $15/tile
	Tile Installation	Estimate	14058	14510	452	See SG Marble Estimate For Details
	New Shower Enclosures	Estimate	2872	2825	-47	
	Mirrors & Glass Shelves	Estimate	1747	1762	15	4 Mirrors & 1 Shelf
	Plumbing	Estimate	3930	3900	-30	
	Plumbing Fixtures & Bathrooms Accessories	Allowance	4750	4623	-127	$400/sink, $250/faucet, $600/shower valve, $1000 for towel bars
	Electrical	Estimate	3890	3810	-80	See Volt Electrical Estimate
	Paint	Estimate	2240	2263	23	1- Coat Includes Primer and Paint
	SUBTOTAL		**47512**	**49078**	**1566**	
	TOTAL		**90552**	**91850**	**1298**	

Figure 5-5-1 A typical house remodeling cost spreadsheet

Chapter 5-6 Foreign Currency Exchange

An international traveler has to make several decisions when converting his or her currency to another country's currency. Should he change the US dollars to foreign currency at the airport before boarding the plane? Should he take along traveler's checks, international postal money orders, or cash along? If he is traveling to several foreign countries, what should he do with the left over currency of one country when leaving that country and going to another one?

In this example, let us consider an international traveler who is going to go from US to Japan, from Japan to South Korea, and from South Korea to Hong Kong and then return home. If his airplane is arriving to a particular country in the middle of the night or during the weekend, all the banks and exchange bureaus can be closed. He can change US dollars at the US airport before boarding the plane, but the commissions and exchange rates can be very unfavorable. Let us say that he is traveling with cash even if traveler checks or international postal money orders have better exchange rates.

In his first stop, Tokyo, he goes to a bank at the airport and observes the exchange board shown in Figure 5-6-1.

	Selling YEN		Buying YEN	
	Traveler's Check	Cash	Traveler's Check	Cash
USD, $	91.25	93.05	89.25	87.25
100 South Korean WON	-	9.46	-	6.46
Hong Kong, $	-	14.06	-	9.2

Table 5-6-1 A bank's exchange board in Tokyo

The bank is also charging a 10% commission for the currency exchange. He also looked up the exchange rate between the US $ and the Japanese Yen at the foreign exchange market (forex) for the day and found out that 1 US$=90.15 Japanese Yen. He also observed that this exchange rate is pretty much fixed during his short stay in Japan. With the above information a simple spreadsheet can be generated, Table 5-6-2, to see how much he is going to lose to commissions and currency exchange rates during his foreign exchange transactions in Tokyo.

US$ - Cash To Be Exchanged	10% Commission (0.1x1000), $	Bought Japanese Yen (900x87.25)	Spent Japanese Yen During His Stay	Actual US$ Spent During His Stay (56,000/90.15)
1,000	100	78,525	56,000	621.19
Left Over Yen Before Leaving Japan (78,225-56,000)	**Sell Yen To Get Back US$ Before Leaving (0.9x22,525/93.05)**	**US$ Out Of Pocket In Japan (1,000-217.87)**	**US$ Lost In Currency Exchange & Commissions (782.13-621.19)**	**% Lost In Currency Exchange & Commissions (160.94x100/782.13)**
22,525	217.87	782.13	160.94	20.6

Table 5-6-2 Foreign exchange calculations for Japan

In Japan, he lost 20.6% of his money for currency exchanges and commissions.

His next stop is Seoul, South Korea. He goes to a bank at the airport and observes the exchange board shown in Figure 5-6-3.

	Selling WON		Buying WON	
	Traveler's Check	Cash	Traveler's Check	Cash
USD, $	1147.624	1170.262	1122.471	1097.318
Japanese Yen	-	13.331	-	11.822
Hong Kong, $	-	154.834	-	137.341

Table 5-6-3 A bank's exchange board in Seoul

The bank in Seoul is charging 8% commission for the currency exchange. He also looked up the exchange rate between the US $ and the South Korean Won at the foreign exchange market (forex) for the day and found out that 1 US$=1133.79 South Korean Won. He also observed that this exchange rate is pretty much fixed during his short stay in South Korea. With the above information a similar spreadsheet to Table 5-6-2 can be generated, Table 5-6-4, to see how much he is going to lose to commissions and currency exchange rates during his foreign exchange transactions in Seoul.

US$ - Cash To Be Exchanged	8% Commission (0.08x1000), $	Bought South Korean Won (920x1097.318)	Spent South Korean Won During His Stay	Actual US$ Spent During His Stay (750,000/1133.79)
1,000	80	1,009,532.56	750,000	661.50
Left Over Won Before Leaving South Korea (1,009,532.56-750,000)	**Sell Won To Get Back US$ Before Leaving (0.92x259,532.56/ 1170.262)**	**US$ Out Of Pocket In South Korea (1,000-204.03)**	**US$ Lost In Currency Exchange & Commissions (795.97-661.50)**	**% Lost In Currency Exchange & Commissions (134.47x100/795.97)**
259,532.56	204.03	795.97	134.47	16.9

Table 5-6-4 Foreign exchange calculations for South Korea

In South Korea, he lost 16.9% of his money for currency exchanges and commissions.
If he would have kept his Yen and directly converted them to Won in South Korea instead of converting back to US$ in Japan, it would have been more advantageous for him. Left over 22,525 Yen converted to Won in South Korea gives him 244,987.3 Wons, 0.92x22,525x11.822=244,987.3, which is equivalent to converting 242.68 US$ into Wons which is 11.4% more than what he got in Japan, namely 217.9 US$.

His next stop is Hong Kong, China. He goes to a bank at the airport and observes the exchange board shown in Figure 5-6-5.

	Selling HK$		**Buying HK$**	
	Traveler's Check	Cash	Traveler's Check	Cash
USD, $	7.855	8.010	7.683	7.510
100 Japanese Yen	-	8.952	-	8.264
1000 South Korean Won	-	7.118	-	6.571

Table 5-6-5 A bank's exchange board in Hong Kong

The bank in Hong Kong is charging 6% commission for the currency exchange. He also looked up the exchange rate between the US $ and the Hong Kong $ at the foreign exchange market (forex) for the day and found out that 1 US$=7.76 Hong Kong $. He also observed that this exchange rate is pretty much fixed during his short stay in Hong Kong. With the above information a similar spreadsheet to Table 5-6-2 can be generated, Table 5-6-6, to see how much he is going to lose to commissions and currency exchange rates during his foreign exchange transactions in Hong Kong.

US$ - Cash To Be Exchanged	6% Commission (0.06x1000), $	Bought HK$ (940x7.510)	Spent HK$ During His Stay	Actual US$ Spent During His Stay (5,400/7.76)
1,000	60	7,059.40	5,400	695.88
Left Over HK$ Before Leaving Hong Kong (7,059.40-5,400)	**Sell HK$ To Get Back US$ Before Leaving (0.94x1,659.40/ 8.010)**	**US$ Out Of Pocket In Hong Kong (1,000-194.74)**	**US$ Lost In Currency Exchange & Commissions (805.26-695.88)**	**% Lost In Currency Exchange & Commissions (109.38x100/805 .26)**
1,659.40	194.74	805.26	109.38	13.6

Table 5-6-6 Foreign exchange calculations for Hong Kong

In Hong Kong, he lost 13.6% of his money for currency exchanges and commissions. If he would have kept his Won when leaving South Korea and directly converted them to HK$ in Hong Kong instead of converting back to US$ in South Korea, it would have been more advantageous for him.

Left over 259,532.56 Won converted to HK$ in Hong Kong gives him 1,603.07 HK$, 0.94x259,532.56x(6.571/1,000)=1,603.07, which is equivalent to converting 227.08 US$ into HK$ which is 11.3% more than what he got in South Korea, namely 204.03 US$.

During this 3-stop international trip, he lost a combined total of 17% of his out of pocket money to currency exchange and commissions as shown in Table 5-6-7, if he converted left over currencies to US$ before he left each country.

Total US$ Exchanged	Total Converted Back To US$	Total US$ Out Of Pocket	Total Actual US$ Spent
1,000+1,000+ 1,000=3,000	217.87+204.03+ 194.74=616.64	3,000-616.64= 2,383.36	621.19+661.50+ 695.88=1,978.57
Percentage lost to currency exchange and commissions =(2,383.36-1,978.57)x100/2,383.36 =17% or 404.79 US$			

Table 5-6-7 Total losses to currency exchange and commissions

References

[1] Paul Richard S. and Ernest F. Haeussler, Jr., “Introductory Mathematical Analysis: For Students of Business and Economics”, Reston Publishing Company, Prentice-Hall, Reston, Virginia, 1973.

[2] US National Weather Service, “Wind Chill Web Page”, at http://www.nws.noaa.gov/om/windchill/

[3] Gilbert Strang, “Linear Algebra and its Applications”, Harcourt Brace Jovanovich, Publishers, San Diego, 1988.

[4] International Standards Organization, “ISO 7243 – Hot Environments – Estimation of the heat stress on working man, based on the WBGT-index (wet bulb globe temperature-index)”, Geneva, 2003.

[5] Shames, Irving H., “Mechanics of Fluids”, McGraw-Hill, New York, 1962.

[6] Parish O. Owen and Terrill W. Putnam, “Equations For Determination Of Humidity From Dewpoint And Psychrometric Data”, NASA Technical Note D-8401, 1977.

[7] Cengel Yunus A. and Michael A. Boles, “Thermodynamics - An Engineering Approach”, Third Edition, McGraw-Hill, Inc., Hightstown, New Jersey, 1998.

[8] Levin, R. I., “Statistics for Management”, Prentice Hall, New Jersey, 1987.

[9] Selby Samuel M. and Brian Girling, “CRC Standard Mathematical Tables”, Fourteenth Edition, The Chemical Rubber Co., Cleveland, Ohio, 1965.

[10] Kunz Kaiser S., “Numerical Analysis”, McGraw Hill, New York, 1957.

[11] Ashar, Kanu G., “Magnetic Disk Drive Technology”, IEEE Press, New York, 1997.

[12] Wiebelt J. A., “Engineering Radiation Heat Transfer”, Holt, Rinehart and Winston, New York, 1966.

[13] Boyce William E. and Richard C. DiPrima, “Elementary Differential Equations and Boundary Value Problems”, John Wiley & Sons, Inc., New York, 1997.

[14] Food and Drug Administration, Department of Health and Human Services, “Determination of SPF Value”, 21 CFR (Code of Federal Regulations) 352.73, December 2005.

[15] Mode, Elmer B., “Elements of Statistics”, Third Edition, Prentice-Hall, Englewood Cliffs, New Jersey, 1961.

[16] National Institute of Standards and Technology, “NIST/SEMATECH e-Handbook of Statistical Methods”, 2003.

[17] Wheeler Donald J. and David S. Chambers, “Understanding Statistical Process Control”, 2nd Edition, SPC Press, Knoxville, 1992.

[18] Diamond, William J., “Practical Experiment Designs for Engineers and Scientists”, Van Nostrand Reinhold Company, New York, 1981.

Index

Q

R

S

T

U

V

W

Made in the USA
Monee, IL
08 November 2026

67383073R00162